Second Edition

Creeping Bentgrass Management

Second Edition

Creeping Bentgrass Management

Peter H. Dernoeden

CRC Press
Taylor & Francis Group
Boca Raton London New York

CRC Press is an imprint of the
Taylor & Francis Group, an **informa** business

CRC Press
Taylor & Francis Group
6000 Broken Sound Parkway NW, Suite 300
Boca Raton, FL 33487-2742

First issued in paperback 2017

© 2013 by Taylor & Francis Group, LLC
CRC Press is an imprint of Taylor & Francis Group, an Informa business

No claim to original U.S. Government works

ISBN-13: 978-1-4665-0992-4 (hbk)
ISBN-13: 978-1-138-07690-7 (pbk)

Library of Congress Cataloging-in-Publication Data

Dernoeden, Peter H.
    Creeping bentgrass management / Peter H. Dernoeden.
       p. cm.
    "A CRC title."
    Includes bibliographical references and index.
    ISBN 978-1-4665-0992-4 (hardcover : acid-free paper)
    1. Creeping bentgrass. 2. Turfgrasses--Diseases and pests--Control. 3. Turfgrasses--Effect of stress on. 4. Turf management. I. Title.

SB608.T87D468 2012
584'.9--dc23   2012013490

Visit the Taylor & Francis Web site at
http://www.taylorandfrancis.com

and the CRC Press Web site at
http://www.crcpress.com

Dedication

This book is dedicated to my family for their patience and love; to my mentors

Dr. Jack Butler at Colorado State University and Dr. Noel Jackson at the

University of Rhode Island for the opportunity to study and for the knowledge

they imparted; to my graduate students for their devotion to academics and

research and for all that they have taught me; and to golf course superintendents

whom I admire and respect for their perseverance and exceptional work ethic.

Contents

The Author

Peter H. Dernoeden is Professor of Turfgrass Science in the Department of Plant Science and Landscape Architecture and joined the University of Maryland in 1980. He earned his B.S. and M.S. degrees at Colorado State University and his Ph.D. from the University of Rhode Island. He served in the U.S. Army Field Artillery from 1970 to 1973 in West Germany. Dr. Dernoeden's appointment includes research and extension components and he teaches a course in pest management strategies for turfgrasses. His research and extension programs involve turfgrass pathology, weed science, and the management of turfgrasses for low maintenance sites. Dr. Dernoeden discovered *Ophiosphaerella agrostis*, the incitant of dead spot disease, and he and his graduate students elucidated many biological aspects of brown patch, dead spot, dollar spot, Pythium-induced root dysfunction, red thread, spring dead spot, summer patch, and take-all patch. He studied and reported on ways to reduce fungicide usage based on soil fertility, irrigation management, and soil microbial interactions. Dr. Dernoeden's weed science program focused on selective control of annual and perennial grass weeds. With graduate students, a growing degree model to predict crabgrass emergence was developed and the emergence patterns of annual bluegrass were elucidated. Dr. Dernoeden pioneered the use of fine-leaf fescue species for use in naturalized areas on golf courses and other low maintenance sites.

Dr. Dernoeden initiated a lab service for turfgrass disease diagnostics in 1980. The lab has proved an invaluable teaching tool since he and his graduate students conduct all diagnoses. It was from his experience in providing a diagnostic service to golf course superintendents in the mid-Atlantic region that it became abundantly clear to him that the majority of summertime problems in golf turf, and in particular creeping bentgrass and annual bluegrass grown on golf greens, often were stress rather than disease related. He recognized that the decline of golf greens in summer was driven by high temperature stress in combination with other factors including poor growing environments, excessively wet soils, mechanical injury, improper irrigation practices, mineral and organic layers in the rootzone, the use and misuse of pesticides and plant growth regulators, pathogens, and other interrelated problems. This monograph is built around the creeping bentgrass decline complex in which abiotic and biotic factors combine to debilitate and damage golf turf.

Dr. Dernoeden is an author of more than 100 scientific journal articles, most of which were coauthored with his graduate students. He is co-author of *Compendium of Turfgrass Diseases*, 3rd ed. (American Phytopathological Society, 2005) and *Managing Turfgrass Pests* (CRC Press, 1994)

Dr. Dernoeden was elected Fellow of the American Society of Agronomy and the Crop Science Society of America (CSSA) in 2003 and 2007. He received the Fred V. Grau Turfgrass Science Award from the Turfgrass Science Division of CSSA in 2007. He also was recipient of the Outstanding Researcher Award from the Northeastern Weed Science Society (2007) and The Dean Gordon Cairns Award for Distinguished Creative Work in Agriculture from the College of Agriculture and Natural Resources at the University of Maryland (2010). In 2012, Dr. Dernoeden was selected to receive the Colonel John Morley Distinguished Service Award from the Golf Course Superintendents Association of America.

Preface

Creeping bentgrass declines in summer and loses its competitiveness with annual bluegrass, blue-green algae (aka cyanobacteria), moss, and pathogens as a result of numerous factors. *Complex* is a word used to describe a decline or death of plants by two or more causal agents, including both biotic (living) and abiotic (nonliving) agents. The cause of premature aging or senescence, and ultimately loss of stand density (i.e., the summer bentgrass decline complex), is a function of numerous and often interrelated factors such as high air or soil temperatures, drought, excessively wet or waterlogged soils, dense thatch or mat (i.e., organic) layers, shade, compaction, mechanical and other physical stresses, improper management practices, the use and misuse of chemicals, and pests. The identification of factors responsible for summer bentgrass decline is made extremely difficult because there generally is no single cause. Golf course superintendents often turn to diagnostic laboratories to help identify a possible pathogen involvement in the decline, especially turf grown on golf greens. Turf samples from greens sent in the summer to diagnostic laboratories often are found to be negative for a primary pathogen. When presented with a negative biotic diagnosis, superintendents often are perplexed, and sometimes are disappointed when a chemical treatment for the decline cannot be recommended. Ultimately, most creeping bentgrass and annual bluegrass problems can be linked to environmental stress factors, poor growing environments, mechanical injury and other abiotic stresses, improper management practices, and sometimes the use of certain pesticides. The goal of this monograph is to assist golf course superintendents, professional turfgrass managers, agronomists, diagnosticians, sales and industry representatives, students, golf course officials, and others interested in golf course management practices in recognizing the complicated interrelationships of the many potential stress factors that contribute to the summer bentgrass decline complex. Once the causal factors of the decline are pinpointed, this monograph can be used to guide golf course managers to those cultural and chemical measures that help alleviate the stress or malady and assist in rejuvenation of the turf.

This is the only monograph devoted to describing the nature of the summer decline complex and the management of creeping bentgrass on golf courses. Annual bluegrass is commonly found growing with creeping bentgrass on golf courses, but emphasis is placed herein on improving the competiveness of the creeping bentgrass in mixed stands. Regardless, in a variety of situations many of the cultural practices discussed are applicable to promoting the health of both turfgrass species. In this second edition, annual bluegrass problems and their management are given a much expanded consideration. Some topics of detailed description include selection of creeping bentgrass cultivars; cultural practices (e.g., mowing, fertility, irrigation, topdressing, coring and other aeration methods, etc.); approaches to limiting mechanical or physical stresses; and recent advances in the nature and management of selected diseases and soil-related maladies of creeping bentgrass.

Some of the diseases and maladies discussed include *Pythium*-incited root dysfunction; anthracnose basal rot; dead spot; dollar spot; brown patch and other *Rhizoctonia* diseases; take-all; yellow tuft; yellow spot; fairy ring; localized dry spots; plant parasitic nematodes; blue-green algae (aka cyanobacteria), moss, and black-layer; etiolated tiller syndrome and causes for yellow or chlorotic creeping bentgrass. It is recognized that there are no absolutes in turfgrass pathology, and focus is placed on common disease symptoms, dominant predisposing conditions, most likely hosts, and current cultural and chemical management strategies for each disease. Snow mold or winter diseases such as Microdochium patch (aka Fusarium patch or pink snow mold) and Typhula blight (aka gray snow mold) are not reviewed, but an excellent overview of these and other winter diseases can be found in the *Compendium of Turfgrass Diseases* and other resources. An overview of advances in biological disease control and the nature and proper use of fungicides is provided.

A weed control section is included which outlines the use of herbicides for the control of annual grass, broadleaf, and other weeds in creeping bentgrass. The weed section also includes a discussion on using diseases specific to annual bluegrass in mixed stands with creeping bentgrass and a description on how to deal with the problem of soil residues of persistent herbicides. The final sections deal with the selection and use of plant growth regulators for annual bluegrass suppression and seedhead control; use of fumigants, herbicides, and plant growth regulators for renovation purposes; and selected invertebrate pests of creeping bentgrass.

Acknowledgments

First Edition

I am grateful to the following individuals for their assistance in the preparation and review of this monograph: Darin Bevard, senior agronomist, U.S. Golf Association (USGA) Mid-Atlantic region; Keith Happ, senior agronomist, USGA Mid-Atlantic region; and Stephen Potter, certified golf course superintendent (CGCS). Portions were edited by James B. Beard. I am especially grateful to Stanley Zontek, USGA Mid-Atlantic region director, for his many comments and suggestions and for sharing his decades of experience in the management of creeping bentgrass in the Northeastern, Mid-Atlantic, and Mid-Continent regions of the United States.

Second Edition

I greatly appreciate the invaluable reviewing assistance rendered by Cale Bigelow, Purdue University; Michael Fidanza, Pennsylvania State University; John Kaminski, Pennsylvania State University; Steven McDonald, Turfgrass Disease Solutions, LLC; Derek Settle, Chicago District Golf Association; Patricia Vittum, University of Massachusetts; and Stanley Zontek, Mid-Atlantic region director, U.S. Golf Association (USGA) Green Section. Their experience and insightful suggestions and edits have substantially contributed to this effort. I thank Darin Bevard and Keith Happ, senior agronomists, USGA Mid-Atlantic region; Nathaniel Mitkowski, University of Rhode Island; and Lane Tredway, North Carolina State University for providing technical information. I learned a great deal about golf course management from Stephen Potter, certified golf course superintendent (CGCS), Woodholme Country Club, and Stanley Zontek, and thank them for their friendship.

1

The Nature of Summer Stresses

Weather extremes can damage creeping bentgrass (*Agrostis stolonifera*; synonym = *Agrostis palustris*) and annual bluegrass (*Poa annua*) turf at almost any time of year. Except for crown hydration, desiccation, and direct ice kill in winter (i.e., anoxia), summer stresses are among the most challenging problems for golf course superintendents. Golf greens in particular face many potential problems during the summer. Golf greens may be simultaneously injured by a combination of biotic (i.e., living stress agents such as fungal pathogens) and abiotic (i.e., nonliving stresses such as heat, shade, water-saturated soil, drought, and mechanical injury) stresses that produce similar symptoms. This combination of factors (i.e., summer bentgrass decline complex) is what makes diagnosing problems of creeping bentgrass greens, tees, and fairways so difficult.

Several important summer diseases such as brown patch (*Rhizoctonia solani*), dollar spot (*Sclerotinia homoeocarpa*), and Pythium blight (*Pythium* spp.) initially produce distinctive circular spots or patches. Diseases such as anthracnose (*Colletotrichum cereale*; synonym *C. graminicola*), Pythium-incited root diseases (*Pythium* ssp.), and the feeding activity of plant parasitic nematodes generally appear as a nonuniform thinning and possibly death of turf. Abiotic stresses that also cause nonuniform thinning include indirect heat stress; poor air and water drainage, anaerobic soil, and black-layer; wet wilt and scald; drought and hydrophobic soils; and shade and other elements related to poor growing environments. More turf on greens in humid regions is damaged in summer by excessively wet or waterlogged soil, rather than by drought or disease. When two or more abiotic or biotic stresses deleteriously affect a stand of turf at the same time, the decline or malady may be referred to as a *complex*. Environmental stresses often predispose turf to more aggressive activity from pathogens. For example, dead spot (*Ophiosphaerella agrostis*) is much more severe in sunny sites when soils are dry, and anthracnose is more damaging in turf grown with low nitrogen (i.e., <2 lb N/1000 ft^2/year; <98 kg N/ha/year) inputs combined with low mowing. The discussion that follows will focus on both the nature and cultural management of summer stresses. Implementing many of the cultural practices discussed, however, may require increases in expenditures for labor and equipment. Due to the importance of putting greens to the game of golf, as well as the reputation of the golf course, these increased costs should be considered money well spent. Although some of the options described will not accommodate all budgets, any suggested practice or combination of options should help to improve creeping bentgrass vigor, quality, and performance.

Wilt and Drought Stress

Creeping bentgrass wilts rapidly in summer, especially on days when the sky is bright and clear and humidity is low. Similarly, during periods of heat stress and in relatively dry rootzones, bentgrass can rapidly wilt. When turf wilts in response to dry soil, low humidity, or both, along with windy air conditions, it develops a smoke-gray or bluish-purple color and is subjected to foot printing (Figures 1.1, 1.2, and 1.3). As discussed below, thatch-mat

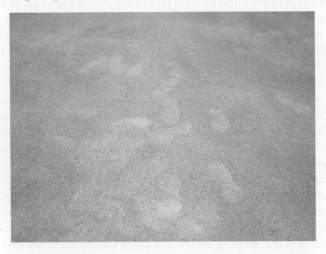

FIGURE 1.1
Footprinting due to wilt on a creeping bentgrass golf green.

FIGURE 1.2
Bluish-purple color in turfgrasses is a symptom of wilt. One clone stands out as more wilt tolerant. (Photo courtesy of S. McDonald.)

FIGURE 1.3
Blue-gray color of wilt on green and surrounds.

layers can dry rapidly in summer and induce drought dormancy in creeping bentgrass. Wilt stress can be alleviated by syringing (Figure 1.4). Syringing involves applying a thin film of water on leaves, without delivering so much water that the underlying thatch, mat, and soil become wet. The evaporation of water cools leaves, allowing stomates (i.e., pores on leaf surfaces) to open. Syringing allows turf to survive that day or until morning when appropriate irrigation programs can be scheduled to replenish water throughout the rootzone (see Chapter 2 for more information). Drought-stressed or wilted bentgrass and annual bluegrass, which is not properly irrigated in a

FIGURE 1.4
Syringing alleviates wilt and involves applying a light film of water to the canopy.

timely manner in summer, will turn tan or brown and enter a dormant state (Figure 1.5). When detected early and properly addressed, creeping bent-grass generally recovers fairly quickly from wilt or even short-term drought dormancy. Annual bluegrass, however, wilts rapidly and dies rapidly (i.e., checks out) on a hot and windy day (Figure 1.6). When properly irrigated and otherwise managed, creeping bentgrass becomes a very tough and resilient grass that can effectively compete with annual bluegrass in the summer.

FIGURE 1.5
Portions of a creeping bentgrass fairway that are drought dormant due to lack of good cover-age by the irrigation system and lack of labor to hand water wilt-prone areas.

FIGURE 1.6
Annual bluegrass wilts and can die rapidly (i.e., "checks out") on hot and dry days.

Heat Stress

A common cause for turf senescing (i.e., aging prematurely) and thinning on greens in summer is heat stress (i.e., physiological impairment), especially when soils are either very dry or saturated with water. Heat stress injury occurs in random patterns but tends to be more severe on high areas and southwest-facing exposures as well as in areas where water drains. Symptoms of heat stress can include slow growth and low clipping yields, chlorosis (i.e., yellowing), and browning and thinning of the stand. The pattern of turf injury could be confused with disease. This is particularly true for creeping bentgrass stands that are less than 1 year old.

Cool-season grasses begin to experience heat stress when air and soil temperatures consistently exceed 86°F (30°C). Stress to creeping bentgrass roots, however, can begin at soil temperatures as low as 75°F (23°C). There are two types of heat stress: direct and indirect (Fry and Huang, 2004). As air and soil temperatures rise above 86°F for extended periods, turf can suffer from *indirect heat stress*. This stress is not immediately fatal but can lead to deterioration of stand density and turf loss. Indirect heat stress often causes a chlorosis or yellowing and thinning of the stand. Direct heat stress occurs in response to a rapid increase in temperatures in excess of 104°F (40°C) for a relatively short period. The most common type of direct heat stress is scald, which is described below. Plant physiologists often use the term *supraoptimal* to describe high-temperature stress (Fry and Huang, 2004). The term *supraoptimal temperature stress* technically refers to any temperature above the optimum for growth but implies a significant amount of heat stress. High-temperature stress in summer results in an increase in respiration and a decrease in carbohydrate production (i.e., photosynthesis). Carbohydrates are an essential source of energy and are used to sustain shoot and root growth. An imbalance in respiration (i.e., converting carbohydrates to energy) that causes a reduction in photosynthesis (capturing sunlight and producing carbohydrates) can result in a rapid depletion of carbohydrates. Significant depletion of carbohydrates weakens the ability of plants to heal wounds and repair themselves, particularly on greens where the grass is cut very low.

Soil Temperature and Rooting

Creeping bentgrass produces a larger and deeper root mass than annual bluegrass growing in sand-based rootzones (Lyons et al., 2011). Soil temperature is more critical to root longevity than air temperature and thus has a dramatic effect on the root systems of grasses in summer. Root mortality naturally occurs in both annual bluegrass and creeping bentgrass in summer in response to higher soil temperatures. Although high-density cultivars (e.g., A-4) experience root mortality in summer, they are able to maintain a larger root mass at deeper depths in summer compared to annual bluegrass and conventional bentgrass cultivars (Lyons et al., 2011). Very high soil temperatures cause dysfunction of cool-season turfgrass roots emanating from the crown and possibly death. Root death occurs before death of aboveground leaves and sheaths. Thus, it is not unusual for the foliage to appear reasonably healthy, while large portions of the root system are being lost. For example, creeping bentgrass and annual bluegrass plants growing on golf greens normally have pearly white roots that extend 4 to 6 inches (10 to 15 cm) or deeper in soil in late spring. Often, however, roots turn brown and are primarily restricted to the upper 0.5 to 2 inches (1.2 to 5 cm) of soil by mid-summer. By late summer, even in open and well-drained sand-based greens, as much as 50% to 60% of a creeping bentgrass root system is likely to be found in the upper 2.6 inches (6.5 cm) of soil (Fu and Dernoeden, 2009b). During sunny and hot weather, soil temperatures generally will be a few degrees higher than ambient air temperature, especially in slow or poorly drained greens. Rapidly draining sand-based rootzones, however, will have soil temperatures similar or slightly higher than ambient (Fu and Dernoeden, 2009b). Regardless, temperatures in thatch-mat layers of sand-based greens can be several degrees higher than air temperature on sunny and hot days. These warmer soils often are not subject to much night cooling during hot summer months in some regions (especially transition and southern regions in the United States). The optimum temperature for root growth of cool-season grasses ranges from 50 to 70°F (10 to 21°C) (Beard, 1973; Fry and Huang, 2004). New root initiation ceases at a soil temperature of 80°F (27°C), and the natural aging process (i.e., senescence) of the existing root system begins. When soil temperatures exceed 86°F (>30°C, i.e., indirect heat stress) for prolonged periods, root hairs die and roots begin to turn brown and fail to function properly (i.e., root dysfunction). Loss of roots probably results in part from an inhibition of carbon supply because photosynthesis (carbon-producing process) declines as temperatures increase in summer and respiration increases (carbon-consuming process) (Huang and Liu, 2003). Plants that lose a major portion of their root system in summer, however, sometimes begin to generate new roots, despite high soil temperatures. Roots that regenerate in the summer are most often found growing in aeration holes. On sunny days when air temperatures exceed 90°F (32°C), the temperature in the upper 2 inches (5 cm) of wet soil can range from 95 to 100°F (35 to 38°C). Turf grown in hot and wet soils in summer often turns chlorotic and roots die rapidly, especially annual bluegrass (Figure 1.7). Thus, it is the combination of air and soil heat stress and wet soils that do not cool sufficiently at night which can be especially lethal to turf.

FIGURE 1.7
Hot and wet soils cause chlorosis and rapid death of roots in summer.

Wet Soils in Summer

Excess water kills more grass on golf courses in summer than too little. Excessive irrigation or heavy rain and thunderstorm activity, particularly on poorly drained soils during summer can lead to problems such as scald and wet wilt. Roots grow through pore spaces in soil. When water saturates soil, air is displaced from pore spaces. Roots need oxygen to survive, and when pore spaces are filled with water there is not enough soil oxygen to physiologically sustain the root system. Heat stress injury to the root system is accelerated in hot and wet soils during summer. Wet soils accumulate heat slowly but retain more heat for longer periods, particularly on golf greens or native soils where water does not drain rapidly. Heat retention is more pronounced in fine-textured native soil compared to sand. As previously noted, water saturation is less likely to occur in sand-based rootzones that drain rapidly (>12 in/hr; >30 cm/hr).

As golf greens age, organic matter accumulates in the upper portion of the profile which slows water and air infiltration. Poorly constructed greens built with highly variable or improperly selected sand particle sizes (i.e., excess fines), sod layers where washed sod was not used, and fine-textured native soil "push-up" greens are most likely to have low water infiltration and percolation rates. Even push-up greens that have received decades of sand topdressing can become saturated when water perches at the interface between sand and underlying native soil. Plants can condition themselves to tolerate high temperatures, but in most cases large portions of root systems die in response to sustained supraoptimal soil heating, low soil oxygen levels, and elevated carbon dioxide levels. Bentgrass can survive many days of water inundation during cool to cold periods from autumn to early spring. This is due in part to cold water holding more oxygen than warm water and because air temperatures are much cooler. When water fills pores or inundates turf during high-temperature stress periods, the result is injury ranging from chlorosis to direct kill of plants. Scald and wet wilt are two damaging phenomena associated with water-saturated soils in summer.

Scald

When water puddles due to heavy rain, excess irrigation, or poor surface or internal drainage and inundates plants during sunny and hot weather, turf can be killed or severely damaged in just a few hours (i.e., direct heat stress). Basically, standing water rapidly absorbs lethal levels of heat from the sun on hot and sunny days. This phenomenon is called *scald*, and it occurs in response to rapid heat buildup (>104°F; >40°C) in a few hours, which causes protein denaturation followed by plant death (Figures 1.8 and 1.9) (Beard, 1973). Water that inundates plants displaces oxygen in soil and around plants. Hence, oxygen depletion (hypoxia) also plays a key role in plant death during a scald event. Standing water should be manually removed (i.e., squeegee) from greens as soon as possible in the summer (Figure 1.10). The term *scald*, however, is a misnomer because turf is not really scalded or burned but dies from a combination of direct heat stress and oxygen deprivation. The injury pattern may be random, but damage is most severe in low areas where water puddles. Scalded tissues become bleached and matted and can have a texture similar to papier-mâché (Figure 1.11). Annual bluegrass

FIGURE 1.8
Scald on a creeping bentgrass tee and fairway.

FIGURE 1.9
Water inundating low areas following a heavy rain on a hot day can result in scald.

FIGURE 1.10
Squeegee water off greens before the sun returns. (Photo courtesy of J. Haskins.)

is more susceptible to scald than creeping bentgrass, but both species are killed rapidly under the right conditions. It is extremely important to manually remove (i.e., squeegee) standing water (i.e., puddles) or casual water off golf greens following thunderstorms and prior to the return of sunshine on hot summer days. Areas chronically prone to water puddling need to be addressed by installing drains or possibly recontouring to ensure rapid and effective surface water drainage (Figure 1.12).

FIGURE 1.11
Scalded tissues become matted and have a papier-mâché texture.

FIGURE 1.12
A chronically wet area containing casual water caused a scald-like event in summer. This problem was resolved by installing a drain.

Wet Wilt

Wet wilt occurs when there is adequate soil moisture yet roots cannot absorb water fast enough to meet the transpirational (i.e., the natural cooling process whereby water moves from roots to shoots and evaporates through open stomates on leaves) needs of a plant. This can occur in plants with a limited or shallow root system on sunny, warm-to-hot days when there is low humidity and windy conditions. Warm, dry, and windy conditions cause stomates to close, which stops water uptake and transpiration from occurring. When stomates close during dry and windy periods, an internal water deficit may occur which can be lethal to plants, despite the presence of adequate soil moisture.

Wet wilt also occurs during hot and humid periods when soils are waterlogged (Dernoeden, 2006). Water in or on the surface of thatch or mat absorbs heat from the sun. Heat in water is transferred to the soil and stored. This is why the upper 1 to 2 inches (2.5 to 5 cm) of soil can have temperatures exceeding 95°F (>35°C) on sunny days, when ambient temperatures are about 90°F (32°C). Due to a combination of heat stress from high air and soil temperatures, low soil oxygen levels, and stomatal closure, roots are unable to absorb water. Turf grown on greens, tees, fairways, and even in roughs can be killed in just a few hours. On affected bentgrass greens, turf initially appears brownish and water soaked, and turf thins out in irregular patterns (Figure 1.13). Damage often is more severe in surface water drainage patterns (Figure 1.14). The combination of heat stress, waterlogged soils (and poor gas exchange), and mechanical injury from mowing causes leaves to collapse. In some instances the turf almost literally "cooks and suffocates" in hot and wet soils.

Fans can be especially beneficial for cooling and drying purposes during wet-wilt events. Fans should be run 24 hours a day during wet-wilt events to improve air circulation. Air movement provided by fans reduces boundary layers of water vapor around leaves and helps to speed cooling of plants and soil as well as drying of soil. Leaf and sheath cooling promotes the opening of stomates. Once stomates open, water can move from soil into roots and shoots and out stomates to assist in plant cooling and drying of soil. Syringing is extremely important during hot and dry periods.

FIGURE 1.13
Wet wilt in creeping bentgrass appears brownish and water soaked.

FIGURE 1.14
Wet-wilt injury often follows surface water drainage patterns.

Syringing also can be helpful during periods of wet wilt, but only if water evaporates off leaves between syringes. Syringing should be done with a handheld hose rather than with the overhead irrigation system to avoid applying too much water and making the situation worse. If water does not evaporate (e.g., on high-humidity days) fairly rapidly, stop syringing because more harm than good may occur. Syringing greens in concert with operating fans can be especially effective in promoting evaporation and transpirational cooling as well as rootzone cooling (Guertal et al., 2005).

Roots need oxygen to survive, especially in waterlogged soils. Thus, improving soil aeration also is needed. Cultural practices that promote rapid soil water and air exchange are necessary to maintain plant health. Greens affected by wet wilt should be solid tine aerated or spiked to improve gas exchange and speed drying, but only when it is possible to operate equipment safely on putting surfaces (Figure 1.15). Aerating or spiking in the heat of the day during a wet-wilt event can cause massive damage to putting greens. These operations are best performed in the evening or very early in the morning when cooler air temperatures prevail. Where rooting is shallow, extreme care should be taken to avoid sod lifting or surface disruption. In some cases heavy equipment should be avoided and localized areas cored by hand with a pitchfork or similar device. It is very difficult to separate damage occurring due to wet wilt versus indirect or direct heat kill (i.e., scald) as these conditions can be interrelated. While imperfectly understood, the wet-wilt phenomenon appears to bridge the definitions of indirect and direct heat stress.

FIGURE 1.15
Green damaged by spiking in late morning during a wet-wilt event.

Wet Organic Layers

Water management is critical to maintaining turf vigor in summer, especially on chronically wet greens. In some cases turf managers can control the amount of water applied to greens. In other situations, such as rainy weather, they cannot. A scald-like condition also can occur where there is a significant thatch or mat layer. Thatch is a layer of organic matter that develops on the surface of turfs. On golf greens, and in some cases tees and fairways, sand topdressing, vertical cutting, and coring are used to mix sand with organic matter, and this mixture is referred to as mat. *Thatch-mat* will be used here to refer to organic layers in creeping bentgrass turfs that are topdressed with sand. Keep in mind that in the absence of topdressing and mixing of sand in the organic matter, you are dealing with thatch layers. A more detailed discussion of thatch-mat appears in Chapter 2.

Sunny, hot days that immediately follow a heavy rain event can cause excessive heating in a water-saturated thatch-mat layer (including layers ≤0.5 inches; ≤1.25 cm), which results in either indirect heat stress or wet wilt. Green plants in core aeration holes are a good indicator that the surrounding tan or brown-colored plants were injured by supraoptimal heating due to excessively wet thatch or thatch-mat layers (Figure 1.16). The green plants in aerifier holes grow in the absence of any significant organic layer so that water drains downward more freely and soil gas (i.e., oxygen, carbon dioxide, others) levels are in better balance. Excessive or improper irrigation during hot summer periods, which causes water to accumulate in low areas,

FIGURE 1.16
Green plants in coring holes are an indicator that the thatch-mat layer became very wet or very dry on a hot day.

also can result in scald. In less severe cases involving wet soils, and hot and humid conditions, turf may turn yellow (i.e. chlorotic). Chlorotic leaves may be in a spindly, succulent, or even flaccid state. Chlorotic annual bluegrass growing on greens often loses its root system rapidly when soils are excessively wet during sunny and warm to hot summer periods. Spindly and chlorotic bentgrass and annual bluegrass canopies are quite susceptible to mechanical injury. When turfgrasses lose their natural green color, this is a clear indicator that a problem exists. See Chapter 3 for more information about chlorosis and its causes.

Regular sand topdressing is the key cultural approach to managing organic layers and mitigating the effects of chronic or excessive surface layer wetness. Insuring the proper dilution of sand with organic matter in thatch-mat layers improves the growing environment where stems and roots live. As discussed below, this requires an aggressive combination of sand topdressing, vertical cutting, and coring on a routine basis. In the short-term, fans, spiking, and solid tine coring are employed to speed organic layer drying and to promote aeration during wet periods.

Dry Organic Layers

Brown grass on greens with green plants in coring holes in summer also can be the result of the thatch-mat layer rapidly drying out on a hot day, especially during windy periods. Turf under wilt stress initially appears purplish or smokish-gray in color. In a matter of minutes tissues can reach the permanent wilting point and turn brown. This brown creeping bent-grass turf initially is dormant and not dead and can be revitalized by good syringing and/or hand-watering practices. Conversely, annual bluegrass may die (i.e., "checks out") suddenly during dry summer periods (Figure 1.6). Coring to the thatch-mat depth with narrow diameter tines in conjunction with syringing can help stimulate a more rapid recovery. Greens that go long periods without proper coring and topdressing develop thatch layers that can become hydrophobic (Figure 1.17). Use of a soil wetting agent or surfactant will be required if the thatch-mat layer is hydrophobic. On areas with a chronic history of hydrophobic problems, programs involving wet-ting agents should begin prior to the onset of drought symptoms (i.e., mid-spring). Hydrophobic thatch layers are extremely difficult to rewet once they dry in the summer.

There is a human tendency to overwater wilted creeping bentgrass on hot summer days. In all too many cases, brown and wilted bentgrass is kept too wet on hot summer days, and plants can be killed by heat stress mechanisms similar to that associated with wet wilt or scald. The key is careful water management and avoiding saturating the thatch-mat layer during hot and humid periods in summer. This involves careful syringing as discussed in detail in Chapter 2. Damage or death of brown bentgrass turf caused by wilt followed by excessive wetness on hot and sunny days often results in matting and bleaching of dead tissue. These tissues are rapidly invaded by secondary organisms like *Curvularia* spp., *Cephalosporium* sp., and *Leptosphaerulina* spp.

FIGURE 1.17
This green was not properly cored and topdressed for several years, and the thatch layer (insert) became hydrophobic.

The presence of the aforementioned organisms (aka senectophytes) on collapsed or bruised tissues indicates to the diagnostician that the turf was damaged or killed by abiotic factors and not a primary pathogen. Senectophytes are discussed in more detail in Chapter 3.

Sand–Soil Layers

There are other types of layers found in golf greens and most negatively impact turf health. Most notably, layering of different soil types, different sand topdressings, sand or soil layers from unwashed sod, sand topdressing overtop of native soil or "push-up" greens, and organic layers between sand layers from heavy and infrequent topdressing (Figure 1.18). In each case, the downward movement of water and air is impeded and may become perched (i.e., a perched water table). When water encounters a layer of a distinctly different soil texture it begins to move laterally or upwards rather than downward. Once the entire area above the restricted layer becomes saturated, water may then begin to move downward. This perching of water can last hours or days. The result usually is poor root development between layers (Figure 1.19). The more layers there are, the longer water remains perched. In some cases, water between layers fills all the pore spaces leading to an anaerobic (i.e., absence of oxygen) condition known as black-layer (see Chapter 3). Roots cannot survive in the absence of oxygen in pore spaces, and gases released in the black-layer are lethal to roots. Moss, blue-green algae, and disease are

FIGURE 1.18
Layers of varying types of sand topdressing cause water to become trapped or perched between layers.

FIGURE 1.19
Different soil types in this sod layer caused water to perch and restricted rooting below.

other consequences of layering and perched water tables. It is imperative to avoid varying the particle sizes of sand topdressing from one year to the next (see Chapter 2). Topdressing mixes and sands must be tested periodically to ensure that they have the proper physical properties for use on golf greens. Assuming a sand-based green has been properly constructed with appropriate materials, the topdressing selected, or something similar, must be used consistently without switching textures at any time. Topdressing lightly and frequently averts development of a topdressing/thatch layer. Using washed sod or custom growing sod on the greens mixture it is to be planted on also avoids damaging layers. Additional information about organic and inorganic layers and their management can be found in Chapter 2.

Shade and Poor Air Circulation

Almost every golf course has one or more greens that are low lying, shrouded by trees and brush, or surrounded by mounds or hills (Figure 1.20). These "pocket greens" are generally among the poorest in appearance and putting quality, particularly during summer months. Because annual bluegrass is more photosynthetically efficient in shade, it invariably outcompetes creeping bentgrass and dominates in shaded environments. Shade filters out much of the photosynthetically active wavelengths before they reach the turf canopy (Danneberger, 1993). Hence, the primary problem presented by shade is simply insufficient sunlight needed for photosynthesis, which results in

FIGURE 1.20
Pocket greens shrouded in trees and brush and poor air circulation usually are fraught with problems in the summer.

poor root development and generally weaker plants more prone to mechanical injury. Other factors in the shade microenvironment such as very high humidity, long leaf wetness durations, poor air circulation, and chronic soil wetness contribute to the problem.

Light, and in particular morning sunlight in the summer, is essential for growth and normal development of turfgrasses. Morning sunlight in the summer helps to speed drying of the canopy and promotes the opening of stomates, thus triggering transpiration as described below. In summer, relatively cool morning temperatures also are more conducive to photosynthesis (carbohydrate producing); whereas, as temperatures rise later in the day they promote respiration (carbohydrate consuming). Simply stated, sunlight reduction results in lower photosynthetic activity and therefore lower carbohydrate production. Lower plant energy (i.e., carbohydrates) levels and other factors contribute to poor root development in shaded turf and leaves are more succulent. Succulent leaves are more easily injured by wear and environmental stress, and they are more easily penetrated by fungal pathogens. Shade also results in less tillering and therefore inferior density, and leaves of shaded plants become longer and thinner. Because turf growing in shade is less dense, even small losses in plant populations or cover can result in a significant reduction in playability and aesthetic quality. The shade microenvironment promotes disease because leaf wetness durations are longer, and there is a higher relative humidity due to less air movement. Furthermore, soils in shaded environments retain water for longer periods, which results in less oxygen and more carbon dioxide in the rootzone, further contributing to poor root development and survival.

Evapotranspiration is a term that combines evaporation of water from the canopy and thatch-mat with the loss of water vapor through stomates from the transpiration process in plants. In the summertime, when air temperatures rise, turf on shaded greens exhibits lower evapotranspiration rates. This is due to higher relative humidity levels and longer leaf wetness duration periods in shade. Because of low light intensity, wet leaves, and high relative humidity, stomates do not open on leaves or they open for only short periods. When stomates are not open, plants cannot properly cool themselves (i.e., transpiration), and there is restricted exchange of gases such as oxygen and carbon dioxide in leaves. Hence, during summer stress periods, plants on shaded greens are more adversely affected by high-temperature stress than turf grown in the full sun. Heat buildup due to impaired transpiration combined with poor rooting, oxygen deprivation in wet soils, and succulent leaves all result in decline of putting surfaces, even in the absence of disease. These same factors, however, also promote disease and as such turf grown on shaded greens is generally weaker and is much more vulnerable to disease, heat stress, and mechanical injury.

A major step in the remedy of shade stress is simply to remove as many trees and as much underbrush as possible (Figure 1.21). It is extremely important that greens receive as much direct sunlight as possible. Ideally, greens should receive 4 to 5 hours of morning sunlight and 9 to 10 hours of sunlight per day between spring and autumn. There are expert arborists who can survey trees around pocket greens and make recommendations on which trees are critical for removal. This service can be expensive but well worth the cost in the long-term health of creeping bentgrass greens. Morning sunlight is recognized as being most important to grass. Morning sun promotes

FIGURE 1.21
Tree and brush removal are necessary for improving the growing environment of pocket greens.

more rapid drying of the canopy and opening of stomates. Plants on greens that receive morning sun, but afternoon shade, usually produce better root systems than plants exposed to morning shade and afternoon sun. The optimum temperature for photosynthesis ranges from 60 to 75°F (16 to 24°C). Temperatures in summer are cooler in the morning, which would promote more efficient photosynthesis as well as evapotranspiration. By mid-morning, summer temperatures are likely to be above optimum for photosynthesis. By early afternoon, respiration increases and stomates are more likely to close in response to increasing levels of heat stress.

Improving air movement is strongly recommended for shaded or pocketed greens. The movement of air over plant surfaces promotes drying of leaf surfaces, lowers relative humidity in the canopy, and cools plants. Air movement provided by fans reduces leaf wetness duration and helps to dry surfaces thus reducing disease pressure and improving wear tolerance of golf greens. The cooling effect opens stomates, which promotes transpiration and gas exchange with the atmosphere. The natural cooling effect provided by transpiration improves plant vigor and helps to reduce soil wetness. Fans help to improve air circulation; therefore, they promote transpiration and evaporation. Large oscillating fans positioned close to the ground provide more effective cooling and drying than fans mounted above the putting surface (Figure 1.22). Fans in pocketed areas should be positioned within 20 feet (6 m) of the green and mounted 10 feet (3 m) or lower to maximize air movement across the green (Moeller and Chapin, 2011). Fans should be angled so that air moves across and not above the turf canopy. During fan installation, place irrigation marking flags throughout the green at about 3 inches (7.5 cm) above the canopy and adjust fan angle according to flag movement. Fans

FIGURE 1.22
Fans are effective in cooling and speeding drying of the canopy and underlying soil.

should be positioned so that air will drain into an open or cleared area. Fans should be allowed to run continuously during periods of excessive heat and humidity. Some golfers as well as homeowners in golf course subdivisions find fans to be noisy and unsightly. If fans cannot be used 24 hours a day, their operation from dusk to dawn will provide important benefits in cooling, drying, and promoting root health and longevity.

Mowing height should be increased and nitrogen fertility reduced on shaded turf. Increasing mowing height will increase leaf surface area and the ability of plants to produce carbohydrates. Nitrogen stimulates leaf extension rate and promotes tissue succulence. Research indicates that the total amount of nitrogen applied to fully shaded greens should be about half of that applied to greens grown in full sunlight (Bell and Danneberger, 1999). Finally, the plant growth regulator trinexapac-ethyl has been shown to improve turf quality in shaded environments (Steinke and Stier, 2003). It is theorized that gibberellic acid (GA) levels increase under low light intensity, causing leaves to grow more rapidly, and leaves become "leggy" and weak over time. Trinexapac-ethyl blocks GA production and slows leaf growth as well as carbohydrate depletion, thus helping to maintain higher turf quality in shaded environments. See Chapter 8 for more information about using plant growth regulators.

Chronically wet soils in shaded environments also must be addressed. It is best to avoid irrigating chronically wet greens, particularly in summer. Syringing by hand should replace using the overhead irrigation system when soils are wet. Spiking and small-diameter solid tine aeration of wet, shaded greens improves gas exchange and promotes soil drying (Figure 1.23). Normally, solid tines are used in summer or when soils are wet. Coring with narrow diameter tines in moist, but not excessively wet, soils causes

FIGURE 1.23
Solid tine spiking in summer promotes surface drying and improves soil aeration.

little disruption to the putting surface while promoting a more healthy turf. Subsurface air movement equipment attached to drainage pipes can remove some water from high sand content greens appropriately equipped for these devices. Variation in elevations on greens complicates soil hydraulics in the rootzone, and these devices are not always effective in removing water from the entire putting green complex. Damaged or plugged drainage lines may need to be replaced or a newer system installed. In extreme cases, greens must be rebuilt using accepted methods. It always is important to keep thatch and mat layers to a minimum (i.e., <0.5 in; <1.25 cm), especially where soils are chronically wet.

Prolonged Overcast Weather

Creeping bentgrass and annual bluegrass plants replace most if not their entire root system annually, with most root initiation occurring in the autumn (Beard, 1973). Creeping bentgrass and annual bluegrass, however, produce a majority (i.e., both root number and length) of their root systems in the spring. Prolonged periods of overcast and rainy weather in mid-to-late spring can contribute to poor root development. This condition is not unlike the debilitating effects of chronic shade. That is, there is a reduction of photosynthetic activity and there may be periods of soil oxygen deprivation, both of which would retard root development. Sunny and hot weather following an extended rainy period can result in indirect heat stress, chlorosis, wet wilt, scald, and black-layer. After temperatures moderate in late summer or early autumn, excessively wet and long periods of overcast weather can result in elevated stems, a proneness to scalping, delayed root development, and chlorosis in creeping bentgrass. Conversely, extremely dry conditions in autumn can retard root initiation and elongation prior to winter. The problem can be even more acute if a drought should recur the following spring. This can lead to problems with autumn or early spring weed control programs, because poorly rooted creeping bentgrass and annual bluegrass generally are less tolerant of herbicides.

Summary

- Proper water management is critical to maintaining quality creeping bentgrass in the summer.
- Too much, rather than too little, water kills more grass in summer.

- Soil and air temperatures exceeding 86°F (30°C) for prolonged periods cause indirect heat stress.
- Scald is a form of direct heat stress, which occurs when plants are inundated with water on hot and sunny days.
- Wet wilt in summer occurs during periods of heat stress, high humidity, and when soils are waterlogged.
- Excessively wet or dry thatch-mat layers can cause turf to die rapidly in summer, especially on golf greens. Green plants in coring holes distributed throughout a dormant or dead turf are a good indicator that the thatch-mat layer became either very wet or very dry during a period of heat stress.
- Water management methods to promote soil drying in summer include the following:
    - Turn off the overhead irrigation system and syringe or hand water as appropriate.
    - Install and run fans continuously until soils are dry.
    - Spike, core, and use other forms of aeration.
- Creeping bentgrass cannot compete with annual bluegrass in shaded environments.
- Turfgrass plants in shaded environments generally are poorly rooted and have lower carbohydrate reserves, fewer tillers, and more succulent leaves.
- Soils on shaded greens remain wet for longer periods, which can result in supraoptimal soil temperatures in summer and lower oxygen and higher carbon dioxide levels in the rootzone.
- Shade management strategies include the following:
    - Promote full sun by removing trees and brush
    - Use fans to improve air circulation
    - Improve surface and subsurface drainage
    - Increase mowing height and reduce nitrogen fertility versus full sun greens
    - Avoid frequent programmed nightly overhead irrigation
    - Syringe or hand water as needed to promote soil drying and transpiration
    - Frequently apply a plant growth regulator like trinexapac-ethyl to improve summer quality of turf grown in shaded environments
    - Spike or solid tine when greens are excessively wet, especially in summer
    - Control blue-green algae (aka cyanobacteria) and thatch-mat

Bibliography

Beard, J.B. 1973. *Turfgrass: Science and Culture*. Prentice Hall, Englewood Cliffs, NJ.

Beard, J.B., and W.H. Daniel. 1965. Effect of temperature and cutting on the growth of creeping bentgrass (*Agrostis palustris* Huds.) roots. *Agron. J.* 57:249–250.

Beard, J.B., and W.H. Daniel. 1966. Relationship of creeping bentgrass (*Agrostis palustris* Huds.) root growth to environmental factors in the field. *Agron. J.* 58:337–339.

Bell, G., and K. Danneberger. 1999. Managing creeping bentgrass in shade. *Golf Course Management* 67(10):56–60.

Danneberger, T.K. 1993. *Turfgrass Ecology and Management*. Franzak & Foster, Cleveland, OH.

Dernoeden, P.H. 2006. Understanding wet wilt. *U.S. Golf Association Green Section Record* 44(2):7–9.

Fry, J., and B. Huang. 2004. *Applied Turfgrass Science and Physiology*. Wiley, Hoboken, NJ.

Fu, J., and P.H. Dernoeden. 2008. Carbohydrate metabolism in creeping bentgrass as influenced by two summer irrigation practices. *J. Amer. Soc. Hort. Sci.* 133:678–683.

Fu, J., and P.H. Dernoeden. 2009a. Carbohydrate level, photosynthesis, and respiration in creeping bentgrass as influenced by spring and summer coring. *J. Amer. Soc. Hort. Sci.* 134:41–47.

Fu, J., and P.H. Dernoeden. 2009b. Creeping bentgrass putting green response to two summer irrigation practices: Rooting and soil temperature. *Crop Sci.* 49:1063–1070.

Fu, J., and P.H. Dernoeden. 2009c. Creeping bentgrass putting green responses to two irrigation practices: Quality, chlorophyll, canopy temperature and thatch-mat. *Crop Sci.* 49:1071–1078.

Guertal, E.A., E. van Santen, and D.Y. Han. 2005. Fan and syringe application for cooling bentgrass greens. *Crop Sci.* 45:245–250.

Huang, B., and X. Liu. 2003. Summer root decline: Production and mortality for four cultivars of bentgrass. *Crop Sci.* 43:258–265.

Huang, B., X. Liu, and J.D. Fry. 1998a. Effects of high temperature and poor soil aeration on root growth and viability of creeping bentgrass. *Crop Sci.* 38:1618–1622.

Huang, B., X. Liu, and J.D. Fry. 1998b. Shoot physiological response of two bentgrass cultivars to high temperature and poor soil aeration. *Crop Sci.* 38:1219–1224.

Lyons, E.M., P.J. Landschoot, and D.R. Huff. 2011. Root distribution and tiller densities of creeping bentgrass cultivars and greens-type annual bluegrass cultivars in a putting green. *HortScience* 46:1411–1417.

Moeller, A., and B. Chapin. 2011. Using turf fans in the northeast. *U.S. Golf Association Green Section Record* 49:31 (August 5), online.

Steinke, K., and J.C. Stier. 2003. Nitrogen selection and growth regulator application for improving shaded turf performance. *Crop Sci.* 43:1399–1406.

Xu, Q., and B. Huang. 2001. Lowering soil temperature improves creeping bentgrass grown under heat stress. *Crop Sci.* 41:1878–1883.

2

Cultural Practices for Summer Stress Management

Symptoms associated with summer decline in creeping bentgrass are nondescript and normally appear as a loss of green color, vigor, and stand density. Root systems often are shallow and brown in color with few white roots visible. On golf greens it is not unusual to see circular yellow or darker-green clones. Some clones may be dense while others are thin or dying. Clonal thinning or dying in discreet circular patches is often confused with disease activity (Figure 2.1). This is because many creeping bentgrass diseases are associated with distinctive color changes and often appear in a spot or circular pattern. When the application of fungicides does not reverse observed symptoms, golf course superintendents usually turn to plant diagnostic clinics to help sort out the potential causes of turf decline or loss on greens. Foliar diseases are relatively easy to diagnose by turfgrass pathologists. Plant parasitic nematode and root pathogen (e.g., *Pythium* species in roots) interactions, however, are more complicated and time consuming to identify. The latter two types of disease-causing agents are almost always found in association with turfgrass roots. The difficult question to answer is are they there as secondary pathogens because the turf is in decline due to some other stress or are they the actual cause of the problem? Problems also arise for diagnosticians when populations of primary pathogens are in the low to moderate range and are mixed with less virulent pathogens or senectophytes (i.e., saprophytes or weak pathogens). To determine the relative importance of these microbes, pathologists need to gather information on cultural and potential stress factors (i.e., golf green location, shade, mowing height, soil moisture level, recent coring or grooming practices, recent fertilizer and pesticide applications, and cultivar grown). Many times, root pathogen activity may be superficial, but in concert with other abiotic stress factors, may become the proverbial "straw that breaks the camel's back" and the grass declines. In most situations, a pathologist will discuss or recommend on a green-by-green basis the potential benefits of applying a plant protection chemical, nutrient, biostimulant, or combination of products. Invariably, pathologists and agronomists will recommend that increasing height of cut, reducing mowing frequency, roll to replace mowing periodically, solid tine aeration to vent wet surfaces, less grooming, and judicious water management must accompany any chemical inputs to alleviate the overall summer stress complex. Simply applying biostimulants, fungicides, and other plant protection

FIGURE 2.1
Two clonal patches (light green and dark green) on a golf green with the dark-green clone in summer decline.

chemicals, however, seldom solves or cures more basic cultural problems. In this section, cultural practices are reviewed, and their potential negative effects in summer are described.

Creeping Bentgrass Cultivars

'Penncross' was the primary creeping bentgrass cultivar grown on greens, tees, and fairways for decades. Penncross is still preferred by some golf course superintendents because of its familiarity as well as its reputation as a resilient and aggressive grass. In particular, Penncross often is grown on tees because of its ability to rapidly recover from divots and traffic. Penncross is still commonly grown in blends with other cultivars on fairways. Beginning about 1990, many new cultivars were released for use on golf greens including the following: 'Pennlinks' and 'Penneagle' followed by 'Southshore', 'Crenshaw', 'Providence', 'SR 1020', 'SR 1119', 'Cato', 'L-93', 'Procup', 'Putter', 'Backspin', 'Princeville', 'Century', 'Trueline', 'Viper', 'Memorial', and others. Crenshaw was widely accepted throughout the south and transition zone because of its improved heat and shade tolerance, but it is very susceptible to dollar spot and greens slowly in spring in northern regions. Crenshaw also is quite susceptible to Microdochium patch (i.e., pink snow mold or Fusarium patch), and the poor winter color and other disease concerns with this cultivar limit its suitability in more northern regions of the United States and Canada. Crenshaw has largely been replaced by other better-performing cultivars in some regions.

In the late 1990s, very fine-textured and high-density cultivars such as 'Penn A-1', 'Penn A-2', 'Penn A-4', 'Penn G-2', and 'Penn G-6' were released for use primarily on golf greens. 'Declaration', 'Tyee', 'Alpha', 'T-1', 'LS-44', and '007' are newer moderate- to high-density cultivars and others will no doubt follow. While high-density cultivars used on greens are aggressive, they may not produce many stolons. Some cultivars are prone to ball-mark injury and slow recovery, take-all patch, and Pythium root dysfunction; however, these problems decline as greens age and their management level tends to become similar to conventional cultivars. In general, high-density cultivars respond favorably to more intense management on greens due to their increased density, when compared to Penncross and other "conventional" cultivars. They can be mowed very low (i.e., <0.125 in; <3.1 mm) and thus provide improved green speed and putting quality. They respond well to double cutting on a more frequent basis, sand topdressing every 2 to 3 weeks, and frequent coring or vertical mowing during the growing season. Although higher inputs of coring, sand topdressing, and vertical cutting remain necessary for maintaining high-density cultivars, these same grooming inputs are now commonly practiced on most golf courses, regardless of cultivar (including annual bluegrass golf greens).

All cool-season grasses will suffer some level of shoot loss during stressful summer months. One advantage of the newer, high-density bentgrasses is that although shoot density declines in summer, they retain substantially greater numbers of shoots than older cultivars. The changes in seasonal shoot densities of Penncross, L-93, and Penn A-4 were monitored (Moeller et al., 2008). In this study all three cultivars exhibited substantially lower shoot densities in late summer compared to late spring levels. Even when A-4 exhibited its lowest shoot density it still was greater than Penncross at its highest density, while L-93 was intermediate. Furthermore, high-density cultivars generally produce a larger and deeper root system when compared to conventional cultivars (Lyons et al., 2011). High-density cultivars grown on greens also look and play better.

Not all high-density cultivars are appropriate for all facilities or all areas of use (i.e., greens, tees, fairway, collars, and approaches). We are in an era of "niche" cultivars, and proper selection is paramount to cultural success and long-term persistence of the turf. For example, a high-density cultivar may perform well on putting surfaces but not on collars or in approaches. Cultivars should be selected based on local or regional performance, aggressiveness, and the level and intensity of play. The new generation of high-density cultivars, however, may require a significant increase in budgeting, due to their more labor-intensive maintenance requirements associated with thatch-mat management and promoting green speed. These newer cultivars offer golfers enhanced green speed and quality, while offering turf managers the next generation of grasses perhaps more adapted to their needs.

Dollar spot is one of the most economically important diseases of creeping bentgrass and consideration should be given to this disease when cultivars are selected, especially for large acreage fairways. The cultivars Backspin, Century, Crenshaw, Penn A-4, Southshore, and SR 1020 are highly susceptible to dollar spot; whereas, L-93, Memorial, 007, and especially Declaration have improved dollar spot resistance. Different cultivars are more suitable for different environments. As noted above, Crenshaw was developed for improved heat tolerance but has limitations when grown in northern climates. Field data generated at the closest agricultural land grant university can be the best source of information on cultivar performance. Visiting neighboring golf courses where newer cultivars are being grown is another good source of information. Regional U.S. Golf Association agronomists, extension specialists, and consultants can provide valuable insights. Unbiased information for cultivar evaluations also can be found on the Web at the National Turfgrass Evaluation Program Web site (www.ntep.org).

In addition to numerous single cultivars being introduced, several commercial suppliers blend cultivars of similar color, density, and texture to provide golf course managers with improved bentgrass seed that possess a broader genetic base (e.g., improved heat or drought stress tolerance, improved dollar spot resistance, etc.). Before blending cultivars yourself, it is crucial to carefully evaluate color, leaf texture, and shoot density to ensure your selections are compatible. For example, Penn A-4 should not be blended with Penncross; however, Penn A-1 and A-4 are compatible. When buying large amounts of expensive bentgrass seed, insist on an independent seed test and a sample size of 50 grams or larger. Even "blue tag" seed lots can contain significant levels of roughstalk bluegrass (*Poa trivalis*) and other noxious weed seeds, and a larger sample size greatly increases the probability of detecting weeds.

Mowing

In the early history of turfgrass management, all greens were mowed by hand. The earliest mowers were push mowers (without motors) with high-speed reels. Today, triplex mowers are commonly used because of the speed with which all greens can be mowed each morning with minimal labor versus hand mowing. Frequently, the outer perimeter or cleanup pass of greens shows wear or thinning, which is called the *triplex ring*. The double cutting that takes place in this location and the turning of mowers in curved areas, as well as wheels, rollers, and groomer attachments traversing the same daily route, cause concentrated soil compaction, abrasion, and mechanical injury (Figure 2.2). Periodically skipping the cleanup or perimeter cut alleviates the problem when turf is not actively growing. During hot and humid periods, the perimeter cut should be skipped every other day. Where loss of density occurs, use of a walk-behind greens mower also may

FIGURE 2.2
Concentrated mower traffic when greens are too wet causes mechanical injury, especially in the cleanup pass. Note goosegrass encroachment.

become necessary. Hand mowing is commonly used on courses with higher operating budgets. This extra expenditure on machinery and manpower can greatly benefit turf during the heat of summer. Even though the wear associated with the triplex ring is discerned easily, a similar yet less conspicuous wear injury can occur throughout the putting surface, regardless of the type of mower. Triplex mowers make it possible to culture creeping bentgrass fairways in warm and humid regions due to their lighter weight and maneuverability compared to the conventional reel mowers of the past (Figure 2.3). Five-gang mowers reduce labor and mowing time, but they are heavier and can cause wear injury where they turn in approaches, fairways, or roughs. Every golf course must ultimately decide which mowing process is best for their unique situation.

By far the most destructive practice for creeping bentgrass is low (<0.125 in; <3.1 mm), daily mowing during hot and humid weather when turf shows signs of wilt and stress, and during excessively wet periods. Mowing when water partially inundates the thatch-mat surface during hot periods in summer is extremely damaging. Superintendents use the term *pushing water* to describe mowing in the presence of excess water (i.e., casual water) on the surface (Figure 2.4). When "casual water" is present on golf greens they should not be mowed; let them drain and dry out.

Mowing greens during hot, humid, and wet periods with grooved rollers also contributes significantly to turf damage. Turf collapses rapidly and injury is usually most severe in low-lying areas and surface water drainage patterns. Leaves and sheaths that collapse rapidly during periods of heat stress and in the presence of excess water often turn white and mat and may have a papier mâché water-soaked appearance and texture (Figure 1.9).

FIGURE 2.3
Lighter-weight mowing units made it possible to grow creeping bentgrass fairways in stressful environments.

FIGURE 2.4
Tee damaged by a mower pushing water. (Photo courtesy of J. Kaminski.)

Low mowing heights, however, generally are preferred to maintain surface smoothness in high-density cultivars. These newer bentgrass cultivars seem to tolerate lower summer mowing better than other conventional cultivars. Turf injured by mowing during hot and relatively dry periods usually will appear reddish-brown for a few days, and in more severe cases leaves turn tan and stand density declines. Anthracnose may appear on drought-stressed grass, especially annual bluegrass grown on greens.

Whenever greens are showing signs of injury from either biotic or abiotic causes, the height of cut should be increased and mowing frequency decreased. This procedure is key to survival. For example, it was shown that creeping bentgrass maintained at a cutting height of 0.160 inch (4 mm) had decreased root mortality compared to turf maintained at 0.120 inch (3 mm), thus showing the importance of raising mowing height on the health of the root system (Liu and Huang, 2002). Weak and thinning greens should be walk-mowed with lightweight mowers, and mowing should be reduced to no more than four or five times per week (Figure 2.5). Simply stated, less mowing is best in stress situations. During hot and humid periods a mowing height at or above 0.140 inches (3.5 mm) is recommended. Increasing mowing height increases leaf area and helps moderate canopy temperature, which improves the ability of plants to produce photosynthate (i.e., sugars) and to naturally cool via transpiration. An increase in canopy height may help to slightly moderate soil temperature and alleviate stress in the rootzone. It is important to replace grooved rollers with solid rollers and disengage or remove grooming devices (i.e., brushes and verticutters) once summer stress conditions begin to slow turf growth and recuperative potential (Figures 2.6 and 2.7). During especially hot and humid periods it is prudent to use walk-behind greens mowers equipped with solid rollers. Even though walk-mowing may be impractical for all greens, it certainly would be most beneficial to greens that are weakened and showing a loss in turf density. Never mow greens in the morning following a major rain event or whenever there is casual water on putting surfaces during periods of heat stress. Greens suffering from heat stress, wet wilt, or anaerobic conditions must be carefully managed, and reducing mower stress is essential to their survival.

Spring and autumn bring additional mowing issues when labor is reduced and there are more frequent rain events. In particular, long periods or rainy weather that keeps mowers inside means that the grass goes uncut for days

FIGURE 2.5
Use lightweight mowers to cut summer stressed greens.

FIGURE 2.6
Replace grooved with solid rollers during summer stress periods.

FIGURE 2.7
Creeping bentgrass damaged by mowing this green with grooved roller during a wet-wilt event.

at a time. If the weather forecast indicates that an extended period of rain is coming, it is recommended that a plant growth regulator be applied to reduce foliar growth. In most regions, golf greens are on some type of growth regulator program. Goals vary, but a stand of grass that is growing more slowly offers managers the valuable option of mowing less frequently, while maintaining reasonable green speeds and limiting the potential for scalping. Once mowers can return to the golf course, and the grass is very high, mowing heights need to be adjusted higher to avoid scalping and heavy clipping deposits.

Scalping and Puffy Greens

Scalping is the removal of excessive green shoot tissue, which leaves turf appearing tan or brown (Figure 2.8). Sometimes crowns and other stems are damaged, which will result in a longer recovery period. Along with the browning of tissue there is physiological damage (i.e., the grass is shocked) because the photosynthetic area is reduced, and in some cases temporarily eliminated and stored carbohydrates will be needed to regenerate new shoots from axillary buds. Scalping may occur when the mower is out of adjustment; if mowing height is drastically reduced; when mowers sink in soft, wet grass and soil; or if the bed knife is worn too thin. A common cause for scalping is when rainy weather conditions make it difficult to mow for long periods. Scalping also is promoted by excessive shoot growth and by the presence of thick thatch layers that hold water. Excessive shoot growth can result from the application of large amounts of nitrogen or when organic nitrogen releases fast in response to warm and wet soil. Scalping greens often is due to the presence of excessive thatch layers, which can be linked to inadequate coring, vertical cutting, and sand topdressing programs and possibly improper use of nitrogen. Conversely, dense mat layers containing sand provide stability and protection against scalping. It is also possible that nitrogen released by microbial degradation of organic matter (i.e., mineralization) during warm and wet periods results in rapid release of nitrogen. The stimulated and rapidly growing turf is scalped as a result of excessive shoot production in response to the released nitrogen (Figure 2.9).

During periods of heat stress and high humidity in summer when creeping bentgrass becomes too wet from either excessive rain or irrigation, the

FIGURE 2.8
Pocket green severely scalped in early summer.

FIGURE 2.9
Scalping due to nitrogen fertilizer overlap and subsequent stimulated growth.

FIGURE 2.10
Scalping puffy creeping bentgrass on an approach. Note aerial roots in thatch (inset).

thatch-mat and plant stems swell. Stems may be lifted or become elevated as a result of swelling. Individual plants may develop elongated stems (i.e., crowns) with new, white roots in summer, which are mostly restricted to the thatch-mat layer (Figure 2.10). These "aerial roots" appear to be pushing crowns upwards as if to seek better exposure to oxygen in the thatch-mat layer, and they are not unlike the adventitious roots produced by corn plants. This swelling of the thatch-mat or stems is referred to as being "puffy." Creeping bentgrass grown on greens, tees, and fairways can become puffy, but it is most problematic on greens and approaches. Mowers sink into swollen thatch-mat and remove an excessive amount of plant tissue thus scalping the turf. Grooming and light sand topdressing in summer helps to control puffiness. However, when greens become puffy, the mowing height should be increased about 0.06 inches (1.5 mm) and grooved rollers should be replaced with solid rollers. Every attempt should be made to avoid scalping, especially in summer on greens.

Solid rollers are preferred in summer mowing because they roll the grass down and even the weight distribution of the mowing head. Conversely, grooved rollers tend to sink into the canopy and promote mechanical damage and scalping. Mowing in the presence of dew can be helpful because the moisture lubricates the bed knife, thus reducing drag or pulling of the mower. Syringing dry turf prior to mowing can help lubricate the bed knife and reduce mower drag. If weather conditions are excessively wet and humid, it may be best to skip mowing for a day or two. Syringing and hand watering should replace use of the overhead irrigation system until the thatch-mat and soil become dry. Use of a gibberellic acid–inhibiting plant growth regulator, such as trinexapac-ethyl, flurprimidol, or paclobutrazol, to slow growth will help to mitigate the problem. Conversely, more than two applications of the plant growth regulator ethephon can cause creeping bentgrass stems to elongate and become scalped. The fungicide mancozeb applied on a 10- to 14-day interval during summer stress periods was shown to partially reduce the damage associated with scalping puffy greens, but the mechanism is unknown (Dernoeden and Fu, 2008). Light and frequent spoonfeeding (0.1 to 0.2 lb N/1000 ft^2; 5 to 6 kg N/ha) also promotes recovery from scalping.

Grain and Grooming

Grain refers to a growth habit in which plants grow in a flat or trailing manner (i.e., procumbent growth). This growth habit is most notable in creeping bentgrass, bermudagrass, and other stoloniferous grasses, and develops on greens, tees, and fairways. Within a few years of establishment, grass plants may orient themselves up or down slopes or along the path of the rising and setting sun (Figure 2.11). The reason for this peculiar growth habit is unknown, but it does not appear to be a response to gravity because the direction of the procumbent growth can be variable. Grain is most problematic on greens because its presence can alter the speed and direction of the ball. Once mowing height is lowered to 0.125 inches (3.1 mm) or lower, grain has less of an effect on ball roll. At these low mowing heights, roll of the ball is most affected by the slope of the green and not by the way leaf blades are oriented. It is largely an aesthetic problem elsewhere on the golf course.

FIGURE 2.11
Grain or procumbent growth of creeping bentgrass on a fairway.

Grooming refers to various methods used to manage grain or stemyness, improve mowing quality, and maintain "smoothness" and "trueness" of the putting surface. Some expand the definition of grooming to include practices that promote green speed, such as rolling and sand topdressing. Some approaches to grooming include shallow vertical grooming, brushing or brooming, and use of grooved rollers. Vertical grooming (i.e., superficial vertical cutting) of putting surfaces on a periodical basis (i.e., 10- to 14-day intervals) during the season is effective (Figure 2.12). Vertical groomers are vertical cutters mounted in front of the mowing reels. The blades are designed to penetrate turf and cut

FIGURE 2.12
Light vertical cutting is a form of grooming used to control grain and improve density and green speed.

FIGURE 2.13
There are several types of brushes used to stand up stolons before mowing to reduce grain.

stolons and stand-up shoots and provide a better cut. More aggressive vertical cutting (i.e., vertical aeration) for thatch management is best performed in the spring and autumn, but even vertical grooming should be avoided during prolonged summer stress periods or whenever turf lacks vigor or is in a stressed or weakened state. Grooved rollers are mounted in front of the mower head to create grooves that stand up shoots and help reduce grain. Brushing also is used to manage grain (Figure 2.13). Brushing is accomplished either manually or by brushes mounted in front of the mower head. Brushing to stand up leaves and stolons before they are mowed is not very disruptive to putting surfaces and can be quite effective in reducing grain. Alternating the pattern of mowing on a daily basis also is helpful. Double cutting, when performed in moderation, not only improves green speed but also is helpful in controlling grain.

Greens under stress should not be double-cut, brushed, vertical cut, topdressed, or otherwise abused mechanically during hot and humid weather. Grooved rollers should be replaced with solid rollers. Grooved rollers and vertical cutting or grooming attachments can dig into the putting surface, especially on turns or when greens are wet and soft. These attachments should be lifted, disengaged, or removed during summer stress periods. Aggressive brushing and vertical cutting to control grain, upright growing stems and stolons, or tufting is best performed at times when the turf is actively growing and not when stressed so that the turf can recover rapidly.

Thatch and Mat Layers

Problems associated with thatch, mat, and other layers were discussed as they relate to summer stresses in Chapter 1, but there is more to this subject that needs to be explored. Thatch refers to an organic layer found in turf above the soil surface. This organic layer is composed of living, dead, or partially decomposed stems, roots, and leaves. Mat refers to a surface layer of organic matter (usually living and dead stem tissue and roots) mixed with sand from topdressing. A modest thatch-mat layer (0.25 inch; 6.2 mm) is considered beneficial on greens because it cushions stems from pitch shots and heavy equipment (Figure 2.14). A 0.5 inch (12.5 mm) thatch layer generally is considered desirable for bentgrass fairways. Because of the common practice of sand topdressing, mat and not thatch layers generally are found at the surface of greens, tees, and anywhere a topdressing program is used. As noted previously, because thatch and mat layers are not exclusively found on bentgrass greens, tees, and fairways topdressed with sand, they are referred to collectively as thatch-mat.

When topdressing heavily yet infrequently with sand, numerous distinct alternating organic/sand layers can develop in the upper rootzone profile (Figure 2.15). Each layer potentially can impede the downward movement of water and gases, which eventually causes problems. Due to lower mowing heights, maintaining a thatch-mat depth of less than 0.3 inches (7.5 mm) on greens is desirable. Under low mowing (≤0.125 in; ≤3.1 mm), thatch-mat layers become extremely dense making it very difficult to incorporate sand topdressing. Mat layers that cause water to perch also can be just as detrimental as thatch in the summertime and both need to be aggressively managed. Again, this is another reason for frequent surface aeration, vertical cutting, and introduction of sand from topdressing. Where there is a good mix of sand and fibrous roots in the thatch-mat layer, water and air exchange occurs freely. However, layers of different sand topdressing mixes or dense organic

FIGURE 2.14
A modest thatch-mat layer is desirable because it cushions the surface.

FIGURE 2.15
On top of the greens mix there is a sod layer, and above that are multiple heavy sand topdressing layers.

layers may impede water movement and could contribute to indirect heat stress and possibly death of turf during hot summer periods.

Some thatch-mat is desirable because it provides cushion for equipment, people traffic, and impact of the ball. Thick or deep and dense thatch-mat layers, however, are detrimental (Figure 2.16). Thick and dense thatch-mat layers reduce the environmental stress tolerance of turf, can predispose plants to indirect heat stress or scald in summer, promote disease, harbor insect pests,

FIGURE 2.16
An excessive thatch-mat layer bypassed by a coring hole filled with sand topdressing.

FIGURE 2.17
Stems and roots developing in thick organic layers are more vulnerable to heat and drought stress because they are not insulted by soil.

and can render surfaces more prone to scalping. When a thatch-mat layer becomes thick, the stem-bases of plants develop in the organic layer rather than at or below the soil surface (Figure 2.17). Stems developing in thatch-mat are not insulated from extremes of heat and cold or wet and dry weather conditions. All roots emanate from stems (i.e., crowns and stolons) and both tissue types are more vulnerable to injury from environmental extremes and mechanical injury when they become largely restricted to these organic layers. Furthermore, axillary buds on stems that produce the new shoots, tillers, and roots are jeopardized by virtue of being exposed to environmental extremes in thatch-mat rather than insulated by soil. Hence, the most important plant tissues (i.e., the growing point, auxiliary buds, and roots) are more vulnerable to desiccation during periods of drought, freezing temperatures in winter, and heat stress in summer. A wet thatch-mat in particular contributes to direct heat stress damage during hot, rainy periods or when turf is excessively irrigated (see Chapter 1). As previously described, water in surface organic layers absorbs and transfers heat from the sun, which can build to lethal levels during periods of heat stress. Green plants in core aeration holes, surrounded by brown turf in the summer, are a good indicator of a thatch-mat problem. This is because the coring hole provides an opening unobstructed by organic matter, which allows for better water infiltration and air exchange. The surrounding brown turf was injured because the thatch-mat layer was either too dry or too wet during a period of heat stress.

The organic matter in thatch-mat also provides harborage for insect pests and pathogens. Annual bluegrass weevil (*Listronotus maculicolis*; formerly *Hyperodes* weevil), black cutworms (*Agrotis ipsilon*), sod webworms (several species), black turfgrass Ataenius (*A. spretulus*), and other insect pests find thatch a suitable medium in which to survive (see Chapter 10). Most pathogens survive unfavorable periods as spores, sclerotia, fruiting bodies, or mycelium on or within dead organic matter. Pathogens often live sapro-phytically on dead organic matter in thatch and soil during environmental periods that are unfavorable for infection of plant tissues. In particular, pathogens causing dollar spot (*S. homoeocarpa*) and anthracnose (*C. cereale*) survive in and build up their populations in thatch-mat layers. Wet thatch-mat also provides moist conditions needed by blue-green algae and moss to proliferate. Furthermore, thatch-mat layers can swell (i.e., become puffy) during hot and humid periods, predisposing bentgrass to scalping. The best managed turf has neither too little nor too much thatch-mat.

Fertilizer management is important in minimizing thatch-mat. Most of the annual nitrogen applied to creeping bentgrass usually is in early autumn to promote recovery from summer stress and coring and vertical cutting injury. Spoonfeeding in summer may account for one-third or more of the total amount of nitrogen applied annually. A substantial amount of organic mat-ter develops in the thatch-mat layer between late autumn and early spring when microbial degradation of organic matter is inhibited by low tempera-tures (Fu et al., 2009). Hence, dormant or winter applications of substantial amounts of nitrogen should be avoided on creeping bentgrass. Dormant applications of soluble nitrogen also are more likely to leach and can pro-mote annual bluegrass.

High nitrogen fertility promotes thatch, but too little nitrogen is det-rimental. Hence, a balance is required. Generally, most mature creeping bentgrass golf turfs grown on sand-based rootzones should receive about 3 to 6 lb N/1000 ft^2 annually (150 to 300 kg N/ha). Bentgrass grown on fine-textured native soils should receive 3 to 4 lb N/1000 ft^2 (150 to 200 kg N/ha) annually. Special situations like heavy play, heavy rain in sand-based root-zones, and extended growing seasons also affect nitrogen needs of the turf. During summer, light applications of nitrogen (0.1 to 0.2 lb N/1000 ft^2; 5 to 6 kg N/ha), known as spoonfeeding, should be applied every 1 to 2 weeks. Several studies have shown that natural organic N-sources are not generally better at promoting thatch-mat degradation and soil microbial activity, when compared to synthetic slow-release fertilizers or urea (Davis and Dernoeden, 2002). Some composted sewage sludges that contain wood chips (high C:N ratios) and dehydrated manures can promote thatch buildup and dollar spot. Applying large amounts of organic materials with high C:N ratios can result in a harmful organic layer buildup. Biological products (e.g., molasses and other sugars; humates, acids, and microbial inoculants) have little significant impact on reducing thatch-mat production. As discussed below, irrigating deeply and infrequently also results in less thatch-mat buildup. Keeping bentgrass on the dry side slows growth and thus less organic matter is pro-duced and carbohydrates are conserved (Fu and Dernoeden, 2008).

Management to reduce the thatch-mat layer is essential for maintaining smooth and firm greens. Research has shown that it is difficult to actually physically remove enough organic matter to keep pace with its production (McCarty et al., 2007). Thatch-mat is aggressively managed by a combination of coring, vertical cutting, topdressing, and proper use of fertilizer and water. A major objective of coring, vertical cutting, and sand topdressing is to incorporate sand into the surface organic matter layer. This mixing of sand and organic matter is called *dilution*, and it helps create a better growing environment for plant stems and roots compared to plant growth in a dense organic layer (Fu et al., 2009). Dilution or sand mixing also helps speed decomposition of organic matter.

Sand Topdressing

Sand topdressing is an indispensible cultural practice for maintaining smooth, firm, and healthy greens and other areas on the golf course. As previously noted, a primary objective of topdressing is to dilute surface organic matter with sand. Sand topdressing also aides in surface organic matter control, smoothens and firms the putting surface, and helps greens recover from ball marks. In general, topdressing sand is applied heavily following core aeration in spring and late summer as part of a soil modification or thatch-mat reduction program (Figure 2.18). Infrequent and heavy topdressing should be avoided because it will create a thatch/topdressing layer consisting of bands of organic matter and sand. The banding is referred to as the "Oreo cookie" or "marble cake" effect (Figure 2.19). Layers cause water to perch and restrict rooting. Perched water tables also promote black-layer, blue-green algae, moss, disease, and other problems. At other times during the growing season, light applications of topdressing on a 2- to 4-week

FIGURE 2.18
Greens should be topdressed heavily with sand following coring in spring and again in late summer or autumn.

FIGURE 2.19
The "Oreo cookie" or "marble cake" layer or banding effect from heavy but infrequent top-dressing. (Photo courtesy of D. Bevard.)

FIGURE 2.20
Sand should be watered-in or blown into the canopy following light topdressing. (Photos courtesy of S. McDonald.)

interval (usually every 2 weeks depending on weather conditions) are recommended (Figure 2.20). Light topdressing involves applying 25 to 50 lb dry sand topdressing per 1000 ft² (1221 to 2442 kg/ha) in each application. Light topdressing may be applied several times each month of the golfing season as long as turf is actively growing.

Heavy topdressing in summer that smothers the canopy in sand can bake or suffocate bentgrass in the summer (Figures 2.21, 2.22, and 2.23). Light

FIGURE 2.21
Heavy sand topdressing that inundates the canopy can be injurious if done in warm weather.

FIGURE 2.22
Heat stress and creeping bentgrass injury from sand around divot mix on a tee.

applications of topdressing can cause abrasion, which may lead to a loss of turf density or aesthetic quality during summer stress periods. In mixed-species greens, consisting of annual bluegrass and creeping bentgrass, the annual bluegrass is more likely to be abraded in summer. Topdressing and the brushing associated with working sand into the canopy can cause abrasion of shoots, which may appear as a yellowing or bronzing within a few days (Figure 2.24). Bronzing of shoots may not develop for 3 or more days after topdressing. During summer stress periods, some golf course managers

FIGURE 2.23
Green damaged by topdressing sand that was allowed to sit for over an hour on a hot day.
(Photo courtesy of S. McDonald.)

FIGURE 2.24
Abrasion from brushing in sand topdressing can cause creeping bentgrass to develop a yellow
or bronze color within a few days.

will topdress only dry turf and wash sand off the canopy by lightly irrigating rather than brushing. Light topdressing when greens are chlorotic, weak, or otherwise stressed in the heat of summer should be delayed until turf is actively growing and can recover rapidly from abrasion. Syringe dry or wilted greens before and after topdressing to reduce the effects of this type of abrasion. Spoonfeeding in summer promotes bentgrass recovery from mechanical injury and environmental stress–related problems.

Sand topdressing, including light and frequent, wears and dulls mower blades rapidly. Mowing with dull or damaged blades causes chronic and sometimes severe mechanical damage and therefore clubs should own their own grinders. Sand from topdressing collars can be very abrasive and is a major contributor to the rapid decline of collars in summer. Causes for collar decline are discussed in detail in Chapter 3. Moderately heavy topdressing greens in late autumn after fungicides for snow mold have been applied is performed in cold climates, because sand provides some protection from winter desiccation.

Vertical cutting prior to topdressing helps move sand into thatch-mat. Vertical cutting after topdressing, however, can cause severe injury to putting surfaces. Injecting sand (e.g., DryJect®) is another approach to introducing sand (Figure 2.25). Sand injection is an attractive alternative for golf course superintendents because it does not involve the level of injury, disruption of play, and cleanup associated with coring or vertical cutting. Injecting sand can improve infiltration and percolation, improve soil aeration, and dilute organic matter. Injecting sand, however, may have negative attributes if incompatible sands are utilized. Sand injection should not be used as a replacement or substitute for traditional coring until more research is conducted to prove otherwise.

Topdressing should consist of straight sand, high sand, or sand/soil/peat topdressing similar to the greens mix. It generally is recommended to use topdressing sand that has the same or similar properties as the greens mix. With the advent of lower mowing heights and planting of high-density cultivars on greens, it may become difficult to brush in or otherwise incorporate

FIGURE 2.25
Sand injection creates a bypass through a sod layer, mixes sand with organic matter, and firms surfaces.

sand. In some situations the topdressing material contains coarse particles, similar to the sand used in the original greens mix, and cannot be easily incorporated below the canopy and into the thatch-mat layer. In such cases finer topdressing sands may be needed, but superintendents are cautioned to obtain expert assistance from soil testing laboratories to ensure a suitable material is utilized. Whether the choice is straight sand or something similar to the rootzone mix, it is important to stay consistent with your sand selection. Alternating among different types of topdressing will create undesirable layers and perched water tables (see Chapter 1). Topdressing mixes should be periodically tested for uniformity of particle size and shape and infiltration properties. Poor topdressing mixes consisting of variable particle sizes ranging from fine to coarse can pack and seal the surface of greens. Silica sands are preferred to calcareous (calcium-based) sands. Calcareous sands may become unstable if treated with sulfur for long periods. The U.S. Golf Association (USGA) has recommended particle sizes for putting green construction and maintenance (USGA Green Section Staff, 2004). They specify that at least 60% of the sand by weight should be in the medium to coarse size factions (0.25 to 0.5 mm) and no more than 20% fines (<0.15 mm). Shape of sand particles is another important factor. Angular rather than round (i.e., spherical) or subround sands should be utilized for golf green construction and topdressing (Li et al., 2009).

As previously noted, it is difficult to incorporate sand from topdressing due to lower mowing heights and the planting of high-density cultivars. This has led to a trend in using finer sands to topdress greens. Agronomists, however, are concerned that use of finer sands overtop a coarse sand may result in packing of the variable sand particle sizes and thus plug the surface. Research is underway to address this concern. Until more research-based information is forthcoming, superintendents who are using finer sands should be coring with wide-diameter tines no less than twice annually and should fill the holes to the surface with coarse sands similar to, if not identical to, the construction mix. It is hoped that this practice will help offset potential problems that may be associated with the use of finer sands in topdressing programs.

Sand topdressing creeping bentgrass fairways is growing in popularity (Figure 2.26). Once considered a luxury, topdressing fairways has become more common due to golfer demands for excellent playing conditions. Topdressing makes fairways more firm and fast and is employed to reduce wetness on chronically wet fairways (Zontek, 2005). Topdressing helps to control or dilute organic matter in the thatch-mat layer and thus improves the growing environment for roots. Topdressing also makes it easier to leach excess salts, but contrary to popular belief it does not appear to reduce earthworm activity. Perhaps after a decade or more of sand topdressing some decline in earthworm activity would occur. Sandy castings will dry more rapidly and eventually become easier to reincorporate by dragging.

FIGURE 2.26
Topdressing with sand helps to keep wet fairways firm and dry. (Photo courtesy of D. Bevard.)

The amount and frequency of topdressing fairways with sand is largely a budget issue. During the initial year of a fairway program an inch (2.5 cm) of topdressing should be applied and continued in successive years until a 5-inch layer (12.5 cm) of sand has been built (Whitlark, 2008). Fairway top-dressing rates range from 25 to 50 tons of sand per acre per year (56,070 to 121,140 kg/ha/yr). Topdressing sand generally is applied to fairways about twice annually, but aggressive superintendents topdress more frequently. Heavy and deep sand from infrequent topdressing will result in alternating bands of organic matter in fairways. These layers potentially pose problems, but probably are not as detrimental as a thick layer of undiluted thatch. Light and frequent topdressing is preferred, and some managers apply sand to fairways four to six or more times annually at rates ranging from 5 to 10 tons per acre (11,215 to 22,430 kg/ha) per application. This approach negates thick alternating layers of sand and organic matter. Early spring and late autumn are the preferred times for topdressing bentgrass fairways and topdressing is best coordinated in concert with core or solid tine aeration and vertical cutting operations. If budgets allow, light topdressing can be performed in the summer as long as bentgrass is actively growing. High-quality sands like those used on greens are not needed, but the sand should be free of gravel and should contain very little clay (<3%) and silt (<5%) (Whitlark, 2008; Skorulski et al., 2010). It is unknown what the consequences of suspending or eliminating a fairway sand topdressing program would be. It is known, however, that topdressing creeping bentgrass fairways with sand promotes localized dry spots, which can lead to increased costs for wetting agents and labor to syringe or hand water dry spots.

Coring and Other Aeration Practices

In turfgrass management the term *cultivation* refers to working the soil and thatch without destroying the turf (Turgeon, 2008). Coring is a cultivation technique often referred to as *core aeration* or *aeration*. The process of coring is deplored by golfers because it negatively affects surface smoothness for a period of time, but it is absolutely critical to maintaining healthy and firm putting surfaces. Coring is used to alleviate compaction, to improve air and water infiltration (i.e., drainage), to diminish rootzone layering, to reduce or help dilute thatch with sand, and to promote root growth and longevity. Basically, coring holes provide a bypass channel through the thatch-mat layer and other layers and provide for improved air and water movement through the rootzone.

Although recommendations can vary, it generally is recommended that greens be cored with wide-diameter tines (0.50 to 0.65 inch; 12.5 to 16.2 mm) in the spring and late summer and with narrow diameter tines during the summer, weather conditions permitting (Figure 2.27). Aggressive coring should only be performed when turf is actively growing. Wide-diameter coring holes are filled to the surface with sand to maintain a continuous channel into the rootzone where air (oxygen) and water can enter unobstructed by layers. To promote turf recovery and closure of holes, nitrogen is applied just prior to coring and then on a 5- to 7-day interval thereafter at rates ranging from 0.10 to 0.15 lb N/1000 ft² (5 to 7.5 kg N/ha). Greens can be periodically cored throughout summer and lightly topdressed, and sand from the cores is broken up and brushed into holes. Filling holes in summer normally is not done to avoid undue abrasion injury to greens. Spoonfeeding and careful

FIGURE 2.27
Use wide-diameter tines in spring and autumn and fill holes to the surface with sand topdressing.

postcoring irrigation and syringing are required to ensure turf around holes does not desiccate. If the surface is under stress from localized dry spot this should be remedied prior to coring or serious desiccation may result. Wetting agents can be applied 7 days in advance of coring to maximize product effectiveness and rootzone hydration and to ensure rapid recovery.

Coring in late summer can be damaging to greens that have become poorly rooted. Coring weakly rooted greens can cause sod to lift, which often requires several weeks of recovery time. Managers therefore are cautioned to stop coring if sod lifts. Coring during periods of high-temperature stress in late summer also can contribute to more injury than normally would be expected, which extends recovery time significantly. Hence, coring should be avoided until turf is actively growing, as measured by better color and an increase in clippings that are removed as heat or drought stress abate and nighttime hours become longer in late summer. Many superintendents are confronted with full event calendars in late summer and cannot always core at this time. However, delaying coring to mid-to-late autumn invariably promotes annual bluegrass invasion as discussed in Chapter 7. Sand injection, a relatively new technique, is used in cases where coring cannot be performed at preferred times. Sand injection in lieu of traditional coring is less disruptive but should not be relied upon as a replacement for traditional coring. Some superintendents core and then inject sand and fill remaining open holes with topdressing. Sand injection into soil also is a firming process useful on greens and approaches.

During the golf season, greens may have to be "opened-up" to promote soil aeration and drying, to facilitate water penetration into hydrophobic soils, to alleviate compaction, or for overseeding purposes. Quadratine, needle-tine, and water injection aeration can promote root growth and root longevity. During the summer, a less-invasive, small-diameter tine is used such as 0.25 to 0.38 inch (6.2 to 9.5 mm) solid or hollow tines (Figure 2.28). Deep, needle-tine aeration treatments can be performed throughout the summer as long as turf is not under stress. This type of aeration is very beneficial and does not significantly affect playability, especially when greens are rolled following tining. While the debate continues, hollow tines generally are preferred when turf is healthy and actively growing. Solid tines, however, are best for aerifying wet or weekly rooted greens in the summer because they cause less injury and disruption to the putting surface. Spoonfeeding in conjunction with summer coring promotes turfgrass recovery.

Continuous hollow or solid tine aeration may lead to a subsurface, hard-pan layer or compaction zone if tines of the same length are used continuously for several years. Therefore it helps to occasionally alternate hollow and solid tines, and to vary aeration depth. Sometimes it is best to use long tines (>6 in; >15 cm) like those provided by the Verti-drain®/Soil Reliever® to loosen hard-pans and alleviate subsurface compaction (Figure 2.29). There are vibratory coring units that also can be used for creating fissures in the rootzone that alleviate compaction and promote drainage. Drill and fill is a more aggressive approach used

FIGURE 2.28
During the season less-invasive, small-diameter tine coring should be performed periodically.

FIGURE 2.29
Deep verti-drain tining is performed to alleviate subsurface compaction, but the operation should cease if it causes lifting of the sod as shown in the upper right photo.

to modify a greens mix and breakthrough layers. This approach often is used as a last measure before considering rebuilding poorly drained or soft greens built with round or subround sand particles. Regardless of depth or type of unit or tine, coring must be stopped if it causes lifting of the sod during summer (Figure 2.30). Lifting of sod is a sure sign of shallow rooting. Soil should be moist but not saturated prior to aeration. For emergency overseeding, an adjustable height coring unit, which produces numerous, closely spaced, shallow (<0.5 in; <12 mm) holes, may be recommended. Following aeration, greens

FIGURE 2.30
Sod lifting during coring is a sure sign of weak root systems and causes additional injury that extends recovery time. (Photo courtesy of S. McDonald.)

should be syringed if dry or wilted and rolled prior to mowing. Despite high soil temperatures and the wounding of healthy plants, the creation of aeration channels can promote some root growth in summer, which may counterbalance in part the roots lost by the process. It again is noted, however, that weak greens should not be mechanically injured by coring or grooming during summer stress periods. When golf greens are excessively wet, solid tine aeration or venting is the best practice.

Water injection aeration is beneficial and usually is performed two to three times per month in summer (Figure 2.31). The weight of hydro-injection machines and the deposition of rootzone sand (i.e., *blow back*, which indicates the machine may be traveling too slowly over the green) can cause wear injury to golf greens that are frequently water injected. On hot and dry days, syringe greens prior to water injection aeration. Spiking, another form of solid tine aeration, usually is restricted to once per week. Additional spiking may be prudent during excessively wet periods if it does not cause undue injury. Spiking should be performed during early morning or evening hours on days when high temperatures (>86°F; >30°C) are expected. Spiking greens suffering from drought or wet wilt during the heat of the day can cause extreme damage to greens (Figure 1.15). In small, localized wet spots or isolated dry spots, it may be necessary to use a pitchfork or similar handheld tool occasionally to rapidly create openings to allow water to drain and to allow air to enter soil. Sometimes these old mechanical practices can still be used effectively.

Tees should be cored and topdressed similar to greens, but due to budget and time restraints these practices often are performed less frequently. Some superintendent's core and vertical cut deeply prior to incorporating sand in tees. Due to the large acreages involved, fairways generally are cored only

FIGURE 2.31
Water injection aeration promotes rooting and root longevity in summer.

once annually with wide-diameter tines (Figure 2.32). Fairway coring can be done either in spring or late summer or autumn to best accommodate golf schedules. As previously noted, some topdress fairways to help provide drier and more firm playing conditions. If this practice is followed, use of solid rather than hollow tines for fairway aeration is considered the best practice.

Studies have shown that coring in the absence of sand topdressing and vertical cutting can actually promote organic matter production and thatch buildup. Murphy et al. (1993) found that total organic matter content as well as thickness of the thatch-mat layer increased with hollow tine

FIGURE 2.32
Creeping bentgrass fairways should be cored at least once annually with wide-diameter tines.

coring; however, the concentration of organic matter (i.e., dilution with sand) decreased (Murphy et al., 1993). McCarty et al. (2007) also reported that coring four times annually had no effect on creeping bentgrass thatch-mat depth. Similarly, Fu et al. (2009) found that coring alone does not effectively control thatch-mat. In the aforementioned study, three coring regimes were assessed as follows: spring-only coring with heavy topdressing (0.5 in; 1.25 cm diameter tines), spring plus three summer corings, and brushing in sand brought to the surface by coring (0.25 in; 0.62 cm diameter tines). There was also a noncored control. At the end of the second year, spring-only and spring plus summer cored plots developed 66% and 89%, respectively, thicker thatch-mat layers compared to noncored bentgrass. Total organic matter content (weight loss-on-ignition) in thatch-mat layers, however, generally was similar among all three regimes in both years. This indicated that organic matter was diluted by inclusion of sand from topdressing and reincorporation of sand from cores. Thus organic matter concentration (i.e., dilution with sand) in the thatch-mat layer was much lower in plots of both coring regimes compared to noncored plots. This dilution of the organic matter improves the growing environment around stems, where all roots emanate. Thatch-mat can only be effectively managed by combining coring, vertical cutting, sand topdressing, and grooming (McCarty et al., 2007).

Coring is disruptive and causes mechanical injury to turf surfaces in the short term, but coring ultimately results in improved quality by enhancing soil aeration and promoting rooting in the long term. It should be noted that coring during the first spring and summer following new seeding should be avoided. Research showed that coring the summer following establishment reduces quality and rooting (Fu et al., 2009). In a more mature stand in the second summer of the study, greater total root number was observed in spring plus summer cored plots, compared to spring-only and noncored plots. Thus spring plus summer coring of mature creeping bentgrass promoted root growth and longevity, but coring during the first summer of establishment injured roots. In short, spring and summer coring were harmful to the immature green, and there were no benefits to counterbalance the first summer negatives. One very important reason for coring is to alleviate compaction and improve aeration and water movement into and through a rootzone. In a new, sand-based green built to USGA particle size specifications, poor drainage and aeration should not be an issue for many years. Hence, there really should be no compelling need to core the first summer. Because grow-in involves very high inputs of nitrogen, thatch formation will be a concern. If thatch levels become excessive (>0.75 in; >19 mm), then a superintendent may consider less-invasive coring and light sand topdressing. That is, coring tines should be cut or set to the depth of the thatch layer and sand from cores should be reincorporated by brushing. Less deep and invasive coring would be expected to have less impact on young root systems as well as leaves, sheaths, and stems. Thus turf recovery is more rapid versus deeper coring an immature green the first summer of establishment.

Vertical Cutting

Deep vertical cutting creates channels through dense organic mats, facilitates entry of sand topdressing materials, helps to control thatch-mat and grain, and disrupts layers. Deep vertical cutting (sometimes called vertical aeration or Graden® vertical cutting) removes large amounts of organic matter and improves aeration and rooting into the channels (Figure 2.33). Some units inject sand that helps to dilute surface organic matter. Using thick blades, however, creates lines or a "corduroy" appearance that may reduce aesthetic quality for months (Figure 2.34). Deep vertical cutting (i.e., >0.5 in; >12 mm)

FIGURE 2.33
Deep vertical cutting improves aeration and thus rooting. (Photo courtesy of S. Potter.)

FIGURE 2.34
Vertical cutting with thick blades can cause a corduroy effect and invite criticism.

FIGURE 2.35
Vertical cutting creeping bentgrass fairways will remove large amounts of organic matter. (Photos courtesy of S. McDonald.)

should only be performed in the autumn or spring when turf can rapidly recover. In summer, deep vertical cutting disrupts playability, damages turf, reduces aesthetic quality, and invites golfer criticism. Less-invasive, light vertical cutting (i.e., vertical grooming) to cut stolons and stems, however, can be performed whenever turf is actively growing and not under environmental stress to reduce grain and improve smoothness of playing surfaces. Moderate or deep vertical cutting in early summer prior to periods of heat or other environmental stresses can cause a decline in turf density, which progresses over time. This progression in intensity often is confused with disease. Aggressive vertical cutting of greens and fairways is best performed in the autumn when temperatures are cooler and turf is growing vigorously (Figure 2.35).

Rolling

Rolling greens has been performed for many decades to firm and smooth surfaces. The first rollers were pushed manually by hand but evolved to the heavier, motorized sidewinder-type roller (Figures 2.36 and 2.37). Rolling is best used to improve green speed in summer during periods when mowing height is increased or when mowing frequency is decreased. Rolling to increase smoothness and green speed is nondestructive if not overdone (no more than three times per week is recommended). Rolling is quite useful for promoting a smooth surface and avoiding scalping following aeration, spiking, verti-draining, and other disruptive practices. Some superintendents

FIGURE 2.36
Rolling is performed to smooth putting surfaces and increase green speed and is performed after spiking or coring.

FIGURE 2.37
Sidewinder rollers are used to improve green speed.

roll during rain-free periods in early spring to improve smoothness of greens that became bumpy during winter due to frost heaving. Rolling is used frequently to maintain green speed during tournaments. Rolling should not be performed when soils are wet or when topdressing sand is on the canopy. Rolling can be overdone. It is best used for special purposes, especially where a quick increase in green speed is desired without changing the way greens are being maintained.

Spot or target rolling a small portion of the green around the hole helps reduce concentrated mechanical injury and reduces the time it takes to roll greens. Avoid making more than five passes on either side of the hole, and avoid pivoting sidewinder rollers on collars. For unknown reasons, rolling has been shown to reduce the severity of dollar spot and anthracnose (Nikolai, 2002). Devices are available which enable turf managers to monitor and maintain firmness for tournament play.

Irrigation Practices

Despite decades of informative research on water management, imposing proper irrigation methodologies during the summer remains an "art." Different irrigation practices may have to be adopted for stands dominated by annual bluegrass. Annual bluegrass greens grown in more northern regions where heat stress is not chronically severe respond well to light and frequent irrigation. In warm to hot and humid climates, however, annual bluegrass can rapidly decline or "check-out" (i.e., die or go dormant) when subjected to wilt conditions or when grown in hot and wet soils (Figure 2.38). Balancing the water requirements of annual bluegrass with potential heat stress injury in the summer is difficult. Regardless of turf species, wet conditions often occur on low-lying and downfall areas where water flows via gravity through the surface drainage pattern of greens. This excess wetness can be exacerbated by poor distribution of water from the overhead irrigation system. Distribution

FIGURE 2.38
Annual bluegrass wilts and can die suddenly ("checks out") in the summer.

of water delivery from irrigation systems should be checked periodically using the simple catch-can method. Uniformly distribute cans or pie tins over the putting surface, run the system for several minutes, and then measure the amount of water in cans. If necessary, adjust irrigation head rotation and nozzle type to promote more even water distribution.

Too much water in summer can be much more harmful than too little (Figures 2.39 and 2.40). When soils are excessively wet during summer, irrigation of greens using the overhead system should be avoided until the soil has

FIGURE 2.39
Too much rather than too little water kills more grass in summer. (Photo courtesy of J. Haskins.)

FIGURE 2.40
Mowing excessively wet greens in summer can be extremely damaging. Note casual water.

dried. During hot, windy, and dry periods, hand syringing during the late morning, mid-afternoon, and late afternoon hours may become necessary to avoid wilt or drought injury. In situations where nearly all greens need water, it will be necessary to use the overhead irrigation system. Care must be taken to not overwater, especially low areas of greens that already may be too wet. Thus the weakest greens or those where water from the irrigation system runs-off high spots and accumulates in swales or low spots should be syringed or watered by hand to ensure that excess water is not applied. Promoting dry soils on greens when plant root systems are restricted to the upper 1 to 2 inches (2.5 to 5 cm) of soil generally will require frequent hand syringing during periods of heat stress.

Deep and Infrequent versus Light and Frequent

Careful water management is critical to growing quality creeping bentgrass during summer stress periods and for providing firm and smooth putting surfaces. In summer, golf course superintendents often use daily irrigation combined with hand watering and syringing practices when managing creeping bentgrass. Frequent or excessive irrigation, however, not only increases costs associated with water consumption but can reduce environmental stress tolerance; predispose turf to injury from mechanical stresses; and promote blue-green algae, moss, and diseases.

Two contrasting practices that can be used to maintain golf greens are light and frequent (LF) and deep and infrequent (DI) irrigation. Light and frequent irrigation involves maintaining soil at field capacity; whereas, DI irrigation is imposed at the first sign of leaf wilt (Fry and Huang, 2004). Deep and infrequent irrigation at the time turf shows signs of wilt generally is recommended in summer. Moderate drought does not greatly affect the summer performance of creeping bentgrass monostands.

In a field study, several Providence creeping bentgrass performance and physiological factors were measured as influenced by LF versus DI irrigation (Fu and Dernoeden, 2008, 2009a, 2009b). The LF plots were irrigated daily to moisten the upper 1.6 to 2.4 inches (4 to 6 cm) of soil, while DI plots were irrigated at leaf wilt to wet soil to a depth of ≥9.5 inches (≥24 cm). It was found that creeping bentgrass quality generally was better in LF- than DI-irrigated bentgrass, and DI-irrigated bentgrass had higher canopy temperatures. The negative effects of DI irrigation on visual quality were greatest in the first summer following establishment and less objectionable in the second year in mature turf. In terms of plant response, DI-irrigated bentgrass had lower chlorophyll levels in leaf and sheath tissue in the summer following establishment but developed better color and quality and had higher chlorophyll levels by summer of the second year. This indicated that the turf was adapting to wilt stress or was a function of turf maturity. Creeping bentgrass subjected to DI irrigation developed a less thick thatch-mat layer, which contained 23% less organic matter than was found in LF-irrigated plots.

Slightly more than twice the amount of water was applied to the LF- versus DI-irrigated plots. Deep and infrequently irrigated creeping bentgrass produced a greater number of roots, longer root lengths, and larger root surface areas than LF-irrigated turf throughout most of the rootzone. When compared to data collected in the first year of establishment, the 2-year-old turf had 55% and 32% fewer roots in LF- and DI-irrigated bentgrass, respectively, by the end of the second year. There were similar reductions in root length and root surface area between years in both irrigation regimes. It was concluded that allowing young creeping bentgrass greens to wilt prior to irrigating during the first summer of establishment should be avoided. The DI method, however, did result in a larger root system, less thatch-mat production, and improved chlorophyll levels overtime. The DI-irrigated turf would be expected to promote a more drought stress–tolerant turf with fewer pest problems, and would result in firm surfaces and increased green speed. Additional benefits of this strategy include less water consumption and lower facility energy costs associated with pumping less water. Thus the research supports the old practical adage that mature turf should be irrigated deeply and infrequently at wilt to the depth of the root system.

Avoiding irrigation during cool and moist periods in spring promotes deeper rooting and is the first step in preconditioning turf for summer stresses. Creeping bentgrass seems to perform best in summer if it is stress-conditioned by allowing plants to enter early stages of wilt prior to initiating a regular irrigation program. This practice reduces tissue succulence and "hardens" or "preconditions" plants to better withstand hot and dry conditions. Automatic, light daily irrigation should be avoided prior to the summer stress period. Too much water promotes tissue succulence, wet soils, scald, blue-green algae, moss, disease, black-layer, and reduced stress tolerance. It therefore is best to allow soil to dry-down throughout the rootzone before irrigating deeply and less frequently in summer. When soil is too dry to support plant vigor, the irrigation system should be allowed to deliver enough water to saturate the entire depth of the rootzone. This is an especially important practice on sand-based greens or where water quality is poor, because greens will need to be periodically flushed or purged of soluble salts. Deep irrigation is best timed at predawn, so that any excess or gravitational water can drain before mid-morning and prior to wear imposed by heavy maintenance equipment. Predawn, compared to night irrigation, also helps to avoid prolonged periods of leaf wetness. This is important because long leaf wetness periods at night are a primary factor in increasing the activity of some diseases. Some managers, however, prefer to irrigate at dusk to provide a longer period for plants to recover from stress of the previous day. This is an acceptable practice assuming that disease is not a threat or a preventive fungicide program is in place. Whenever possible, soil should again be allowed to dry-down. Some wilting, however, often develops on greens during hot weather despite adequate soil moisture due to high evapotranspiration rates. These wilted areas should be syringed or hand watered as described below.

Ridges, slopes, and localized dry spots require special attention including use of soil wetting agents, and frequent hand watering or syringing during hot and dry periods to avoid wilt and to keep turf green and alive.

Irrigating nightly for a set period of time on a regular basis can create problems. By watering daily for a set time the upper soil profile remains constantly wet, thus discouraging deep rooting while promoting root loss during periods of heat stress. Furthermore, this practice keeps thatch-mat layers saturated promoting blue-green algae, moss, black-layer, scald (where water puddles), large divots and ball marks, scalping and other mechanical damage, and generally less than optimum playing (i.e., soft) conditions. Excluding annual bluegrass, it is beneficial to keep cool-season turfgrasses as dry as possible during summer. In summary, deep and infrequent irrigation will improve the environmental stress tolerance of turf, helps to discourage pests, minimizes problems associated with large divots and deep ball marks, and enables creeping bentgrass to compete more effectively with annual bluegrass and other weeds.

The previous discussion on irrigation pertains to creeping bentgrass and most other turfgrass species. Greens dominated by annual bluegrass, however, generally require more light and frequent irrigation in summer. Unique irrigation programs may have to be specially designed for individual courses, and sometimes for just a few selected greens (e.g., low lying, shaded, or pocketed). The best practice is to frequently monitor soil moisture with a knife, soil probe, or soil moisture sensor and to visually assess greens for wilt several times daily during hot and dry weather. Many turf managers are using handheld or buried soil moisture sensors that provide quantitative measurements of volumetric soil moisture in the upper 0 to 3 inch (0 to 7.5 cm) soil profile. Individual moisture thresholds for different soils and sites can be assigned to assist with irrigation programming and hand watering to ensure water is placed where it is needed. There still is no substitute for visually assessing dew patterns during early morning hours. This is key to determining daily irrigation, hand watering, or syringing requirements of golf greens during the summer.

Syringing, Fans, and Hand Watering

Syringing and hand watering (also known as manual watering) are essential cultural practices for managing bentgrass areas prone to wilt as well as localized dry spots and fairy rings. Syringing and hand watering are very different practices, and the nature of these differences is discussed below.

Syringing

Turf under drought stress develops a bluish-purple color and is subjected to foot printing (Figures 1.1 and 1.2). The purpose of syringing is to alleviate wilt and high-temperature stress without creating or exacerbating an

FIGURE 2.41
"Chasing wilt" with a hose to syringe and alleviate wilt and heat stress.

existing wet soil or thatch-mat condition on hot summer days. The primary goal of syringing is to rehydrate plants and alleviate wilt stress. Greens are most often syringed to alleviate wilt rather than to abate high-temperature stress in the canopy (Figure 2.41). Fans are more effective in cooling the canopy than syringing in the absence of wilt. Syringing involves applying a light thin film of water on leaves without delivering so much water that the underlying thatch-mat or soil becomes wet. The evaporation of water cools leaves, allowing stomates (i.e., pores on the leaf surface) to open. Assuming adequate soil moisture, the opening of stomates induces the natural movement of water from soil into roots, through the plant via the transpiration stream (i.e., via xylem elements in the vascular system), and water vapor emerges from stomata thereby cooling plants. This process is called *transpirational cooling.*

Wilting of turf in the presence of adequate soil moisture is referred to as "wet wilt" and can occur on windy days when relative humidity is low or when soils are waterlogged during hot and humid periods (see Chapter 1). In drier soils, syringing allows turf to survive that day or until a time when appropriate (i.e., deep) irrigation programs can be scheduled. During periods of high temperature and low relative humidity, and particularly on windy days, syringing may be required three or more times daily. The critical periods generally are between 11 A.M. (11:00 hours) and 4 P.M. (16:00 hours). Syringing, however, is less effective during periods of high humidity when water cannot quickly evaporate. Regardless, it is prudent to lightly syringe wilted greens during periods of high humidity. Greens, however, should not be syringed repeatedly unless all of the water has evaporated and leaves are dry. On most greens there is a fine line between applying too much or too little water. Syringing is one of the unique "arts" of greenkeeping.

Syringing is best performed manually using a handheld hose equipped with a misting nozzle; however, if this is impractical the overhead irrigation system can be used. Generally, a full rotation of the irrigation heads (1 to 2 minutes) is enough to supply a light film of water to the canopy. This practice should be monitored carefully because many irrigation heads are designed to deliver moderate water volumes quickly and with large droplets to reduce wind drift. Thus relying regularly on overhead irrigation for syringing may result in unwanted excess water being applied. Furthermore, many in-ground irrigation systems around greens lack individual head control. Lack of head control may result in excess water being applied, particularly on downslope areas or in the middle of the green where water from irrigation heads overlaps.

Hand syringing is preferred to using the automatic irrigation system because there is more control on where water is needed most. To properly syringe by hand the nozzle should be kept horizontal (i.e., not directed downward) and the operation should take less than 3 to 5 minutes to cover an average-sized (i.e., about 6000 ft^2; 550 m^2) green. In many situations it may be necessary to syringe only localized wilted areas (i.e., "hot spots") rather than the entire surface of greens. It cannot be emphasized enough that it is important not to over-wet the thatch-mat layer and underlying soil during periods of heat stress. Excessive water in the thatch-mat layer and soil on hot and sunny days can lead to high-temperature stress around stems and roots, which may cause yellowing of turf, loss of vigor, and possibly a scald-like condition that can be lethal to turf. There is a natural tendency for new and inexperienced employees to apply too much water while syringing greens during the heat of summer (Figure 2.42). Equipping hoses with syringing nozzles that deliver only 4 to 6 gallons (15 to 23 L) of water per minute may

FIGURE 2.42
New employees should be shown the proper method of syringing.

be a good idea where there is a tendency for employees to apply too much water while syringing. Additionally, handheld soil moisture probes assist in determining soil moisture needs.

Fans for Cooling

Syringing is an effective tool for alleviating wilt and rehydration of plants, but it will only cool the canopy temporarily. Fans can be highly effective in cooling the canopy. More importantly, the combined use of fans and careful syringing have a large impact on reducing canopy and soil temperatures and help to promote root survival in summer (Figure 2.43) (Guertal et al., 2005). If water does not evaporate between syringes the canopy is not cooled. Hence, during periods of heat stress and high humidity or wet wilt, only a fan is likely to provide for sufficient cooling and drying of a golf green. Where permanently mounted fans are not available, the use of portable fans or a buffalo blower (i.e., large blowers for removal of leaves and cores) in conjunction with syringing may provide enough air movement to promote evaporative loss and keep turf healthy. Water temperature (warm versus cold) does not influence the effectiveness of syringing on lowering canopy temperature (Danneberger and Gardner, 2004).

Hand Watering

The practice of hand watering is different than syringing. The intent of hand watering is to wet underlying soil and replenish soil moisture in a specific localized area. The advantage of hand watering is that it allows the manager

FIGURE 2.43
Use of fans and syringing together is effective in cooling the canopy and soil.

to place water where it is needed without overwatering areas that contain sufficient soil moisture. Hand watering involves applying enough water to rewet dry soil areas to rootzone depth. The need for hand watering normally occurs in chronic "hot spots" such as mounds or high spots (usually due to great variation in the depth of the rootzone mix), south-facing slopes, or where wicking of water occurs on collars at the interface between the sand rootzone and adjacent native soil. This normally is accomplished by concentrating water onto small areas. Water is applied using the same showerhead nozzle used to syringe. The nozzle is used in a gentle back-and-forth pattern until the turf surface begins to appear "glassy" (Figure 2.44). It is important not to "fire-hose" the area causing water to run off and puddle in low areas. Once an area has become glassy, the applicator should move onto another area to allow water in the glassy areas to completely penetrate the turf canopy. Applying repeated, small amounts of water will eventually provide sufficient hydraulic head to move water out of thatch-mat and into underlying soil. Do not just wet the thatch-mat, but instead probe the soil frequently to ensure that the rootzone has become sufficiently hydrated. By just wetting the thatch-mat, turf becomes prone to scalping and creates an environment that promotes supraoptimal heating, blue-green algae, and other problems. The ability of a given rootzone soil to be wetted will depend on its physical properties (i.e., texture, compaction, infiltration and percolation rates, and organic matter levels). Turf grown on sand-based rootzones generally requires more frequent hand watering than fine-textured native soil because it holds much less water. It is best to hand water in the evening as air temperature is falling; however, this may be impractical. The next best time for hand watering is early in the morning so that excess water has time

FIGURE 2.44
Slopes and ridges often need to be hand watered.

to drain before traffic is imposed by maintenance equipment and the heat of the day sets in.

Where soils are rendered hydrophobic by fairy rings and localized dry spots, affected areas will require spiking or coring to create openings for water to move into soil. For localized dry spots, most water repellency occurs in the top 1 to 2 inches (2.5 to 5 cm) of soil, but the problem can be as deep as 6 inches (15 cm) (Karnok and Tucker, 2002). Hence, soil wetting agents usually are required to assist with water penetration through the hydrophobic zone of repellency. The same is true for alleviating drought damage caused by fairy rings. Powered water injection devices (e.g., water fork or water injection aeration) also are useful for wetting hydrophobic soils. See Chapter 4 for more information on managing localized dry spots and fairy rings.

Summer Fertilization and Soil pH

An in-depth discussion of soil fertility, fertilizers, and soil and tissue testing is beyond the scope of this book. Carrow et al. (2001) and Christians (2007) provide in-depth coverage of soil fertility management. In native soils, phosphorus (P) and potassium (K) levels at or above 40 lb P and 500 lb K per acre (45 kg P and 561 kg K/ha), respectively, are considered adequate for turfgrass growth (Christians, 2007). In sand-based rootzones, where cation exchange capacity (i.e., nutrient holding capacity of soil) is low, adequate levels of P and K are not usually found and thus adjustments need to be made based on soil test recommendations.

Nitrogen (N) is the most important nutrient used in turfgrass management. Basically, N has the greatest elemental effect on turfgrass growth, green color, and recovery. Most mature creeping bentgrass golf turfs grown on sand-based rootzones should receive about 3 to 6 lb N/1000 ft^2 annually (147 to 294 kg N/ha). Bentgrass grown on fine-textured native soils should receive about 3 to 4 lb N/1000 ft^2 (147 to 196 kg N/ha) annually. Much higher amounts of nitrogen are used during establishment to promote turf coverage and shoot density. During summer, light applications of nitrogen (0.1 to 0.2 lb N/1000 ft^2; 5 to 10 kg N/ha), known as *spoonfeeding* or *foliar-feeding*, should be applied every 1 to 2 weeks. The terms *spoonfeeding* and *foliar-feeding* are often used interchangeably because 37% to 56% of N applied in small amounts (0.1 lb N/1000 ft^2; 5 kg N/ha) in liquid form to foliage is absorbed in about 8 hours (Steigler et al., 2011). Even when applied as a liquid to the canopy, half or more of the N eventually is taken up by roots.

While most N usually is applied in late summer and autumn to creeping bentgrass, about one-third of the annual total is supplied by spoonfeeding. Without overstimulation of grass, spoonfeeding in summer promotes

FIGURE 2.45
Spoonfeeding stimulates the bluish creeping bentgrass to outcompete wilted annual bluegrass in summer.

creeping bentgrass vigor, thus enabling it to more effectively compete with annual bluegrass and to more rapidly recover from divots, ball marks, and mechanical damage (Figure 2.45). Use of water-soluble N sources periodically also has been shown to provide some suppression of anthracnose, dollar spot, and blue-green algae. Spoonfeeding is performed by mixing a soluble N source in water and spraying the material, because it is not possible to otherwise uniformly apply such low amounts of N.

The primary water-soluble N sources include urea, ammonium nitrate, ammonium sulfate, potassium nitrate, and calcium nitrate. Urea is the preferred N source used in summer spoonfeeding programs because it has a lower burn potential than most other water-soluble N sources. Urea also is less expensive and more readily available than most other N sources. Ammonium sulfate has the greatest potential to cause leaf injury or tip "burn" and should be lightly watered-in immediately (Figure 2.46). Calcium and potassium nitrate can cause a tip burn but are less potentially injurious than ammonium sulfate. Avoid using spreaders to apply granular forms of water-soluble N during periods of heat stress and humidity as they can cause speckling and a general foliar burn.

Natural organic fertilizers are good N sources and they are safe. Application of natural organic N sources using a spreader may not be as practical or as effective as low rate (<0.25 lb N/1000 ft²; <12.5 kg N/ha), spray applications of water-soluble N (e.g., urea) in promoting summertime vigor. Some natural organic fertilizers contain water-soluble N from ammonium sources, which can burn greens in the summer. Thus it is prudent to carefully read all the N sources listed on fertilizer bags. The rates of natural organic fertilizer normally should not exceed 0.25 to 0.33 lb N/1000 ft² (12.5 to 16.5 kg N/ha) in

FIGURE 2.46
Ammonium sulfate burn to a golf green due to an application error.

a single application in spring and summer. Because natural organic fertilizers rely on soil microbes to release N, their response is greatly affected by soil temperature. This reliance on microbes makes their expected response sometimes less consistent and unpredictable than water-soluble N. Thus there may be slow or rapid growth provided by natural organics depending on prevailing soil and air temperatures.

Use of a nonburning, complete fertilizer (i.e., N + P + K) usually is recommended if soil phosphorus and potassium levels are low or when fertilizing seedlings. Potassium assists with wear and stress tolerance. Phosphorus is especially critical at seeding but should only be applied to mature turf according to soil test recommendations. Phosphorus can promote blue-green algae on golf greens, and its use should be avoided in summer. More periodic applications of potassium, along with micronutrients like manganese, magnesium, and iron should be considered for high sand rootzones. This is because nutrient levels normally are lower in sand-based rootzones that have a lower ability to hold nutrients (i.e., lower cation exchange capacities). It is popular to apply calcium to sand-based greens, but research has not demonstrated any benefits from using calcium. Calcium (calcium sulfate or gypsum), however, is used to displace sodium in high salt soils. The aforementioned calcium sources also are used where water quality is poor (i.e., high in salts) and in low rainfall areas where salts accumulate in rootzones due to lack of flushing rains.

Humic acids can promote rooting and improve nutrient uptake, but most studies suggest that they have a significant effect only if incorporated into soil prior to establishment, rather than being applied overtop of mature turf (Kaminski et al., 2004). Biostimulants may be beneficial, but most supporting evidence has been forthcoming in laboratory and greenhouse rather than field studies. Biostimulants containing gibberellic acid (GA) stimulate growth and are used to improve density and vigor of "stressed-out" greens. Biostimulants containing GA should not be used continuously after density of greens has been restored. Excessive use of GA-containing materials may cause wider leaf blades, leaf etiolation, grain, and chlorosis, and may defeat plant growth regulator programs. Cytokinins are plant hormones that regulate cell division, root and shoot growth, and chlorophyll synthesis. Cytokinins in roots regulate leaf senescence and regulate plant tolerance to heat stress by promoting root growth in bentgrass. However, most field studies conducted on creeping bentgrass sand-based greens show little or no visual foliar response to these products when applied alone (i.e., without iron or N). Spoonfeeding with water-soluble N is perhaps the most powerful tool in maintaining color and vigor of greens in summer (Settle and Dernoeden, 2009). Both humic acid and biostimulant products should be viewed only as supplements to a sound nutritional program based on adequate elemental (N-P-K) fertilizer. These products are not designed to substitute for basic nutrition.

Soil pH in the range of 6 to 7 is preferred. A pH in this range allows plants to absorb nutrients more efficiently. For several pest management programs, such as for suppression of annual bluegrass, earthworms, sod webworm, summer patch, take-all patch, and dead spot, a soil acidification program may be recommended. In these programs, acidification using ammonium sulfate rather than sulfur or other chemicals is preferred. Generally, 3 to 5 lb N/1000 ft^2/year (147 to 244 kg N/ha/year) from ammonium sulfate is recommended. Bentgrasses perform well in acid soils ranging from a pH of 5.3 to 6.0, but lower pH's should be avoided. In the summertime, ammonium sulfate can be applied through a sprayer at 0.10 to 0.125 lb N/1000 ft^2 (5 to 6 kg N/ha) in spoonfeeding programs. Ammonium sulfate, however, can scorch, burn, or otherwise injure tissue and should be lightly syringed off leaves immediately following application, even when air temperatures are below 80°F (27°C). Use of wetting agents in tank-mixes with ammonium sulfate can cause severe discoloration; thus they should be applied separately. Moderate to high amounts (>0.5 lb N/1000 ft^2 or 25 kg N/ha in a single application) of ammonium sulfate applied over a short period of time can cause pH to drop rapidly in sand-based rootzones. When this occurs creeping bentgrass may develop a gray color or a gray and yellow mottling (Figure 2.47). Small amounts of limestone or calcium carbonate (5 to 10 lb/1000 ft^2; 244 to 488 kg/ha) increase pH and restore green color; however, the process may take a few weeks to have its beneficial effect. Hence, it is best to apply ammonium sulfate to bentgrass grown on sand-based rootzones in small increments of ≤0.20 lb N/1000 ft^2 (≤10 kg N/ha) in any single application. Excluding perhaps the Pacific Northwest United States and Canada, acidification with elemental sulfur is discouraged because it can be phytotoxic. Calcium and potassium nitrate increase soil pH over time, which can promote some diseases such as take-all patch and dead spot.

FIGURE 2.47
Yellow-gray mottling in creeping bentgrass due to a rapid drop in pH can occur with ammonium sulfate use on sand-based rootzones.

Surface and Subsurface Drainage

Agronomists universally agree that without good drainage you cannot grow quality turf. New sand-based greens have infiltration rates in excess of 20 inches (50 cm) per hour. As greens age, organic matter increases not only in the form of a thatch-mat layer, but throughout the profile (Kerek et al., 2002). The source of this organic matter in the profile is from root systems past and present as well as algae and invertebrates like insects and earthworms. Organic matter increases the water-holding capacity of soil. Thus drainage slows as organic matter increases in the rootzone profile. Researchers attempted to quantify a critical organic matter content threshold (i.e., percentage by weight) where the negative effects of excess organic matter occur and results varied between 3% and 8%. Generally, problems begin when organic matter exceeds 3% to 4% throughout the rootzone profile. Poorly drained creeping bentgrass areas, and especially golf greens, are fraught with problems including scald, wet-wilt, blue-green algae, moss, black-layer, direct and indirect heat stress, and other stress- and disease-related problems. Where drainage is poor, it is not possible to provide a quality putting surface because traffic from players and equipment eventually will cause extensive turf damage (Figure 2.48). Furthermore, during rainy periods, poorly drained golf courses must be closed for much longer periods, which reduces revenue and aggravates golfers. Ensuring effective surface water drainage is essential and should be accomplished during the construction

FIGURE 2.48
Wear around the cup on greens is exacerbated by wet surfaces.

phase. Unfortunately, this often is not the case, and expensive measures eventually have to be taken to correct surface drainage problems.

Excessively wet areas on golf greens will require special attention. This may involve something as simple as creating holes in the surface with a pitchfork or solid tine aeration unit in chronically wet spots during summer or as drastic as trenching through greens to install supplemental drainage lines. Adding coarse-textured sand into channels will improve lateral drainage. Several methods are available ranging from the low-cost installation of thin (<2 in; <5 cm) channels of coarse sand (i.e., slit drainage) to the more expensive practice of installing subsurface drainage pipes via sand channeling. Installing slit drains throughout a green can be done rapidly but causes a bumpy surface that takes considerable time and sand topdressing to rectify (Figure 2.49).

Sand channeling involves removal of narrow strips of sod, trenching, installing pipe, and backfilling with coarse materials and reinstalling the sod (Figure 2.50). Once installed, some superintendents use drill and fill aeration between channels to further improve water infiltration and drainage (Figure 2.51). Sand channels are very effective in improving surface water drainage with minimal disruption and can be done rapidly. Another option is called *passive capillary drainage* in which a synthetic "rope" in which an internal stainless steel hollow core is installed in the rootzone about 4 to 8 inches (10 to 20 cm) below the surface. These ropes are led from the green cavity to a low-point water collection area. Basically, water is wicked out of wet areas and travels through the hollow core and ultimately drains. Installation of ropes is minimally disruptive.

Where an existing sand-based rootzone exists which possesses good internal drainage, adding additional subsurface drainage may be appropriate in

FIGURE 2.49
Slit sand drains can be installed rapidly and inexpensively but disrupt the putting surface for
a long period.

FIGURE 2.50
Channel drainage systems are very effective and can be quickly installed.

low-lying areas where pipe or drainage lines do not exist or have been damaged or plugged. The addition of "smile drains" at the lowest point of the
golf green cavity also may aid in facilitating drainage in chronically wet
areas. Plastic liners used between sand-based rootzones and native soil can
act as dams and create anaerobic (i.e., black-layer) conditions. Therefore in
new construction, it is important to ensure drain lines are placed through
liners in low areas or that perforations exist in the barrier at low points. In

FIGURE 2.51
Drill and fill is performed between channels to further improve water drainage.

some cases, liners are either not installed or are removed if it is felt that they impede drainage out of the low side (i.e., false fronts).

Whichever corrective method is chosen, these practices facilitate internal drainage in situations where surface drainage is insufficient to quickly remove excess water. If surface drainage does not already exist, then those issues should also be addressed because installing internal drainage may be a temporary Band-Aid to a larger, more chronic problem. These practices are best accomplished in late autumn or winter when the greens are being used less frequently.

In addition to drainage installation, other technologies are available that may help improve subsurface drainage in golf greens. Mechanically induced subsurface air-movement/drainage (e.g., Precision-Air® or Soil-Air Technology®) devices are designed to take advantage of existing subsurface drainage pipes to remove excess soil water via vacuum removal or to inject air into the rootzone. These systems are attached to the drainage outlet and pull excess water from underneath the green. Research conducted on contoured golf greens, however, showed that these devices may not be very effective for water removal in a sand-based rootzone system. A field study that evaluated one of these systems showed that water content of vacuum-treated plots was decreased only 2% to 3%, when compared to the untreated plots (Bigelow et al., 2001). These rather small reductions in water content may not be agronomically important.

It is important to mention that a properly constructed sand-based rootzone is designed to resist compaction and drain quickly, while retaining a reservoir of moisture as a "perched water table" above the gravel drainage blanket. In theory, these idealized physical properties exist in a very new golf green that has experienced little accumulation of organic matter. In reality, soil physical characteristics change in older greens. As organic matter is deposited over time, irrigation practices associated with maintaining a mature turfgrass stand during summer months may not lend themselves to successful use of a subsurface drainage system. As rootzones mature, two regions of wetness develop. One region of wetness develops at the top of the rootzone, which is related to the matric forces of the thatch-mat and accumulation of moisture associated with light and frequent irrigation cycles. Another wetness zone is located at the sand–gravel interface (i.e., "perched water table"). It is the zone of wetness near the surface that becomes difficult to manage. Due to a discontinuity of water between these two wet zones, a disconnect occurs in capillary continuity between the wet zones, which limits the effectiveness of subsurface vacuum removal machines. Because all of the water is not connected, there will be variation in the matric forces between the highly organic layer and the underlying unaffected rootzone materials. Therefore vacuum machines may be unable to extract water from the thatch-mat layer on greens. This renders them ineffective for removing excess surface moisture, which is a major cause of turf decline during summer stress periods. Hence, carefully controlled supplemental irrigation practices and utilization of surface fans may be the best strategy for promoting a dry, firm putting surface and healthy turf.

In addition to water removal, these devices can be reversed to blow air into the soil at night when ambient temperature is cooler than soil temperature. These systems have the ability to change soil temperature and modify soil gas composition (Bunnell et al., 2004). The changes, however, were rather small (2 to 5°F; 1 to 2°C) when the soil temperatures were >85°F (>29°C). This may be enough cooling to help reduce root respiration and other stresses associated with supraoptimal temperatures. Golf course managers in the Southeastern United States who utilize this method of cooling, however, believe that even small temperature modifications during intense summer stress conditions are helpful. Conversely, newer systems that blow refrigerated air can reduce soil temperature significantly in the rootzone. Cooling the rootzone reduces root respiratory carbon consumption, helps maintain or increase carbohydrate accumulation, and improves turf quality during hot summer periods.

Subsurface air movement systems may be most effective for removing toxic gases like excess carbon dioxide (CO_2), methane (CH_4), or hydrogen sulfide (H_2S) that become more pronounced during warm weather. These gases normally are higher in situations where the existing soil is not well oxygenated. Therefore, either pulling or pushing surface air that contains higher concentrations of oxygen and very low concentrations of harmful gasses into the rootzone may be beneficial. A larger problem, however, probably exists with regard to the internal physical characteristics of the rootzone, which may need to be addressed. Hence, using these devices only may be a temporary remedy for the problem. In any case, subsurface drainage technologies are primarily used where golf green turf is grown on high sand-content mixes. They seem to be best suited for greens located in shade, and where surface-air circulation or where existing internal drainage is poor. These systems are expensive to install and operate.

Small (0.25 in; 6.2 mm) solid coring tines, spiking, water injection aeration, and fans also provide short-term drying and aeration benefits when soils are excessively wet. In many situations, supplemental slit trenches installed through the top 12 inches (30 cm) of soil can provide drainage through wet spots in fairways and green surrounds and work well for draining excess water.

Where problems occur year after year to any particular green, it is important to evaluate all factors affecting the surrounding environment. Surface-air drainage is nearly as important as sound surface and internal water drainage. Enhancing the natural drying of greens by creating an environment conducive to ideal evapotranspiration is achieved by removing low-growing brush and trees and where appropriate installing fans around pocketed green. Finally, there is no substitute for adequate surface drainage. Eliminate any water puddles as soon as possible; this can be accomplished simply by using a squeegee to push standing water away. This will help to avoid scald in summer and retard the development of blue-green algae. If not controlled, blue-green algae can dramatically affect the ability of water or gases to penetrate the surface and actually may function like an unwanted gasket, which seals the surface and suffocates roots.

Common Pesticide Injuries

The potential for a phytotoxic reaction from pesticides increases under conditions of heat stress, high humidity, and low mowing (Figure 2.52). Products formulated as emulsifiable concentrates are potentially most injurious and should be applied with caution during hot weather. Most turf pesticides today are available in safer formulations. Herbicides present a greater potential risk of injury than either fungicides or insecticides. Whenever in doubt, apply pesticides in the evening when temperatures are falling versus the morning when temperatures are rising in the summer.

Some *Pythium*-targeted fungicides can cause a tip burn or scorch golf green turf if applied in the heat of the day. Chloroneb and etridazole are notorious for burning greens when applied during warm and humid periods. Fosetyl-aluminum and phosphites can cause a tip burn when applied in hot weather. High rates and repeated applications of dimethylation (DMI) sterol-inhibiting (SI) fungicides such as triadimefon, propiconazole, fenarimol, tebuconazole, and myclobutanil can elicit a blue-green or brown color and cause thinning in creeping bentgrass greens. Frequent applications of high rates of DMI/SI fungicides in summer can widen and darken bentgrass leaf blades, giving the golf green canopy a "leafy" appearance, especially when used in combination with a plant growth regulator program. Some formulations

FIGURE 2.52
Green burned with nitrogen tank-mixed with a plant growth regulator and applied on a bright summer day.

of triticonazole and triadimefon can injure annual bluegrass. Dry flowable formulations of chlorothalonil may yellow turf temporarily. Tank-mixes of phosphite fertilizers and fungicides with the growth regulator trinexapac-ethyl can injure bentgrass in summer, especially when applied on a hot afternoon. Some insecticides like chlorpyrifos and trichlorfon can scorch greens in hot weather. There have been relatively few reports of phytotoxicity from pyrethroid and other insecticides, but it would be prudent to apply these materials in the evening during periods of hot weather. Increasing the amount of water or spray volume (e.g., 100 gallons/acre; 935 L/ha) also helps to reduce potential phytotoxicity from pesticides.

There is no faster way to kill or injure bentgrass than with the improper use of an herbicide. Most herbicides used on creeping bentgrass will elicit a yellowing and should only be applied in early spring or autumn during cool periods when turf is actively growing. Yellowing from early spring applied preemergence herbicides, however, may not appear for 30 days after application. Granular forms of preemergence herbicides that wash due to heavy rains and accumulate in low areas can cause extensive damage, especially on golf greens. Bispyribac-sodium, fenoxaprop-ethyl, and quinclorac are used during warmer periods in summer on bentgrass tees and fairways to control some grass weeds (none labeled for greens use). Herbicides like chlorsulfuron and metsulfuron are used for control of some grass weeds in the autumn. All five of these herbicides are likely to elicit a yellowing in creeping bentgrass but do not cause severe injury when properly applied. Tank-mixing the aforementioned herbicides with chelated iron partially masks the chlorosis associated with their use. In general, use of broadleaf herbicides should be avoided during hot summer weather. Clopyralid, however, is regarded as being among the safest broadleaf herbicides for use on bentgrass to control white clover in summer. See Chapter 7 for more information on using herbicides on creeping bentgrass.

Some plant growth regulators (PGRs), such as paclobutrazol and flurprimidol, are not applied during stressful summer periods on weak or severely stressed turf. Trinexapac-ethyl generally is more benign than other PGRs and is more likely to be recommended for use on golf greens in the summer. All three can cause discoloration when applied in cooler weather. See Chapter 8 for more information about plant growth regulators. Products that contain copper hydroxide can yellow turf in hot weather, especially annual bluegrass and immature creeping bentgrass.

Summary

- Select creeping bentgrass cultivars based on regional performance, resistance to dollar spot and other diseases, and adaptability to your budget and likely cultural inputs.

- Summer stress management of golf greens begins with increasing mowing height and reducing mowing frequency. Use solid rather than grooved rollers. Do not mow excessively wet or soft greens, especially during periods of heat stress (>86°F; >30°C). Roll more and mow less.

- Green plants in core aeration holes surrounded by tan or brown grass indicate that thatch-mat layers had become either very wet or very dry on a hot summer day.

- Thatch-mat control on golf greens requires routine inputs of coring, sand topdressing, vertical cutting, and proper irrigation and nitrogen fertility management. The organic layer should be uniformly intermixed (i.e., diluted) with sand topdressing (i.e., mat) and not layered.

- Delay sand topdressing, core aeration, and other abrasive cultural practices when greens are in a weakened condition during summer stress periods.

- Thatch-mat layers and stems may swell during hot and humid periods causing puffiness and scalping.

- Grooming is used to control grain and improve trueness of putting surfaces.

- Sand topdress heavily in spring and late summer with sand to fill holes following coring with wide-diameter tines. Topdress with sand lightly and frequently throughout summer.

- Periodic narrow-tine aeration during the summer can be very beneficial—reincorporate sand by brushing in cores and blowing off excess organic matter.

- Aggressive, deep vertical cutting and incorporation of sand is best performed during cool and moist periods of autumn.

- Rolling is best performed on days when mowing is skipped or to maintain consistent green speed throughout the day during tournaments.

- Irrigate creeping bentgrass deeply and infrequently. Hand syringe frequently when turf is showing signs of wilt and avoid water puddling or excess soil wetness. Hand water "hot spots" that are chronically dry.

- Remove trees and brush to improve sunlight penetration and air circulation.

- Install fans to promote air circulation to cool the canopy and speed soil cooling and drying.

- Reduce excessive soil wetness by installing or replacing drainage lines. Keep drainage cleanouts and outfalls open and cleared where water from drain lines exit greens.

- Spiking and solid-tine aeration will provide some short-term aeration (i.e., oxygen) and drainage benefits when soils are excessively wet. Avoid aeration/spiking during the heat of the day, especially when turf is showing signs of wilt or wet wilt.

- Most (≥50%) of the total annual N use should be applied to bentgrass in the late summer and autumn months, but substantial amounts (≥33%) are applied in summer spoonfeeding programs.

- Spoonfeed liquid nitrogen at low levels (0.1 to 0.2 lb N/1000 ft²; 5 to 10 kg N/ha) every 7 to 14 days during summer to promote color and recovery from injury. Use chelated iron (can cause foliar blackening if applied in hot weather) or Epsom salts ($MgSO_4$) to improve turf color without stimulating turf growth.

- Water-soluble nitrogen can burn turf and should be watered-in immediately in the summer, especially when spoonfeeding greens.

- Use caution when applying pesticides (especially those formulated as emulsifiable concentrates) during periods of high-temperature stress. Most herbicides and some fungicides and insecticides applied during summer stress periods will cause yellowing and in some cases burning of turf.

- Trinexapac-ethyl is the preferred plant growth regulator for use during periods of high-temperature stress. The use of PGRs, however, should be avoided on damaged or weak greens because their use may slow recovery.

Bibliography

Beard, J.B. 2002. *Turf Management for Golf Courses*, 2nd ed. Ann Arbor Press, Chelsea, MI.

Bigelow, C.A., D.C. Bowman, D.K. Cassel, and T.W. Ruffy. 2001. Creeping bentgrass response to inorganic soil amendments and mechanically induced subsurface drainage and aeration. *Crop Sci.* 41:797–805.

Bunnell, T.B., L.B. McCarty, and H.S. Hill. 2004. Soil gas, temperature, matric potential, and creeping bentgrass growth response to subsurface air movement on a sand-based golf green. *HortScience* 39:415–419.

Carrow, R.N., D.V. Waddington, and P.E. Rieke. 2001. *Turfgrass Soil Fertility and Chemical Problems: Assessment and Management*. Wiley, Hoboken, NJ.

Christians, N.E. 2007. *Fundamentals of Turfgrass Management*. Wiley, Hoboken, NJ.

Cooper, R.J., C. Liu, and D.S. Fisher. 1998. Influence of humic substances on rooting and nutrient content of creeping bentgrass. *Crop Sci.* 38:1639–1644.

Danneberger, T.F., and D. Gardner. 2004. Syringing can dramatically affect canopy temperature. *Turfgrass Trends* June:63–64.

Davis, J.G., and P.H. Dernoeden. 2002. Dollar spot severity, tissue nitrogen and soil microbial activity in bentgrass as influenced by nitrogen source. *Crop Sci.* 42:480–488.

Dernoeden, P.H., and J. Fu. 2008. Fungicides can mitigate injury and improve creeping bentgrass quality. *Golf Course Management* 76(4):102–106.

Fry, J., and B. Huang. 2004. *Applied Turfgrass Science and Physiology*. Wiley, Hoboken, NJ.

Fu, J., and P.H. Dernoeden. 2008. Carbohydrate metabolism in creeping bentgrass as influenced by two summer irrigation practices. *J. Amer. Soc. Hort. Sci.* 133:678–683.

Fu, J., and P.H. Dernoeden. 2009a. Carbohydrate level, photosynthesis, and respiration in creeping bentgrass as influenced by spring and summer coring. *J. Amer. Soc. Hort. Sci.* 134:41–47.

Fu, J., and P.H. Dernoeden. 2009b. Creeping bentgrass putting green response to two summer irrigation practices: Rooting and soil temperature. *Crop Sci.* 49:1063–1070.

Fu, J., and P.H. Dernoeden. 2009c. Creeping bentgrass putting green responses to two irrigation practices: Quality, chlorophyll, canopy temperature and thatch-mat. *Crop Sci.* 49:1071–1078.

Fu, J., P.H. Dernoeden, and J.A. Murphy. 2009. Creeping bentgrass color and quality, chlorophyll content and thatch-mat accumulation response to summer coring. *Crop Sci.* 49:1079–1087.

Guertal, E.A., E. van Santen, and D.Y. Han. 2005. Fan and syringe application for cooling bentgrass greens. *Crop Sci.* 45:245–250.

Jordan, J.E., R.H. White, D.M. Vietor, T.C. Hale, J.C. Thomas, and M.C. Engelke. 2003. Effect of irrigation frequency on turf quality, shoot density, and root length density of five bentgrass cultivars. *Crop Sci.* 43:282–287.

Kaminski, J.E., C.A. Bigelow, and P.H. Dernoeden. 2004. Soil amendments and fertilizer source effects on creeping bentgrass establishment, soil microbial activity, thatch and disease. *HortScience* 39:620–626.

Karnok, K., and K. Tucker. 2002. Water-repellent soils: Part 1: Where are we now? *Golf Course Management* 70(6):59–62.

Kerek, M., R.A. Drijber, W.L. Powers, R.C. Sherman, R.E. Gaussoin, and A.M. Streich. 2002. Accumulation of microbial biomass with particulate organic matter of aging golf greens. *Agron. J.* 94:455–461.

Li, D., D.D. Minner, and N.E. Christians. 2009. Evaluation of factors contributing to surface stability of sand-based turf. *Agron. J.* 101:1160–1167.

Lui, X., and B. Huang. 2002. Mowing effects on root production, growth, and mortality of creeping bentgrass. *Crop Sci.* 42:1241–1250.

Lyons, E.M., P.J. Landschoot, and D.R. Huff. 2011. Root distribution and tiller densities of creeping bentgrass cultivars and greens-type annual bluegrass cultivars in a putting green. *HortScience* 46:1411–1417.

McCarty, L.B. 2010. *Best Golf Course Management Practices*. Prentice-Hall, Upper Saddle River, NJ.

McCarty, L.B., M.F. Gregg, J.J. Camberato, and H.S. Hill. 2005. Minimizing thatch and mat development in a newly seeded creeping bentgrass golf green. *Crop Sci.* 45:1529–1535.

McCarty, L.B., M.F. Gregg, and J.E. Toler. 2007. Thatch and mat management in an established creeping bentgrass golf green. *Agron. J.* 99:1530–1537.

Moeller, A.C., C.A. Bigelow, J.R. Nemitz, and G.A. Hardebeck. 2008. Bentgrass cultivar and annual nitrogen regime affects seasonal shoot density. *Applied Turfgrass Science*. doi:10.1094/ATS-2008-0421-01-RS.

Murphy, J.A., P.E. Rieke, and A.E. Erickson. 1993. Coring cultivation of a putting green with hollow and solid tines. *Agron. J.* 85:1–9.

Nikolai, T.A. 2002. More light on lightweight rolling. *U.S. Golf Association Green Section Record* 40:9–12.

Settle, D., and P.H. Dernoeden. 2009. Evaluation of cytokinins plant extract biostimulants, iron and nitrogen products for their effects on creeping bentgrass summer quality. *U.S. Golf Association Turfgrass and Environmental Research* 8(1):1–15.

Skorulski, J., J. Henderson, and N. Miller. 2010. Topdressing fairways: More is better. *U.S. Golf Association Green Section Record* 48:14–17.

Steigler, J.C., M.D. Richardson, and D.E. Karcher. 2011. Foliar nitrogen uptake following urea application to putting green turfgrass species. *Crop Sci.* 51:1253–1260.

Turgeon, A.J. 2008. *Turfgrass Management*. Pearson Prentice Hall, Upper Saddle River, NJ.

U.S. Golf Association Green Section Staff. 2004. USGA recommendations for a method of putting green construction (USGA.org/Content.aspx?id=25890).

Whitlark, B. 2008. Christmas wish: Firm and fast fairways. USGA Green Section Southwest regional update (www.usga.org/turf/regional updates/regional updates.asp).

Zontek, S.J. 2005. Fairway topdressing in the Mid-Atlantic region. *U.S. Golf Association Green Section Record* 43(1):18–21.

3

Selected Abiotic and Imperfectly Understood Maladies of Creeping Bentgrass

There are several abiotic maladies of creeping bentgrass that are weather related. Stresses can induce physiological changes in plants that may manifest themselves in the form of color changes in creeping bentgrass and annual bluegrass foliage. Frost may induce a purpling and high-temperature stress a leaf chlorosis (i.e., yellowing). Both of these color changes can be confused with diseases. Mechanical stresses causing bentgrass to decline similarly can be confused with disease. Laboratory diagnosticians help to sort out many of these issues for golf course managers. Turfgrass pathologists can be baffled by the presence of fungi and bacteria that are found in association with chlorosis and decline, but there often is no consistent relationship among symptoms and organisms. Many of these organisms are saprophytes (living or senescent, dying or dead tissues or senectophytes), but their dominance in samples can be good indicators of other stress factors. Then there are symptoms such as etiolated, yellow-green leaves or white, chlorophyll-deficient leaves that mimic disease, but no causal agent can be found. Some examples of the latter include etiolated tiller syndrome and white leaf disease. There also are diseases that are new and little is known about them. Acidovorax bacterial disease fits this category and is considered in this section. Finally, when environmental and mechanical stresses combine with poor growing environments, pathogens, and saprophytes there can be foliar color changes, decline in density, and some or all can contribute to the summer bentgrass decline complex.

Red or Purple Golf Greens in Winter and Spring

The first frosty nights in autumn bring about some pronounced physiological changes in plants. In creeping bentgrass and annual bluegrass, especially on golf greens, leaves may develop a reddish-brown, purple, or blue-gray color. The discoloration may be uniform in some cultivars (especially 'Crenshaw'), but frequently the various shades of purple, red, or blue appear in circular patches (Figure 3.1). These patches constitute different clones within the bentgrass polystand. In annual bluegrass grown on greens during winter,

FIGURE 3.1
Creeping bentgrass clones turning purple in response to low temperatures is a physiological phenomenon.

the perennial biotypes (i.e., *Poa annua* spp. *reptans*) often develop a purple color while the annual types remain green (Figure 3.2). The colors are most prevalent on older greens. Prior to the advent of 'Penncross', many greens were seeded with 'Seaside' and 'South German' bentgrasses. The aforementioned cultivars were genetically variable, and as a result all plants emerging from seed were not true to type. Today, even the most advanced cultivars grown in monostand can segregate and distinctly different colored patches

FIGURE 3.2
Perennial biotypes of annual bluegrass may develop a dark-purple color in response to low temperatures.

may develop. As individual plants grow, the more aggressive biotypes dominate in a natural selection process to produce circular patches (i.e., segregated clones). These patches are not unlike the circular areas of blighted turf associated with some diseases. The purplish discoloration often is misdiagnosed as red leaf spot (*Drechslera erythrospila*) by some golf course superintendents. Red leaf spot, however, is a late spring and summer disease of colonial bentgrass and some cultivars of creeping bentgrass (e.g., 'Penn A-1', vegetative, and South German bentgrasses).

If the purpling is due to frost, the underside of purple leaves will retain their green color. The purpling from frosts often is prominent in Penncross golf greens, but similar color changes appear in older stands of most other bentgrass cultivars. As previously noted, a similar purpling can occur in annual bluegrass, but it is more commonly associated with the perennial rather than annual biotypes. Segregated clones can react differently to the application of some chemicals.

Why do golf green grasses turn red, blue, or purple in response to low night temperatures? The cool to cold temperatures trigger the color responses. During mid to late autumn, many regions experience relatively warm days (65 to 75°F; 18 to 24°C), but cool to cold nights (32 to 55°F; 0 to 13°C). Sunny, bright, and warm days stimulate plants to produce large amounts of sugars (through photosynthesis) in leaves and sheaths. At night sugars must be translocated out of leaves to crowns for storage or use in other physiological processes. When nights are very cool or frosty, sugars are not completely moved out of leaves and they accumulate. There are many types of sugars. Glucose is a common plant sugar and sometimes glucose molecules are chemically bound with anthocyanins (Salisbury and Ross, 1992). Anthocyanins are pigments, and their function in plants is unclear. The word *anthocyanin* is from Greek: anthos means "flower" and kyanos means "dark blue." Anthocyanins provide the red, purple, and blue colors in flowers. Anthocyanins always are present in leaves but normally are masked by the presence of chlorophyll. They are expressed in the foliage of trees during cool and bright weather to provide the spectacular colors in autumn leaves. Hence, creeping bentgrass and annual bluegrass may experience a similar accumulation of sugar, and therefore anthocyanins, following the first cold and frosty nights of autumn. Frost injury may denature the green chlorophyll, thereby exposing the anthocyanin pigment. These colors may intensify and persist throughout winter months and slowly disappear in mid-spring after turf begins active growth.

A similar phenomenon appears to occur on creeping bentgrass and annual bluegrass greens, tees, and fairways in the spring. This usually coincides with unseasonably warm temperatures in late winter or early spring, which stimulates a premature green-up of bentgrass and annual bluegrass. Should night temperatures plummet into the low 20s°F (–7°C), or there are several nights of frost following a premature green-up, creeping bentgrass and annual bluegrass again may develop a reddish-brown, brick-red, or purple color. This condition is aggravated by sand topdressing, brushing, core aeration, foot traffic (especially around cupping and walk-on and walk-off areas), and other grooming practices performed at the time cold temperatures recur following an early green-up. Application of some plant growth regulators, particularly paclobutrazol and flurprimidol, coinciding with frost or dry and windy conditions can cause an objectionable reddish-brown or purplish-blue discoloration in bentgrass. Trinexapac-ethyl can cause yellowing in annual bluegrass when applied in cold weather. Sometimes only sensitive clones turn purple, red, or reddish-brown, but usually the effect of the aforementioned plant growth regulators in cold weather is more generalized.

Reddish-brown lesions may develop on creeping bentgrass or annual bluegrass leaves, but it should be noted that leaf lesions can develop in response to many different kinds of injury mechanisms. The lesions observed on red or purple leaves during cold periods most likely are caused by mechanical injury associated with grooming, sand topdressing, or mowing greens too early in the morning following a frosty night. As previously noted, applications of some plant growth regulators (especially paclobutrazol and flurprimidol) that coincide with frosts can result in discoloration of creeping bentgrass turf. The bentgrass will recover rapidly with the advent of consistently warmer weather (i.e., days >70°F; >21°C and nights >45°F; >7°C). An application of 0.125 to 0.25 lb N/1000 ft² (6 to 12 kg N/ha) from a water-soluble nitrogen source will speed recovery as temperatures rise.

Despite the aforementioned explanation, many persist in believing that the reddish color of greens must be disease or a nutrient deficiency. After all, textbook photographs of red leaf spot disease and phosphorus deficiency depict symptoms similar to the reddening or purpling associated with frosts. To explain this better it may be helpful to review the Helminthosporium diseases and their incitants. Many of the fungi that cause leaf spotting and melting-out diseases of turfgrasses once belonged in the taxonomic genus *Helminthosporium*. Today, these fungi are classified as species of *Drechslera* or *Bipolaris*. Turfgrass pathologists (Smith et al., 1989; Couch, 1995; Smiley et al., 2005; Vargas, 2005) agree that red leaf spot (*D. erythrospila*) is a warm weather disease that develops in late spring or summer. Vincelli and Doney (1995) reported that colonial, brown top, and dryland bentgrasses were very susceptible to red leaf spot, but all 15 cultivars of creeping bentgrass evaluated in their trial showed high levels of resistance. There are other *Drechslera* species known to be associated with bentgrasses, and they include *D. catenaria* and *D. gigantea. Drechslera catenaria* is the most likely species to attack bentgrass in the spring, especially vegetatively propagated (e.g., 'Toronto') creeping bentgrass cultivars. As temperatures rise in mid-spring, *D. dictyoides* can be found producing large numbers of spores on senescent or dead leaves of creeping bentgrass. In most cases, *D. dictyoides* is behaving as a saprophyte (i.e., an organism living on senescent or dead organic matter), because there generally are no lesions on green leaves, and spores are only found on dying or dead leaves. Red leaf spot, however, develops as temperatures rise in early summer in colonial bentgrass and Penn A-1 and possibly other creeping bentgrass cultivars.

Most reports of *Bipolaris* and *Drechslera* diseases in *Agrostis* (i.e., bentgrasses) species involved colonial bentgrass, redtop, or vegetatively propagated creeping bentgrasses (e.g., 'Congressional', 'Old Orchard', Toronto, 'Washington', and other C-series cultivars). Except for Toronto in the Great Lakes region, these grasses are no longer used in the United States. Hence, the relative rarity of *Bipolaris* and *Drechslera* diseases in creeping bentgrasses today is due in part to improved resistance in Penncross as well as a large majority of the cultivars released since Penncross. Furthermore, the widespread usage of broad-spectrum fungicides applied from late autumn to spring for controlling winter diseases on golf greens (e.g., snow molds) also may help explain why leaf spotting and melting-out diseases are uncommon in spring in creeping bentgrass and annual bluegrass.

During summer, *Bipolaris sorokiniana* can attack creeping bentgrass and annual bluegrass, but the melting-out phase is uncommon in both grasses. Occasionally, *B. sorokiniana* attacks only selected clones in older stands of 'Penneagle' and other cultivars. It is not unusual, however, to find a few zonate lesions (i.e., circular to oblong, brownish-purple lesions with or without a tan spot in the center) produced in response to *B. sorokiniana* spore penetration on creeping bentgrass and annual bluegrass leaves during summer stress periods. Colonial bentgrass, however, may exhibit Bipolaris melting-out in the summer, particularly when treated routinely with dimethylation or sterol-inhibitor (i.e., DMI/SI) fungicides; thiophanate; or flutalonil (see Chapter 6 for more information about these fungicides). In the aforementioned situation, the fungicides noted can promote leaf spotting and melting-out, but the phenomenon is uncommon.

Some golf course superintendents reported seeing a positive response from a fungicide applied in the spring to purple, frost-injured greens. This would be more convincing if an untreated strip were left for comparative purposes. Some fungicide combinations (e.g., fosetyl-aluminum + chlorothalonil, green-pigmented mancozeb, and other pigmented fungicides) can induce a green color enhancement in bentgrass, even in winter following a late autumn application. Therefore, in some cases a positive response from a fungicide applied to a frost-discolored green may be due to a "paint" effect or color enhancement elicited by the chemical.

A somewhat similar purpling or blackening of leaves also may be elicited by the following: iron or sulfur applications; ammonium sulfate applications; low soil phosphorus levels; high application rates of some DMI/SI fungicides (especially propiconazole and triadimefon); some plant growth regulators, particularly when applications coincide with frost; and in past years toxicity from arsenic used in annual bluegrass control programs on greens. The foliar color responses induced by the aforementioned chemicals are well known and can occur at any time of year.

Summary of Key Points

- Red or purple creeping bentgrass or annual bluegrass foliage appearing between early winter and spring develops in response to cold temperatures and frost.

- The purpling response is physiological in nature and is most pronounced in clones within a bentgrass polystand and in perennial biotypes of annual bluegrass.

- This physiological response reduces aesthetic quality but is not harmful.

- The turf will recover its normal green color with the advent of consistently warmer temperatures as spring progresses. In spring as temperatures rise, spoonfeeding with water-soluble nitrogen (e.g., urea) will speed the replacement of discolored leaves with green leaves.

- Applications of flurprimidol and paclobutrazol plant growth regulators coinciding with frost can elicit a red, reddish-brown, purplish-blue, or bronze color in creeping bentgrass foliage.

- Except for Penn A-1 and some older or vegetative cultivars, red leaf spot is an uncommon disease of creeping bentgrass, and it is not known to occur in annual bluegrass.

Summer Bentgrass Decline

The decline of cool-season turfgrasses during stressful summer months is not a new problem (Beard and Daniel, 1965). As noted previously in the Introduction and Chapter 1, summer decline usually involves a complex of two or more interacting biotic and or abiotic stresses. An important aspect of decline is the natural loss of root mass as soils warm in early summer. Root decline accelerates in response to low mowing heights and reduction in soil oxygen due to excessive rainfall or over irrigation. The latter commonly occur in summer in greens with poor water infiltration and percolation rates during wet periods. Shade combined with poor air circulation, and poor internal and surface water drainage intensify the problem (Figure 3.3). Huang and coworkers (1998a, 1998b) investigated the root loss phenomenon in field grown creeping bentgrass. They found that there was a greater and more rapid loss of roots in field plots maintained at 0.125 inch (3.2 mm) versus 0.156 inch (4.8 mm) mowing height. Using a minirhizotron, they could detect root loss beginning as soil temperatures rise in early summer. Root loss actually begins close to the crown, because roots emanating from the crown are subjected to higher-temperature stress than roots growing deeper in the soil profile. Huang and coworkers (1998a, 1998b) further demonstrated that greater root loss occurs in overwatered and saturated soil versus

FIGURE 3.3
Shade and excess water team to damage this golf green.

well-watered and well-drained soils. Root mortality was also increased when plants were maintained at higher temperature of 95°F day/77°F night (35/25°C), when compared to a lower temperature of 77°F day/58°F night (25/14°C). The mechanism of accelerated root loss in wet soils is likely due to displacement of air by water and therefore a depletion of soil oxygen and a buildup of carbon dioxide. High soil temperatures (>86°F; >30°C) and lower mowing heights speed the rate of root loss. When soils become saturated during high-temperature stress periods, creeping bentgrass and annual bluegrass often appear chlorotic. In this situation, root loss is rapid and oftentimes annual bluegrass grown on greens almost completely loses its root system in a few days (Figure 1.7).

Low mowing during heat stress, poor growing environments, mowing when there is standing water or otherwise excessively wet thatch-mat, intense grooming, and concentrated traffic promote creeping bentgrass and annual bluegrass summer decline (Figures 3.4, 3.5, and 3.6). Most golf course superintendents likely will agree that their most troublesome greens in summer historically tend to be in low-lying, pocketed, or shaded sites (Figure 3.7). Soils in pocketed greens remain wetter longer, and plants growing on these greens invariably are much more poorly rooted when compared to plants grown on greens in full sun and with good air circulation. Wet soils also retain heat longer than well-drained soils. As previously noted, studies show that root mortality is increased by excessive soil wetness, high temperature, and by reducing mowing height. Hence, root decline in creeping bentgrass is primarily a response to environmental or mechanical stress factors (i.e., abiotic or nonliving stresses). Most of the stress factors responsible for summer bentgrass root decline were reviewed in detail in Chapter 1.

FIGURE 3.4
Summer decline often is caused by mowing too low in poor growing environments during stressful summer conditions. Note healthy yellow-green clones in the forefront.

FIGURE 3.5
Mowing golf greens that are spongy wet or in the presence of casual water are common causes for turf loss in summer.

The "summer bentgrass decline complex" originally was described as a malady caused by an interaction between abiotic stress factors and root pathogens. *Rhizoctonia solani* and *Pythium* species that attack roots were thought to be the pathogens that interact with abiotic stresses to cause the summer bentgrass decline complex. The pathogen connection to this theory was based on a favorable color response and overall improvement of turf quality

FIGURE 3.6
Wear injury around a weekend cup placement is a common cause of turf decline.

FIGURE 3.7
Greens pocketed by trees are commonly damaged in response to the adverse effects of shade, poor air circulation, and in this case poor irrigation practices.

when stressed greens were treated in summer with fosetyl-aluminum alone or tank-mixed with either chlorothalonil or mancozeb. Subsequent studies, however, revealed that *R. solani* rarely attacks roots of even severely blighted cool-season grasses (Zhang and Dernoeden, 1997), and that the *Pythium* spp. are common inhabitants of roots and are most problematic in new constructions (Feng and Dernoeden, 1999). Fosetyl-aluminum and tank-mix partners control Pythium diseases, brown patch, and blue-green algae, but as discussed below they also improve color and mitigate or mask mechanical injury.

Studies conducted on creeping bentgrass greens have demonstrated a consistent improvement in color and overall turfgrass quality with fosetyl-aluminum, especially when tank-mixed with chlorothalonil or mancozeb (Dernoeden and Fu, 2008). The enhanced quality appears in the absence of disease during periods that are either cool and moist or hot, humid and/or dry. The improvement in turf quality usually becomes apparent within a week following spraying and may persist for one to three weeks. The mechanism of this improved bentgrass quality is unknown, but fosetyl-aluminum may promote the production of antioxidants. Field studies, however, have failed to find a link between fosetyl-aluminum use and improved nutrient uptake or enhancement of chlorophyll levels. There also is no apparent improvement in photosynthesis or respiratory activity in plants treated with fosetyl-aluminum or phosphite fungicides. Hence, the improved quality associated with fosetyl-aluminum use may be partially the result of a "paint effect" from the dye or pigment in the product. It was suggested, however, that the pigment in the aforementioned fungicide provides a sunscreen effect by protecting plants from ultraviolet light. Improved quality also may be due in part to blue-green algae (cyanobacteria) control, which is provided by tank-mixing fosetyl-aluminum with either chlorothalonil or mancozeb. Finally, these fungicides may be controlling weak pathogens or senectophytes that accelerate tissue senescence.

Biostimulants are organic materials that can enhance plant growth, improve root growth, and condition turfgrasses to better tolerate environmental stresses (Schmidt et al., 2003). Seaweed extracts and humic acid are the two most common ingredients used in biostimulants, which contain cytokinins and auxins (i.e., hormones). The definition of biostimulants has broadened, and these products may also contain nitrogen and other nutrients, amino and other organic acids, vitamins, microbial inoculants, iron, and other compounds. Cytokinins are extracted from seaweed or kelp, and humic substances can be extracted from soils, peats, and coal. Seaweed or kelp extracts also contain small amounts of nitrogen, phosphorous, potassium, and sodium and trace amounts of calcium, magnesium, sulfur, iron, copper, manganese, zinc, and boron. Humate extracts contain humic acid and fulvic acid, which contain auxin, and they are capable of chelating some inorganic compounds like iron (Schmidt et al., 2003). Humate incorporated into a sand-based rootzone was shown to enhance rooting for a few months following establishment but had no long-term impact on bentgrass root systems (Kaminski et al., 2004). Although growth, rooting, and stress enhancement benefits of biostimulants were measured in controlled laboratory studies, there remains little evidence that they improve the summer stress tolerance of creeping bentgrass in the field. For example, Settle and Dernoeden (2009) assessed four biostimulants containing kelp extracts and other compounds (e.g., terpenoids, vitamins, humic acids), urea (0.15 lb N/1000 ft^2; 7.5 kg N/ha), and several chelated iron products for their visual effects on summer quality on creeping bentgrass greens. The bentgrass was not subjected to long-term periods of wilt stress, and roots and foliar growth were not measured. They found that chelated iron products darkened foliage, but the biostimulants not containing iron yielded little or no visual responses when compared to untreated plots. None of the biostimulants or chelated iron products impacted chlorophyll production as interpreted by normalized difference vegetation index (NDVI) measurements. Urea, however, improved turf color and quality and increased chlorophyll. Urea mitigated injury from scalping, but none of the biostimulants or chelated iron products improved quality of injured turf when applied without urea. Biostimulants may enhance stress tolerance under certain environmental conditions, but spoonfeeding with urea during summer is a powerful yet inexpensive tool in improving bentgrass vigor during stressful periods (Figure 3.8).

In summary, the decline of creeping bentgrass during mid-to-late summer is mostly the result of a combination of abiotic stresses, including one or more of the following: high-temperature stress (indirect and direct), high humidity, shade, excessively wet soils, or mechanical injury from mowing too low or from grooming greens during hot weather when turf is not actively growing. Interestingly, some usually light-green clones seem to better tolerate summer stresses (Figures 3.4 and 3.9). Bentgrass roots will shorten and darken in color as soil temperatures rise in the summer. The best approaches to promoting the survival of declining creeping bentgrass in summer is to increase mowing height, reduce mowing frequency, avoid grooming and other mechanical stresses, spoonfeed, and pay very careful attention to proper water management as described in Chapter 2. As noted above, some fungicides, biostimulants, and other products can be beneficial but are not a

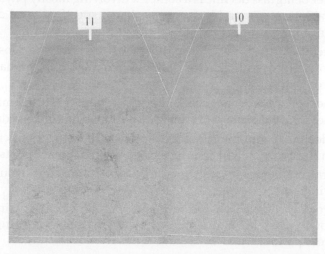

FIGURE 3.8
There is no substitute for a good nitrogen spoonfeeding program in the summer. The plot on the left was treated with a chelated iron and micronutrient product versus urea on the right.

FIGURE 3.9
Creeping bentgrass clones exhibiting improved tolerance of summer stresses on this shaded green. (Photo courtesy of S. McDonald.)

substitute for conservative cultural practices. Decline in creeping bentgrass and annual bluegrass roots will naturally occur throughout the summer until root initiation and regrowth of the root system resumes with the advent of a sharp drop in temperature in autumn (Beard and Daniel, 1966).

Summary of Key Points

- Summer bentgrass decline is a complex involving mostly physiological or abiotic stress problems, rather than a specific fungal disease problem.
- High-temperature stress in the rootzone causes root hair loss, root discoloration, dysfunction, or death of bentgrass and annual bluegrass roots in summer.
- Summer bentgrass decline accelerates when soil or the thatch-mat layer becomes very wet or very dry, and when mowing height is reduced.
- Root decline in summer progresses until a sharp drop in temperature in the autumn, which induces the initiation of new roots.
- Cultural management strategies for summer decline include the following:
  - Avoid excessive soil wetness and mechanical injury, increase mowing height, and reduce mowing frequency during high stress periods.
  - Fosetyl-aluminum tank-mixed with either chlorothalonil or mancozeb and other fungicides often elicits a favorable color response in golf green turf in both warm and cool weather.

- Biostimulants containing cytokinins, humic acids, and possibly other compounds can improve rooting and environmental stress tolerance.
- Spoonfeeding with urea and other water-soluble nitrogen sources (e.g., 0.10 to 0.15 lb N/1000 ft^2 [5 to 7.5 kg N/ha] on 7- to 14-day interval) is a powerful tool in improving color and vigor and mitigating turf injury during summer stress periods.
- Apply a micronutrient product containing nitrogen, and chelated iron or magnesium.
- Promote air circulation, sunlight penetration, and soil drying, especially in the summer.
- Avoid mechanical injury and the application of potentially phytotoxic chemicals in the summer.

Some Causes for Yellow Spots, Speckles, Patches, or Generally Chlorotic Creeping Bentgrass

Throughout the year plant pathologists receive numerous samples of yellowish-green to entirely yellow appearing creeping bentgrass or annual bluegrass from golf greens. Superintendents sometimes suspect yellow tuft, yellow patch, or perhaps some other disease is the problem. Pathogens generally do not cause a uniform chlorosis (i.e., yellowing) throughout large portions of individual golf greens. A generalized yellowing of golf green turf often is due to several interacting environmental, physiological, nutritional, soil, and other factors. For example, annual bluegrass plants often turn yellow in the spring and early summer for physiological reasons when producing seedheads. Creeping bentgrass may turn a yellowish color in response to heat stress, low nitrogen fertility, and hot and wet soils (Figure 3.10).

Biotic Causes of Chlorosis

Acidovorax bacterial disease may cause a generalized chlorosis in creeping bentgrass greens during prolonged periods of heat stress in summer (see below for more information). Plant parasitic nematodes are often implicated as causal agents of generalized chlorosis, but in reality this seldom is the case. *Pythium* spp. that attack roots also can cause chlorosis. Generalized yellowing is associated with anthracnose and summer patch diseases in annual bluegrass during summer months.

Most fungal diseases are associated with spots or circular patch symptoms. Some examples of common diseases associated with yellow spots include yellow tuft and yellow spot. Yellow spot is described below. Basal

FIGURE 3.10
Chlorosis in bentgrass due to excessive soil wetness due to an irrigation head being out of adjustment on the right side of the green.

rot anthracnose in annual bluegrass can appear as yellow or orange speck-les in winter and early spring (Figure 3.11), whereas bacterial wilt in annual bluegrass appears as lime-green and eventually as tan-colored speckles in summer. There are three *Rhizoctonia* spp. that can elicit the formation of yel-low patches or rings in putting greens. Yellow patch (*R. cerealis*) appears in distinctive circular patches or rings that may be yellow or reddish-brown in

FIGURE 3.11
Yellow speckling associated with basal rot anthracnose in annual bluegrass in early spring. The creeping bentgrass is unaffected.

color. *Rhizoctonia zeae* (sheath and leaf spot) and *Waitea circinata* var. *circinata* (brown ring patch; Waitea patch) can produce yellow rings in summer and spring, respectively, on golf greens. Brown ring patch is mostly a disease of annual bluegrass, whereas creeping bentgrass usually remains unaffected in mixed stands on golf greens. Circular and yellow-colored patches in creeping bentgrass also may indicate a potential take-all patch (*Gaeumannomyces graminis* var. *avenae*) problem. A band of yellow plants may appear in the outer peripheral edge of some fairy rings in creeping bentgrass golf greens. The aforementioned diseases are described in detail in Chapter 4.

Hail damage, sod webworms, and black cutworms cause white to tan-colored speckles as described in Chapter 10 (Figure 3.12). Finally, there is an obscure malady known as "white leaf" in creeping bentgrass. In the case of white leaf, leaves lose chlorophyll and have, as the name implies, a white color or there are white leaf tips (Figure 3.13). There is a similar condition in bermudagrass (*Cynodon* spp.) in Japan which is caused by a bacteria-like organism called a *mollicute* (Tani and Beard, 1997). White leaf in creeping bentgrass normally is not injurious, and plants eventually gain their natural green color.

Abiotic Causes of Chlorosis

Creeping bentgrass, annual bluegrass, and other cool-season grasses may develop a chlorosis with the advent of hot and humid conditions in summer. Excessive water from rain or irrigation, soil compaction, poor internal drainage, poor air circulation, anaerobic conditions, heat stress, declining root systems, and shade can cause yellowing during summer. The yellowing is often

FIGURE 3.12
Hail damage can cause speckling similar to sod webworm damage on creeping bentgrass greens.

FIGURE 3.13
White leaf disease in a creeping bentgrass green. (Photo courtesy of C. Robinson.)

caused by a combination of heat stress (i.e., direct or indirect) and oxygen depletion in wet or waterlogged soils, which were discussed in Chapter 1. In summer, the yellowing is largely due to altered physiological processes in response to heat stress, such as lowered photosynthesis and increased respiration leading to early senescence. Low soil oxygen and elevated carbon dioxide levels, impaired transpiration, or possibly an inability of roots to effectively absorb nutrients can also cause turf to turn yellow. When soils are excessively wet during periods of heat stress, it is important to promote soil drying and cooling by restricting irrigation; to judiciously syringe; to avoid mechanical stress; and to improve air circulation with fans or by tree and bush removal. Blue-green algae also should be controlled.

Segregation of biotypes into distinct, circular clonal patches is common in creeping bentgrass greens (Figures 3.4 and 3.9). Each bentgrass cultivar is often the product of several pollen crosses with different parents. It is estimated that a pound (454 grams) of bentgrass seed contains between 5 and 7 million individual seeds. Hence, in a pound of seed there are going to be numerous potential variant types that can eventually form circular patches of turf with colors ranging from yellow-green to dark green. These segregated patches become very conspicuous during prolonged periods of heat stress. Some clones are genetically predisposed to chlorosis in response to heat stress and possibly oxygen depletion in wet soils. The chlorosis in circular, segregated clones intensifies in hot and wet soils (i.e., direct or indirect heat stress) and can be misdiagnosed as disease. Turf may eventually turn brown, thin out, or completely die in these circular, segregated, or clonal patches. Generally, the advent of cool and dry weather conditions results in turf recovery and a return to a normal green color. When thinning and turf loss occurs in distinctly circular patterns, samples should be sent to a diagnostic laboratory to determine if a pathogen is present. Some biotypes of annual bluegrass and creeping bentgrass are naturally yellow-green in color, and their color cannot be darkened significantly by applying nitrogen or iron.

Prolonged periods of cool, overcast, and rainy weather in spring, autumn, or early winter also can induce chlorosis, particularly in low-lying, poorly drained, and shaded sites. Autumn and spring chlorosis is common in creeping bentgrass and other cool-season grasses, and generally coincides with warm day and cool to cold night temperatures. The chlorosis in affected areas often develops in isolated pockets. There is usually no apparent relationship between the appearance of chlorosis and soil condition (e.g., well drained versus wet; pH; etc.). Chlorosis, however, is more common on slopes and in shaded or wet sites, but these factors are not associated with all warm day and cool night–related chlorosis problems. Spring and autumn yellowing of turf may be caused by an interruption in chlorophyll production or retention. Because problems with nonuniform chlorosis often appear during spring and autumn, it may be that environmental conditions are interfering with root uptake of nitrogen, iron, magnesium, or other nutrients required for chlorophyll production. Spring and autumn are characterized by generally warm days and cool nights, and these conditions are ideal for rapid growth of cool-season grasses. It therefore is also possible that cool to cold nights impair the ability of plants to produce sufficient chlorophyll levels in rapidly growing leaf and sheath tissues, and as a result the turf develops a yellow appearance. Although the cause of spring and autumn chlorosis is unknown, it appears to be linked to a combination of environmental, physiological, soil, and nutritional factors. In most situations, the chlorosis that develops in response to abrupt temperature or relative humidity changes or extremes dissipates within a few weeks. Reversing some types of chlorosis between spring and autumn may be achieved by applying a low rate of water-soluble nitrogen or by applying iron or magnesium. Covering chlorotic greens during unusually cold spring or autumn nights may be beneficial.

Nutritional Causes of Chlorosis

The most common nutritional problems leading to chlorosis in turf would include low nitrogen fertility; high nitrogen use in combination with low potassium levels in some soils; and iron or magnesium deficiencies. Even though growth-limiting iron and magnesium deficiencies are relatively uncommon in most soils, even in high sand rootzones, creeping bentgrass usually exhibits a "green-up" response following an application of iron and magnesium. An accumulation of copper from overuse of copper hydroxide can induce iron deficiency, which appears as a chlorosis. Other factors, however, can limit iron or magnesium uptake such as extremes in soil pH (especially iron chlorosis in alkaline soils), soil nutrient imbalances, and possibly cold or wet weather.

Nutritionally related chlorosis can debilitate and weaken plants, especially annual bluegrass and creeping bentgrass growing on sand-based rootzones. Should thinning of the turf become evident, an application of 0.1 to 0.2 lb N/1000 ft^2 (0.5 to 1 kg N/ha) from a quick-release, water-soluble nitrogen source mixed with a micronutrient product may alleviate the condition. Because iron and magnesium are involved in chlorophyll production and elicit a shoot green-up response, they are frequently recommended for ameliorating a chlorotic condition. In reality, applications of iron darken the color of leaves within an hour following its application on a warm summer day, and while it is a constituent of the chlorophyll molecule, it does not stimulate chlorophyll production in turfgrass plants. Iron sulfate (1 to 2 oz/1000 ft^2; 30 to 60 gr/93 m^2) or chelated iron materials are options. Epsom salts or $MgSO_4$ (2 oz/1000 ft^2; 60 gr/93 m^2) are a good source of magnesium. Chelated forms are preferred because the iron and magnesium will remain soluble longer in soil and thus provide a longer-lasting response than sulfate compounds. Finally, it is good to use a complete fertilizer (i.e., N + P + K) the next time large increments of nitrogen (i.e., >0.5 lb N/1000 ft^2; >25 kg N/ha) are to be applied. The application of other micronutrients and biostimulants may be beneficial. If the yellowing is genetic or clonal, a significant green-up in response to nitrogen, iron, magnesium, or other micronutrient applications is unlikely.

Other Causes of Chlorosis

There are other causes for generalized chlorosis including the following: (a) integration of applications of plant growth regulators and biostimulants containing gibberellic acid; (b) application of herbicides and other pesticides, especially those formulated as emulsifiable concentrates, during warm to hot weather which can scorch or yellow turf; (c) other chemical or fertilizer burns; (d) use of extremely high seeding rates that result in large numbers of plants occupying a small space; and (e) possibly viruses and other unknown pathogens. For the turf manager, however, yellow and chlorotic turf is an indicator that something is wrong with the grass, soil, growing environment, or way the turf is being managed.

Summary of Key Points

- Chlorotic (i.e., yellow) turf is an indicator that something is wrong in the growing environment.
- Yellow speckles, spots, and circular patches generally indicate a disease, whereas nonuniform yellowing (i.e., chlorosis) often indicates that one or a combination of environmental, physiological, genetic, nutritional, or soil factors are responsible for the chlorosis.
- Chlorosis can be induced by a nutrient deficiency (especially nitrogen, iron, and magnesium), nutrient imbalances, soil compaction,

extremes in soil pH, environmental stress, an abrupt change in temperature, excessively wet or anaerobic soil (particularly during hot weather), and plant genetics.

- Chlorosis management strategies include:
  - Get a soil and tissue test and/or send samples of chlorotic turf to a diagnostic lab.
  - Apply a micronutrient product containing nitrogen, and chelated iron and/or magnesium.
  - Promote air circulation, sunlight penetration, and soil drying, especially in the summer.
  - Avoid mechanical injury and the application of potentially phytotoxic chemicals in the summer.

Bentgrass Collar Decline

Loss of density and death of creeping bentgrass plants in collars of new or newly renovated golf greens is a common problem (Figure 3.14). The decline often begins as days lengthen and get warmer from late spring through summer. On older greens, turf grown in collars (bentgrass and annual bluegrass) also can experience a significant loss in density during hot and humid weather or when turf is mowed when excessively wet. The mechanism of injury or death of plants on collars is due mostly to mechanical injury

FIGURE 3.14
This less than 1-year-old bentgrass collar was mowed when excessively wet following a rainstorm the previous night.

inflicted by mowers, rollers, and topdressing sand. Collar damage is greater in walk-on and walk-off areas, areas adjacent to bunkers where soil temperatures are higher (and the abrasive effects of sand pitched up from balls played out of bunkers), greens pocketed by trees and brush, where soils are wet, and where the layout of the green complex makes it difficult to avoid turning the mower on the collar (Figures 3.15 and 3.16). Nearly all cases appear to be due to mechanical damage and not disease. Pythium blight (*Pythium* spp.) is the

FIGURE 3.15
Collar damaged where the mower enters a curve. Sand pitched from the bunker further exacerbates the problem.

FIGURE 3.16
Shade and poor air circulation contribute to summer decline of creeping bentgrass collars. (Photo courtesy of S. McDonald.)

greatest disease threat to immature collars in summer because of the large amounts of nitrogen applied during establishment.

In many situations entire collars die or exhibit severe thinning, but the putting green remains in relatively good condition. This phenomenon is opposite of what some would expect. Normally, it is the shorter-cut grass that suffers, but in this situation the opposite can occur. Most injury appears in the spring or summer following an autumn seeding. The problem may persist for several years after initial seeding and eventual reseeding or sodding of new bentgrass cultivars into collars. Use of high nitrogen levels during grow-in produces succulent leaves, which are rendered more prone to mechanical injury. Also, following a grow-in there generally is little thatch to cushion plants from the wear inflicted by mowers and other maintenance equipment. Decline or plant death also occurs on older collars in hot and humid weather and in shaded environments.

Mowing collars when they are subjected to both heat and drought stress often results in a rapid loss of turf (Figure 3.17). Similarly, mowing creeping bentgrass or annual bluegrass collars during very hot and humid weather or when collars are very wet or spongy often causes severe mechanical injury. Pocketed greens are most prone to this type of damage, primarily because they remain wet longer and as such plants have more succulent leaves, fewer tillers, and tend to be poorly rooted. Turf is damaged mechanically when mowers are routinely turned on collars. Collar damage is more common where walk-behind mowers are used. Turning or merely passing the mower over the collar with the back roller moving produces enough friction to injure bentgrass (McDonald, 2009). The twisting of the longer leaves on collars and possibly the crowns, stolons, and roots of plants results in a destructive grinding action. Even if this is not the case, the turning action of the mower (walk-behind or triplex) when the

FIGURE 3.17
Collar was mowed while creeping bentgrass was wilting on a hot day.

collar is being cut causes more twisting and crushing (i.e., a shearing effect or wear) of leaves than occurs on the putting surface. This is probably because the mower cuts in straight lines on greens, whereas collars meander in a circular or serpentine pattern where the torque of a turning mower becomes much more of a factor. A similar phenomenon is common at the outer periphery (i.e., the clean-up pass) of greens that are subjected to the stressful turning action (including wear and compaction) of a triplex mower, while the inner portions of the putting green remain in good condition. Bentgrass growing in collars usually is less dense (i.e., fewer tillers) and very stemy and the amount of stoloniferous tissue greatly exceeds the amount of foliar tissue. The taller, wider-bladed and succulent foliage as well as stolons on collars have a much greater surface area for reels, rollers, groomers, and wheels to inflict injury. Rooting also may be less deep in collars versus the bentgrass growing on the green, which may weaken turf rendering plants more susceptible to mechanical injury.

Topdressing sand and heavy rollers are major contributors to collar decline, which is discussed by McDonald (2009) and summarized below. Topdressing sand is much more damaging to taller (i.e., greater leaf surface area) succulent leaves of the collar canopy as well as exposed stolons once foliar density declines (Figure 3.18). A decline in plant density on collars following topdressing may not appear for several days. Sand topdressing in spring can be destructive to young succulent bentgrass, but topdressing even mature bentgrass in hot weather can be very injurious. No matter how gently the topdressing is applied or brushed in, the abrasive sand (especially during hot-humid weather) is injurious to succulent leaves, crowns, and exposed stolons. The warm summer sun heats the sand particles after topdressing or adjacent to bunkers where sand is pitched onto collars. The increase in

FIGURE 3.18
Angular topdressing sand on the canopy in summer contributes to an increase in mechanical injury to collars. (Photo courtesy of S. McDonald.)

temperature combined with the angular shape of sand causes additional stress and abrasion to the longer leaves of plants growing in collars. It is important to keep sand off collars in the summer and to topdress when the surface is dry to facilitate the downward movement of sand to stem bases and off leaves. After topdressing, blow sand off collars prior to mowing the following day. Collar decline also is more common where sidewinder rollers are used routinely. The pivoting of the sidewinder roller causes friction due to its large surface area and weight. It is important not to allow these rollers on collars, except where they enter and exit the green.

As soils warm in summer, root function becomes less efficient, and the root system may no longer be able to provide sufficient water and nutrients to plants. Collars typically wilt sooner and require more hand watering (McDonald, 2009). The wear injury on collars, however, can mimic drought stress, and thus too much water may be applied to collars making the condition worse. Use of soil wetting agents can reduce the need for hand watering and help reduce additional wear caused by excess water.

Most turf managers who experience this injury initially are likely to conclude it must be a disease. This is not usually the case; however, Pythium blight is a real threat because of the large amounts of nitrogen and water used during the grow-in process. Damage inflicted by the annual bluegrass weevil is generally more severe on the perimeter of greens and collars, and thus these insects can cause collar damage in both annual bluegrass and creeping bentgrass (McDonald, 2009). In summary, the cause of collar decline can involve several interacting factors, but it is largely the result of mechanical injury to either immature and succulent plants or mature plants that are growing less vigorously due to environmental stress factors (e.g., heat stress, waterlogged soils, drought stress, shade, sand abrasion, wear from mowers and rollers, etc.).

Management

Because collar decline is more severe in wet and shaded sites, the first step in its management is to remove trees and brush to improve sunlight penetration. Installing a fan on pocketed greens to promote cooling and evapotranspiration can be very beneficial. Whenever mechanical injury becomes evident, it is important to reduce mowing frequency and to avoid sand topdressing if weather conditions are rainy or hot and humid. Also, avoid any kind of core aeration or grooming practices until cooler weather (i.e., <85°F or <29°C high in the 7-day forecast), and only perform these practices when turf has recovered and is actively growing. Even water injection aeration, when used very frequently, can cause mechanical injury to collars. During periods of heat stress, reduce mowing frequency of collars to once every 3 to 4 days and only mow during the coolest part of the day and when surfaces are dry. Do not mow collars or the putting surface when there is casual water or when soils are waterlogged during hot and humid periods. Never mow young bentgrass in the first year after seeding on rainy days. Use a "floating

FIGURE 3.19
Use lattice, rugs, or boards to protect collars from turning mowers. (Photo courtesy of S. McDonald.)

or flex-head," walk-behind mower to cut weak collars. It is helpful to mow collars and the clean-up in the evening when it is cooler and leaves are dry. Some superintendents lay down rugs, mats, plywood, or plastic lattice on collar areas where mowers cross or turn to reduce mechanical injury (Figure 3.19). A gradual reduction in height of cut to 0.25 inches (6 mm) oftentimes alleviates collar decline. This may be a result of having less leaf surface area to be abraded by sand or worn by mowers and rollers. The lower mowing height also may stimulate tillering or perhaps less foliage provides for a higher and therefore a more favorable root-to-shoot ratio. Spoonfeeding helps worn collars to recover. Use a drop spreader and apply a nonburning natural organic fertilizer to collars at a rate of 0.2 to 0.3 lb N/1000 ft^2 (10 to 15 kg N/ha) per month (McDonald, 2009). Hand syringe frequently when weather conditions warrant and avoid water puddling. Use soil wetting agents for collars that are prone to rapid drying. Use small signs or fences to divert foot traffic away from damaged areas (Figure 3.20). Avoid sand topdressing whenever daily temperatures are >86°F (>30°C) in the 7-day forecast. Keep sand off collars or use blowers to remove sand from leaves. Do not allow sidewinder rollers on collars. Enter greens with rollers using lattice to protect worn collars.

Once creeping bentgrass is actively growing, grooming practices can be resumed to reduce steminess of turf on collars by core aeration, vertical cutting, and sand topdressing. Collars should receive the same management practices for thatch-mat control as the putting surface. When building new or renovating old greens, one consideration may be to maintain the entire bentgrass surface at putting green height. Just before opening, increase the height of cut on perimeter collar areas. Also consider seeding or sodding collars with an aggressive cultivar like Penncross (Figure 3.20). Where collar

FIGURE 3.20
Rope off weak collars and where no longer functional, overseed or sod with an aggressive cultivar.

decline is chronically severe, it may become necessary to overseed with perennial ryegrass (*Lolium perenne*) that better tolerates wear stress on collars and is a very practical solution to this problem in many regions.

Summary of Key Points

- Bentgrass collar decline or death in summer is primarily due to mechanical injury from mowers, rollers, and sand topdressing. This type of injury is common in young bentgrass but also can occur on older bentgrass and annual bluegrass collars, particularly during hot and humid weather in shaded environments.

- Injury is most severe where mowers turn, in walk-on and walk-off areas, in wet and shaded areas, on sloped areas, and in areas adjacent to bunkers.

- Mowing collars during rainy periods or whenever turf is excessively wet or soft can cause severe damage in summer. Similarly, mowing collars that are debilitated by heat and drought stress can result in a rapid loss of turf.

- Bentgrass collar decline management strategies include the following:
  - Divert traffic away from injured areas with rope fences or signs.
  - Turn mowers on mats or lattice and not collars.
  - Mow collars when grass is dry with a lightweight, walk-behind mower.
  - Replace grooved rollers with solid rollers during summer.
  - Reduce mowing frequency to two or three times weekly and mow in cooler evening hours when turf is dry.

- Gradually reduce collar mowing height to 0.25 inches (6.2 mm).
- Install a fan where air circulation is poor (i.e., shaded and pocketed greens) to promote cooling and drying.
- Delay sand topdressing, core aeration, and grooming practices until cooler weather when bentgrass is actively growing.
- Apply 0.2 to 0.3 lb N/1000 ft^2 (10 to 15 kg N/ha) per month to collars from a natural organic fertilizer or other nonburning, slow-release nitrogen source.
- Use blowers to remove sand pitched from bunkers or from topdressing.
- Keep sidewinder rollers off collars.
- Syringe as needed to alleviate wilt, avoid water puddling in summer, and use a soil wetting agent to reduce hand watering.
- In the autumn after bentgrass has recovered, reduce stemyness by core aeration, vertical cutting, and sand topdressing, and maintain collars using the same cultural practices used on greens.
- When all else fails, overseed chronically damaged collars with perennial ryegrass. Unfortunately, perennial ryegrass winter-kills under ice in colder regions.

Black-Layer

Black-layer is a soil physical condition primarily associated with sand-based greens or older greens modified overtime with sand topdressing. Black-layer can develop in native soils and is commonplace in wetlands. Anaerobic conditions in soil are the cause of black-layer. Blue-green algae may play a role in the development of some black-layers by plugging pore spaces or by sealing the surface (Hodges, 1992).

Symptoms usually first appear in low spots as turf develops a yellow, purple-gray, red-brown, or bronzed appearance, and eventually turf thins in irregular patterns (Figure 3.21). Loss of density generally is most severe in low-lying, shaded areas where air circulation is poor. Black-layer detection is easy. Inspection of the soil profile reveals the presence of a surface or subsurface black-layer (Figure 3.21). Examine a plug by tearing it apart and smell the soil for a foul odor. Where layers are restricted to the surface, the thatch-mat may have a black color. Subsurface layers are coal-black in color and may develop several inches (3 to 12 cm) below the surface. Bands of blackened sand usually range from 0.25 to 1 inch (6.2 to 25 mm) in width, but bands greater than 1 inch (>2.5 cm) in width are not uncommon. Under anaerobic conditions, sulfur is converted by sulfate-reducing bacteria to form hydrogen sulfide gas,

FIGURE 3.21
Black-layer appears as nonuniform, reddish or purplish-colored creeping bentgrass that eventually collapses.

which is directly toxic to roots. The foul, sulfurous, rotten-egg odor associated with affected soils is due to the production of hydrogen sulfide gas. The actual blackening is caused by the precipitation of metal sulfides, usually iron, which are black in color (Figure 3.21). Root development generally is poor below a black-layer or between layers of different soil types.

Black-layer develops during periods of excessive rain or because of excessive irrigation, especially on golf greens with poor internal drainage or where soil water perches. The condition is frequently encountered where sand topdressing is layered over a heavier soil type or an organic layer. Layering of different topdressing mixes with variable particle sizes or other physical properties also can cause water to perch. Black-layer can develop on greens where soil from unwashed sod forms a layer effect. These situations can cause water to become trapped in the upper layer, resulting in soil becoming anaerobic. This trapping of water between layers is referred to as a *perched water table*. New sand greens that are intensively irrigated and fertilized during the grow-in period may become anaerobic periodically. In this situation, the blackening may not appear in distinct bands but as speckles of black aggregates or as a general blackening in the profile (Figure 3.22). Following construction, it is common for liners between the green's mix and native soil to act like dams (i.e., false fronts) on the low side of greens. The damming of water by liners often leads to black-layer.

Black-layer can also develop when a superintendent irrigates every night to the same soil depth, which keeps the upper soil profile almost continuously damp. Although waterlogged soils of low oxygen content are invariably associated with black-layer, there are no other physical, chemical, or cultural factors common to all turfs affected by this condition.

FIGURE 3.22
Black precipitate in the rootzone of this new construction is due to excessive irrigation, which
caused an anaerobic condition.

Management

Increasing soil aeration and reducing soil wetness are key approaches to
alleviating anaerobic conditions. Deep core aeration with the verti-drain
or using drill-and-fill machines greatly improves aeration of compacted
soils. More traditional core (quadratine, needletine, etc.) aeration is also
beneficial (Figure 3.23). These operations may have to be performed several

FIGURE 3.23
Improved aeration in coring holes creates a bypass through the black-layer. (Photo courtesy of
D. Bevard.)

times before their overall benefits can be objectively assessed. Installing new drainage lines, sand channeling, and adding drains through liners at the low point of greens usually resolve a black-layer problem. The use of subair vacuum systems may be an option for sand-based greens. During the golf season, use of fans, spiking, and solid or hollow tine aeration will help to temporarily alleviate the condition. Affected soils must be dried out; therefore, greens should be syringed only when there is evidence of wilt. Trees and brush should be removed where necessary to improve air movement and sunlight penetration. It is especially beneficial to ensure morning sunlight exposure of greens. Morning sun speeds drying of the canopy and thatch-mat while stimulating the opening of stomates on leaves and sheaths. Furthermore, it is helpful to control blue-green algae and thatch-mat. Avoid materials containing sulfur (Berndt and Vargas, 2006). Nitrate forms of nitrogen may help to alleviate black-layer (Vargas, 2005), but they can increase the severity of some diseases including summer patch, dead spot, and take-all.

Finally, check your irrigation schedules, especially for perimeter automatic irrigation systems. These systems tend to concentrate water in the middle of greens, where it is needed least. Greens with low spots or located in areas of shade and poor air drainage should be watered much less, if at all, by a perimeter irrigation system. This is especially true in summer. Syringing and hand watering areas prone to wilt are preferred for greens having a black-layer problem. Even on newer, sand-based greens (after grow-in), it is recommended that water be applied deeply and infrequently versus irrigating the same amount of time every night. This consistent irrigation scheduling (i.e., nightly) tends to keep the surface of the green too wet and soft, which increases the potential for black-layer, blue-green algae, moss, disease, scald, wet wilt, chlorosis, and scalping.

Summary of Key Points

- Anaerobic conditions induced by excessive soil wetness are the cause of black-layer.
- The condition is diagnosed by inspecting the soil profile for the presence of black layering or blackened soil aggregates and a foul, sulfurous odor.
- Excessive rain or irrigation and possibly blue-green algae can cause black-layer to develop.
- Black-layer is frequently encountered where different soil types are layered, causing water to perch or not drain properly. This can be the result of using different sand topdressing mixes or where greens have been grassed with unwashed (i.e., soil not removed) sod.
- Management of black-layer is aimed at practices that increase soil aeration and decrease soil wetness, including:

- Core, quadratine, or needletine aeration and in more extreme cases verti-drain or drill-and-fill to improve soil aeration.
- Avoid routine automatic overhead irrigation and rely on syringing to alleviate wilt until the soil profile is dry.
- Install fans and remove trees and brush that restrict air movement and sunlight penetration.
- Install drains through liners during construction.
- Control blue-green algae, thatch, and mat.
- Avoid using products containing elemental sulfur. Oxidized forms of sulfur (i.e., SO_4 found in ammonium sulfate) leach and do not contribute to black-layer problems.

Etiolated Tiller Syndrome

Etiolated tiller syndrome (ETS), also referred to as *mad tiller disease* or *ghost disease*, is commonplace in annual bluegrass, creeping bentgrass, and perennial ryegrass during cool, wet, and overcast periods in spring and autumn, but may remain evident during summer. A similar etiolation of leaves is a prominent symptom of bacterial wilt disease (*Xanthomonas translucens*) in annual bluegrass, which is described in Chapter 7. In creeping bentgrass, ETS may be more common in the summer. The cause of ETS is unknown, but some suggested gibberellin-producing fungi (*Fusarium* spp.) or bacteria (*Xanthomonas* spp.) as possible causal agents (Fidanza et al., 2008). Others suggested that it is an environmentally induced physiological disorder or is caused by plant growth regulators or biostimulants containing gibberellic acid. However, the malady has been observed in turfs where no plant growth regulators or biostimulants were applied. Symptoms of ETS include rapid elongation (i.e., etiolated growth) of the youngest leaf through the bud shoot, which is distinctively lime-green or yellow (Figures 3.24 and 3.25). Except during periods of extreme heat stress, affected plants generally do not die, but leaves may die back from tips. Etiolated leaves are not readily removed by mowing. This phenomenon is worse in approaches, collars, and cleanup areas of greens. Attempts to mow more aggressively to remove etiolated leaves during summer stress periods often result in mechanical damage. There are no documented control measures for ETS. Anecdotally, an application of trinexapac-ethyl in combination with a DMI/SI fungicide (e.g., propiconazole, triadimefon, others) may reduce, but does not eliminate leaf etiolation.

FIGURE 3.24
Etiolated tiller syndrome in creeping bentgrass.

FIGURE 3.25
Leaves exhibit elongated growth and a yellow color when affected by etiolated tiller syndrome. (Photo courtesy of S. McDonald.)

Summary of Key Points

- Etiolated tiller syndrome (ETS) affects several cool-season grasses and has no confirmed cause. ETS may be physiologically induced by unknown environmental and soil conditions or be associated with a gibberellin-producing fungus or bacterium.

- Symptoms of ETS include rapid elongation of the youngest leaf in the bud shoot, yellow or lime-green leaves, and leaves that are not effectively removed by mowing.
- ETS is commonly observed during extended cool, wet, and overcast weather in spring and autumn and is more common in perennial ryegrass and annual bluegrass than creeping bentgrass.
- In creeping bentgrass, ETS may be more common in the summer.
- Plants affected with ETS generally do not die, but attempts to lower mowing height to cut the longer leaves in summer may cause mechanical injury.

Acidovorax Disease

In 2010 a new bacterial disease of creeping bentgrass caused by *Acidovorax avenae* pathovar. *avenae* was reported (Giordano et al., 2010). This disease should not be confused with bacterial wilt of annual bluegrass, which is caused by *Xanthomonas translucens* and is described in Chapter 7. Acidovorax bacterial disease (ABD) is triggered by prolonged periods of high-temperature stress. Symptoms of the disease in creeping bentgrass golf greens, as described in the aforementioned report, include general wilt, decline, and necrosis from the leaf tips down. Bacterial streaming from cut infected leaves is the key diagnostic sign. Other predisposing conditions and symptoms not found in Giordano et al. (2010) are noted below.

Symptoms of ABD are variable and also can include etiolated growth and yellowing of leaves, and generalized chlorosis throughout infected greens with islands of healthy darker green biotypes or possibly bentgrass clones (Figure 3.26). Annual bluegrass growing in affected greens does not develop ABD. Etiolated or elongated bentgrass leaves may curl and twist and are not easily removed by mowing. Chlorosis can be mild or pronounced with gradual turf loss, or there can be rapid and widespread decline (Figure 3.27). In mild cases, affected bentgrass can recover rapidly following a shift to a less stressful weather pattern. The disease may begin in late spring but intensifies in response to prolonged periods of heat stress in summer. Daily air temperatures above 88°F (31°C) and night temperatures above 80°F (26.7°C) appear to be optimal for this disease. In creeping bentgrass golf greens, *A. avenae* also has been associated with prolonged periods of overcast and rainy weather in late summer. There are cases where chlorosis and etiolation have not been associated with *A. avenae*. Therefore, some pathologists have suggested that ABD is a weak or secondary pathogen associated with mechanical damage and intense, prolonged periods of heat and possibly other environmental and mechanical stresses. Presumptive identifications

FIGURE 3.26
Chlorotic and darker green islands of creeping bentgrass are symptoms associated with Acidovorax bacterial disease. (Photo courtesy of S. Zontek.)

FIGURE 3.27
Chlorosis followed by rapid decline are symptoms associated with Acidovorax bacterial disease in creeping bentgrass. (Photo courtesy of S. Zontek.)

for this disease in creeping bentgrass are based on chlorosis and streaming of bacteria from cut leaves. Research is underway to better understand the biology and epidemiology of ABD.

There are no effective chemical control materials for bacterial diseases of turfgrasses. Copper-based and antibiotic products may slow spread of bacterial pathogens; however, these products can be phytotoxic when applied in hot summer weather. Hydrogen dioxide also may help slow spread of bacteria, but it should not be tank-mixed with any other chemicals and used only in potable water. Acibenzolar is a chemical used on some crops to suppress bacterial diseases by stimulating natural plant defense mechanisms (aka "systemic acquired resistance"). Research with acibenzolar shows promise in reducing the severity of several turfgrass diseases and is available in a pre-packaged mixture with chlorothalonil (i.e., Daconil Action®) to suppress ABD. Key cultural approaches to managing bacterial diseases include the following: increase mowing height, reduce mowing frequency, mow in the evening when surfaces are dry and when temperatures are declining, use solid rollers, make sure mower blades are sharp, use a dedicated mower and disinfect (10% to 20% sodium hypochlorite or Clorox® solution) the mower when finished mowing, maintain adequate moisture by syringing often and avoid using the overhead irrigation system if possible, spoonfeed, use fans, avoid use of products containing gibberellic acid, maintain or even tighten fungicide applications during stressful periods, and apply chlorothalonil and mancozeb frequently to keep blue-green algae (cyanobacteria) from developing. Avoid sand topdressing and any cultural practice that may mechanically damage turf.

Summary of Key Points

- Acidovorax bacterial disease (ABD) of creeping bentgrass develops during periods of extreme heat stress, but little is currently known about the disease.

- Symptoms of ABD include general wilt and decline, etiolated and yellow leaf growth, green islands of healthy bentgrass, die-back of tissue from leaf tips, and possible rapid loss of turf.

- ABD is diagnosed based on elongated (i.e., etiolated) growth, yellowing of leaves, and streaming of bacteria from cut leaf tips.

- There is no known chemical control for ABD but research with acibenzolar is encouraging. Reduce stress by increasing mowing height and reducing mowing frequency, and mow when the canopy is dry in the evening when temperatures are moderating.

- Proper water management, use of solid rollers, spoonfeeding, and use of fans are key elements that may enable infected plants to survive.

Senectopathic Disorders

Houston B. Couch (1995) coined the terms *senectophyte* and *senectopathic disorder*. He noted that most turfgrass pathogens infect healthy juvenile and mature tissues, but other microorganisms attack only senescent tissues. The latter organisms have been traditionally referred to as secondary pathogens, facultative saprophytes, or saprobes that live on dead or weakened tissues. Senectophytes are basically microbes that cause disease of senescing tissues, and the disease that develops is a senectopathic disorder.

Senescent tissue is in the natural process of aging and will soon die. Older leaves of turfgrass plants are continuously being replaced by younger leaves. As pointed out by Couch (1995), there are numerous biotic (living) and abiotic (nonliving) factors that can cause a premature senescence of leaves and stems. These would include summer stresses such as high air and soil temperature, and drought. Excessively wet or anaerobic soils, low light intensity or shade, nutrient deficiencies or excesses, and heavy thatch and mat accumulations are other abiotic stress factors that lead to premature senescence. Very low mowing and mechanical injury or chemical phytotoxicity are common management-related factors that induce premature senescence. Fungal pathogens and plant parasitic nematodes are biotic causes of premature senescence.

According to Couch (1995), there are three defined senectopathic disorders: anthracnose (*Colletotrichum cereale*), Curvularia blight (*Curvularia* spp.), and Leptosphaerulina blight (*Leptosphaerulinia australis*). Various species of *Ascochyta*, *Fusarium*, *Cephalosporium gramineum*, *Gaeumannomyces incrustans*, and *Nigrospora sphaerica* are other potential agents of senectopathic disorders. *C. cereale*, however, also behaves as a pathogen when found infecting juvenile and mature plant tissues.

Senectopathic disorders in creeping bentgrass and annual bluegrass are most often encountered in summer. A common predisposing scenario occurs when turf is allowed to wilt severely (i.e., the permanent wilting point) and then it is heavily irrigated in hot and humid weather. Wet soils, as previously discussed, store more heat and have low oxygen levels. Photosynthesis is impaired and respiration increases, and turf may appear chlorotic. In this weakened state, tissues begin to senesce. Some cultural practices in summer can promote premature senescence, including lower mowing heights, while using less nitrogen, and the increased level of sand topdressing and other abrasive practices (e.g., brushing, grooved rollers, verti-cutting, wear, etc.). During senescence, fungi such as *C. cereale*, *L. australis*, and *Curvularia* spp. get a competitive advantage and attack tissues, causing a significant thinning of the stand.

Leptosphaerulina blight is commonplace on most cool-season grasses grown on golf courses. The disease often appears in late spring or summer with the advent of high temperature and drought stress. Soil compaction, hard-pans (i.e., a hard surface or subsurface impermeable soil layer), and hydrophobic conditions also promote Leptosphaerulinia blight. The most common species of *Leptosphaerulinia* found in association with turf is *L. australis*, which is considered a saprobe that is incapable of causing disease (Mitkowski and Browning, 2004) (Figure 3.28). In concert with imperfectly understood complexes with other stresses, *L. australis* can rapidly attack and blight tissues that still contain chlorophyll. The disease may appear as small, reddish-brown

FIGURE 3.28
Leptosphaerulinia sp. fruiting bodies embedded in leaves are an indicator of drought stress.

spots that mimic Pythium blight. The affected plants turn tan, and spots may coalesce to encompass large areas. Drought-stressed turf can be severely blighted by *L. australis* following the onset of rain or heavy irrigation.

Fruiting bodies of *Leptosphaerulinia* spp. are easily found in necrotic leaf tissues. The pathogen enters cut leaf tips and works its way down to sheaths. Once it attacks a weakened turf, it rapidly blights tissues and stand density declines dramatically in just a few days. Leptosphaerulinia blight is frequently associated with other diseases such as anthracnose, Pythium blight, and gray leaf spot (a perennial ryegrass and tall fescue disease). This complex of stresses and pathogens can be particularly devastating.

Anthracnose is commonly found on golf courses, particularly in annual bluegrass and less frequently in creeping bentgrass. The causal agent is *C. cereale*, and it can behave as a senectophyte or as a pathogen (see "Anthracnose Basal Rot and Foliar Blight" section in Chapter 4). It also is a common saprophyte, living on dead tissue in thatch. *Curvularia* blight (*C. lunata* and *C. inaequalis*) has long been recognized as a secondary disease. It causes a leaf spotting and melting-out of annual and Kentucky bluegrass. It can cause melting-out or crown rot in bentgrasses but is more common in colonial than creeping bentgrass. *Curvularia* blight is more of a problem in severely heat stressed and compacted soil sites. *Curvularia* blight is perhaps the least invasive of the three senectopathic disorders in creeping bentgrass but is a common annual bluegrass senectophyte. *Cephalosporium gramineum* causes Cephalosporium stripe in cereal crops. It is more of a "true" saprophyte in turf and is commonly found colonizing mechanically damaged and environmentally stressed tissue with masses of white spores being prevalent (Figure 3.29). It is one of the first colonizers of scalded tissue.

FIGURE 3.29
White spore masses of *Cephalosporium* sp. are a good indicator of mechanical injury. Inset
shows leaves damaged by sand topdressing.

Gaumannomyces incrustans is a worthy candidate senectophyte. Root
decline has been associated with *G. incrustans*; however, research indicates
that this fungus is noninfectious or a secondary contributor to the summer
bentgrass decline complex (Landschoot and Jackson, 1989). This fungus is
an extremely common inhabitant of creeping bentgrass roots in summer.
It typically produces black runner hyphae that can be seen on surfaces or
within bentgrass roots. Sometimes black resting structures can be seen on
or embedded on or in roots, and steles may be blackened (Figure 3.30). There

FIGURE 3.30
Blackened mats and root damage associated with *G. incrustans* on summer stressed bentgrass.

is considerable similarity in the hyphal characteristics of *G. incrustans* and *G. graminis* var. *avenae* (i.e., take-all pathogen). Because of these similarities, some diagnosticians mistakenly diagnose a take-all patch problem.

Management

An application of a broad-spectrum fungicide may slow the foliar activity of senectophytes but at best provides only a short-term benefit. There is seldom a response from a fungicide application to turfs affected with *Leptosphaerulinia* blight or other senectophytes. The lack of a response from fungicides is a good indicator that other stress factors are involved in the decline. The underlying cause for predisposition of plants to premature senescence should be identified and addressed. Proper water management, alleviating soil compaction, reducing thatch-mat, increasing mowing height, avoiding mechanical injury during summer, and using a complete (i.e., N + P + K) fertilizer at appropriate times are key factors in improving turfgrass vigor and minimizing the severity of these disorders. In the case of anthracnose basal rot, however, a more sophisticated fungicide program is often required.

Summary of Key Points

- Senectophytes are secondary pathogens or saprophytes that attack weakened and senescing tissues. The disease that develops as a result of these fungi attacking senescent tissue is called a senectopathic disorder.
- Anthracnose, Curvularia blight, and Leptosphaerulina blight are the most common foliar senectopathic disorders.
- *Gaeumannomyces incrustans* is the most common bentgrass root senectophyte, which can be mistaken as the causal agent of take-all patch (i.e., *G. graminis* var. *avenae*).
- Senectopathic disorders are most common in summer and can be induced by heat and drought stress, excessively wet soil, mechanical injury, or the injurious effects of a primary pathogen.
- The activity of these senectophytes may be arrested by applying a broad-spectrum fungicide, but oftentimes there is little or no positive response. The underlying cultural or environmental factors that induce premature senescence and the weakening of tissue need to be corrected.

Yellow Spot

Yellow spot is an undescribed malady of closely mown turfgrasses found on golf courses. Yellow spot mainly is seen in creeping bentgrass grown in

FIGURE 3.31
Yellow spot in a creeping bentgrass golf green.

sand-based rootzones on greens and tees. Symptoms appear in the summer as yellow spots or patches ranging from 1 to 6 inches (2.5 to 15 cm) in diameter (Figure 3.31). Typically, spots are 2 to 3 inches (5 to 7.5 cm) in diameter or about the size of a tennis ball. Turf within the yellow spots does not die and in most cases does not even thin out. Hence, yellow spot is largely a visual problem. Yellow spot often appears almost overnight in large numbers in the summer during periods of high humidity and high nighttime temperatures (>70°F; >21°C). Yellow spots develop in full sun and shaded environments. They may rapidly disappear with the advent of cooler night temperatures (<18°C; < 65°F) in late August. Sometimes they remain evident as late as early autumn.

The proposed causal agents of yellow spot are cyanobacteria (aka bluegreen algae) in the genera *Oscillatoria* and *Phormidium* (Tredway et al., 2006). The mechanism responsible for the chlorotic appearance induced by cyanobacteria is not by infection. According to Tredway et al. (2006), cyanobacteria filaments migrate at night from thatch to the base of plants and glide up leaf sheaths and onto leaf blades. It was suggested that cyanobacteria induce or elicit a chlorotic response either through the production of a toxin or iron chelating compounds called siderophores. High levels of cyanobacteria must be present, and toxins or siderophores diffuse from the filaments and enter plants either through cut leaf tips or hydathodes. Other researchers, however, failed to find cyanobacteria in association with yellow spot (Dernoeden and Fu, 2008). The yellow spot symptoms mimic the description given for yellow dwarf, a mollicute (bacteria-like organisims bound only by a membrane) disease of bentgrass golf greens in Japan (Tani and Beard, 1997).

Management

Yellow spot is effectively controlled by chorothalonil and mancozeb applied in summer in a preventive program (Gelernter and Stowell, 2000; Dernoeden and Fu, 2008). These same fungicides control blue-green algae on greens and lend evidence that these organisms may in some way be the cause of yellow spot. Curatively, two or three applications of the aforementioned fungicides on a 7-day interval may be required to effectively reduce yellow spot presence. Yellow spot primarily develops in creeping bentgrass subjected to frequent irrigation, but little yellow spot develops if turf is allowed to wilt prior to irrigation (Fu and Dernoeden, 2008). Limited test data indicate that yellow spot may be suppressed by ammonium sulfate applied in spoonfeeding programs on golf greens.

Summary of Key Points

- Yellow spot is a disease of unknown cause but has been linked to blue-green algae.
- Yellow spot is promoted by frequent irrigation during summer when night temperatures and humidity are high.
- Yellow spot is controlled by promoting soil drying and applying either chlorothalonil or mancozeb.

Bibliography

Beard, J.B., and W.H. Daniel. 1965. Effect of temperature and cutting on the growth of creeping bentgrass (*Agrostis palustris* Huds.) roots. *Agron. J.* 57:249–250.

Beard, J.B., and W.H. Daniel. 1966. Relationship of creeping bentgrass (*Agrostis palustris* Huds.) roots growth to environmental factors in the field. *Agron. J.* 58:337–339.

Berndt, W.L., and J.M. Vargas, Jr. 2006. Dissimilatory reduction of sulfur in black layer. *HortScience* 41:815–817.

Couch, H.B. 1995. *Diseases of Turfgrasses*. Krieger, Malabar, FL.

Dernoeden, P.H., and J. Fu. 2008. Fungicides can mitigate injury and improve creeping bentgrass quality. *Golf Course Management* 76(4):102–106.

Feng, Y., and P.H. Dernoeden. 1999. *Pythium* species associated with root dysfunction of creeping bentgrass in Maryland. *Plant Dis.* 83:516–520.

Fidanza, M., J. Gregos, and D. Brickley. 2008. Are etiolated tillers a visual nuisance or something else? *Golfdom* 63(10):60,61,64.

Gelernter, W., and L.J. Stowell. 2000. Cyanobacteria (a.k.a. blue-green algae) *wanted* for causing serious damage to turf. *PACE Insights* (www.pace-ptri.com) 6(8):1–4.

Giordano, P.R., J.M. Vargas, Jr., A.R. Detweiler, N.M. Dykema, and L. Yan. 2010. First report of a bacterial disease on creeping bentgrass (*Agrostis stolonifera*) caused by *Acidovorax* spp. in the United States. *Plant Dis.* 94:922.

Hodges, C.F. 1992. Interaction of cyanobacteria and sulfate-reducing bacteria in sub-surface black layer formation in high-sand content golf greens. *Soil Biol. Biochem.* 24:15–20.

Huang, B., X. Liu, and J.D. Fry. 1998a. Effects of high temperature and poor soil aera-tion on root growth and viability of creeping bentgrass. *Crop Sci.* 38:1618–1622.

Huang, B., X. Liu, and J.D. Fry. 1998b. Shoot physiological response of two bentgrass cultivars to high temperature and poor soil aeration. *Crop Sci.* 38:1219–1224.

Kaminski, J.E., C.A. Bigelow, and P.H. Dernoeden. 2004. Soil amendments and fertil-izer source effects on creeping bentgrass establishment, soil microbial activity, thatch and disease. *HortScience* 39:620–626.

Landschoot, P.J., and N. Jackson. 1989. *Gaeumannomyces incrustans* sp. Nov., a root-infecting hyphopodiate fungus from grass roots in the United States. *Mycol. Res.* 93:55–58.

McDonald, S. 2009. Buttoned-down management for collars. 2009. *Golf Course Management* 77(6):60–65.

Mitkowski, N.A., and M. Browning. 2004. *Leptosphaerulina australis* associated with intensively managed stands of *Poa annua* and *Agrostis palustris*. *Can. J. Plant Pathol.* 26:193–198.

Salisbury, F.B., and C.W. Ross. 1992. *Plant Physiology*, 4th ed. Wadsworth, Belmont, CA.

Schmidt, R.E., E.H. Ervin, and X. Zhang. 2003. Questions and answers about bios-timulants. *Golf Course Management* 71(6):91–94.

Settle, D., and P.H. Dernoeden. 2009. Evaluation of cytokinin plant extract biostimu-lants, iron and nitrogen products for their effects on creeping bentgrass sum-mer quality. *U.S. Golf Association Turfgrass and Environmental Research* Online 8(1):1–15.

Smiley, R.W., P.H. Dernoeden, and B.B. Clarke. 2005. *Compendium of Turfgrass Diseases*, 3rd ed. APS. St. Paul, MN.

Smith, J.D., N. Jackson, and A.R. Woolhouse. 1989. *Fungal Diseases of Amenity Turf Grasses*. E. & F.N. Spon. London and New York.

Tani, T., and J.B. Beard. 1997. *Color Atlas of Turfgrass Diseases*. Ann Arbor Press, Chelsea, MI.

Tredway, L.P., L.J. Stowell, and W.D. Gelernter. 2006. Yellow spot and the potential role of cyanobacteria as turfgrass pathogens. *Golf Course Management* 74(11):83–86.

Vargas, J.M., Jr. 2005. *Management of Turfgrass Diseases*. Wiley, Hoboken, NJ.

Vincelli, P., and J.C. Doney. 1995. Reaction of bentgrasses to *Drechslera* leaf blight. *Biol. Cultural Tests* 10:36.

Zhang, M., and P.H. Dernoeden. 1997. *Rhizoctonia solani* anastomosis groups and other fungi associated with brown patch-affected turfgrasses in Maryland, USA. *Intern. Turfgrass Soc. Res. J.* 8:959–969.

4

Selected Creeping Bentgrass
Diseases and Pathogens

This chapter will focus on common diseases and pathogens of creeping bentgrass, with an emphasis on those most troublesome in golf greens. There are no absolutes in turfgrass pathology. Diseases sometimes occur when not expected if weather conditions shift to favor the pathogen. For example, dollar spot normally appears in mid-to-late spring, but unusually warm night temperatures in early spring or late autumn can trigger the disease out of season. Another example is take-all patch. Take-all patch typically appears in new constructions about 2 years following seeding. The disease may intensify for several years and then naturally decline. There are many cases in which the disease is problematic for only a year or two and others where it persists for 10 years before declining. Diseases can be initiated by a pathogen in spring or even the previous autumn, but symptoms or damage may not become prevalent until the advent of warm and dry weather that stresses plants. These include root diseases like Pythium-induced root dysfunction and take-all patch. Brown patch and Pythium blight generally are most active during warm, wet, or humid periods in the summer, whereas dead spot and fairy ring are most destructive during periods of heat stress and drought. Appreciating the relationship among environmental conditions and season of the year, recognition of symptoms during the early stages of disease greatly facilitates field diagnoses and effective management. Invariably, diseases can and often do become problematic when unexpected. Golf course superintendents are cautioned to carefully monitor weather patterns, check "hot spots" where disease usually first appears, and use growing degree day or weather prediction models where available in an attempt to get ahead of their most chronic disease problems.

Pythium-Induced Root Dysfunction

Pythium species are associated with all turfgrasses and cause seed decay, postemergence damping-off, foliar blight, root and crown rot, and snow blight. Foliar blight is known as Pythium blight, cottony blight, grease spot, and spot blight. This disease is most destructive during hot and humid periods and should not be confused with Pythium-induced root dysfunction (PRD) or Pythium root rot. Unlike foliar blight, Pythium root diseases may occur throughout the entire growing season. These root diseases primarily develop during cool and wet periods of spring, causing a general decline of the turfgrass stand (Couch, 1995). The symptoms of this decline are very difficult to describe, and a positive diagnosis usually requires a laboratory analysis. Creeping bentgrass and annual bluegrass grown on greens are the primary turfgrasses affected.

Pythium species are common inhabitants in roots of creeping bentgrass and other grasses (Figure 4.1). The name *Pythium root rot* was first used by Robert Endo in 1961 to describe damping-off of bentgrass seedlings as well as a root rot of more mature plants in California. Hodges and Coleman (1985) described a Pythium disease of turfgrasses that they named Pythium-induced root dysfunction. The occurrence of Pythium-induced root dysfunction or PRD of creeping bentgrass was first associated with renovated golf greens grown on high sand content mixes. Hodges and Coleman (1985) found that roots of plants infected with *Pythium aristosporum* and *P. arrheno-manes* were not killed, and root lesions or rotting were not observed although plant growth was reduced extensively. Although the names Pythium root rot and PRD can denote different diseases, they often are used interchangeably.

The pathogenicity of *Pythium* species that infect turfgrass roots received intensive study in the 1990s. In 1991, Nelson and Craft reported Pythium root rot in New York and found *P. graminicola, P. aphanidermatum, P. aristosporum, P. torulosum,* and *P. vanterpoolii* were pathogenic to creeping bentgrass seedling roots maintained under both cool 55°F (13°C) and warm 82°F (28°C) temperatures. Based on recovery frequency and virulence studies, Nelson and Craft (1991) concluded that *P. graminicola* was the principal species

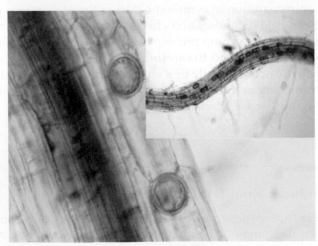

FIGURE 4.1
Pythium oospores are common inhabitants of creeping bentgrass and annual bluegrass roots.

associated with root rot in New York, particularly in annual bluegrass. In North Carolina, Abad et al. (1994) isolated 237 *Pythium* isolates from turfgrasses with symptoms of Pythium root and crown rot. They reported that 29 of 33 *Pythium* spp. isolated from turfgrasses were pathogenic to roots of creeping bentgrass seedlings. Among these species, *P. arrhenomanes*, *P. aristosporum*, *P. aphanidermatum*, *P. graminicola*, *P. myriotylum*, *P. tardicrescens*, *P. vanterpoolii*, and *P. volutum* were highly aggressive. Sixteen of the 29 species were only weakly or nonpathogenic. They concluded that *P. arrhenomanes* was the most important pathogen causing root and crown rot of creeping bentgrass in North Carolina.

Feng and Dernoeden (1999) found eight species of *Pythium* (i.e., *P. aristosporum*, *P. aphanidermatum*, *P. catenulatum*, *P. graminicola*, *P. torulosum*, *P. vanterpoolii*, *P. ultimum* var. *ultimum,* and *P. volutum*) associated with the roots of creeping bentgrass greens in Maryland and adjacent states. *Pythium aristosporum* was the most common and virulent species isolated, whereas *P. torulosum* was weakly pathogenic but was the most common species isolated. Endo (1961), Nelson and Craft (1991), and Abad et al. (1994) also found that *P. torulosum* was among the most frequently recovered species from turfgrasses exhibiting symptoms of root and crown rot. Although several studies have shown *P. torulosum* to be weakly pathogenic or nonpathogenic, it appears that this species can cause root dysfunction when its oospores are found in high populations. Feng and Dernoeden (1999) found that Pythium-induced root dysfunction was primarily found during cool and wet periods of spring and autumn on relatively new golf courses or older greens recently renovated with methyl bromide. The most destructive cases appeared just after seeding in autumn (i.e., October and November in the Mid-Atlantic region) or in the spring following an autumn seeding.

In 2004, young greens in North Carolina were found with circular patches in summer. *Pythium volutum* was isolated from infected roots and was shown to be highly pathogenic to bentgrass (Kerns and Tredway, 2008a, 2008b). Symptoms of PRD incited by *P. volutum* (PRD-Pv) are distinctive and greatly differ from other *Pythium* spp. incitants. Additional information on PRD-Pv is given below.

The severity of Pythium root diseases normally declines in bentgrass over time. Regardless, Pythium root diseases can develop in older creeping bentgrass and annual bluegrass greens that are excessively wet from rain or irrigation or pocketed greens that remain wet for long periods in summer. Apparently, natural antagonists build up in soil to suppress or compete with the pathogen, but this can take from 3 to 5 years or longer. In annual bluegrass greens, however, the disease can be a chronic problem, particularly in colder northern regions. This may be due in part to a greater susceptibility of annual bluegrass to injury from ice or very low temperatures. It is likely that annual bluegrass plants weakened by winter stresses are predisposed to root-attacking *Pythiums* during cool and moist periods of spring. Typically, the wettest greens are affected most and roots are variously impacted. In some cases, oospores (thick-walled spores produced by *Pythium* spp.) are sparse and there is little obvious root damage (Figure 4.1). In other cases, oospores are prevalent, mycelium can be observed within root cells, and the majority of the root system is in severe decline.

Symptoms of PRD

Disease develops primarily in poor growing environments and is most pronounced in low areas or follows surface water drainage patterns. Some injury, however, may be observed throughout the putting surface even in higher, well-drained areas. Pythium-induced root dysfunction (PRD), regardless of pathogen, is mostly found in new, sand-based constructions and in greens less than 3 years old (Feng and Dernoeden, 1999; Kerns and Tredway, 2010). Infection by most *Pythium* spp. associated with PRD occurs during cool and moist periods in the spring or autumn when soil temperatures range from 61° to 75°F (16–24°C), but symptoms may not become evident until summer stress periods. The symptoms of PRD are diverse, and infected roots often appear white and healthy, making it one of the most difficult diseases to diagnose.

In immature creeping bentgrass golf greens, leaves of infected plants may appear yellow and stunted and tend to be narrower than leaves of healthy plants (Figure 4.2). In most situations, there are dark green, perfectly healthy looking plants dispersed throughout pockets of chlorotic and smaller, stunted plants. Mixed within chlorotic areas, small groups of plants turn brown and die. In some cases, especially Penn A-series cultivars, there is a proneness to wilt. Infected plants develop a bluish color associated

FIGURE 4.2
Pythium-induced root dysfunction (PRD) can initially appear as a mix of chlorotic and green plants.

FIGURE 4.3
Wilt symptoms associated with Pythium-induced root dysfunction (PRD) in a creeping bent-grass golf green. (Photo courtesy of D. Bevard.)

with wilt, and in other cases affected turf has a purplish color that gives the appearance of a severe phosphorous deficiency (Figures 4.3 and 4.4). Death of plants usually first occurs at the outer periphery of the greens (i.e., cleanup pass), where *Pythium* is most likely to ingress from adjacent native soil and where there is more mechanical stress from mowing. Death of plants is due to a combination of mechanical injury from mowing stressed turf with dysfunctional root systems. More mechanical injury occurs at the

FIGURE 4.4
Severe cases of Pythium-induced root dysfunction (PRD) may appear as a purplish cast similar to that associated with phosphorous deficiency in creeping bentgrass.

point where the greens mower enters a curved area, where the mower must turn (Figure 4.5). The turning action of mowers twists and grinds leaves and stems placing a lethal stress on infected plants. Eventually, large areas throughout the entire putting surface may die. The disease generally is more severe on shaded or pocketed greens.

In annual bluegrass, leaves appear yellow-brown or reddish-brown in color before plants die. These color symptoms are similar to those associated with the use of some plant growth regulators such as paclobutrazol or flurprimidol. Older annual bluegrass greens that are excessively wet or pocketed may develop a generalized yellowing. As previously noted, PRD is more common in annual bluegrass in colder climates and is less commonly observed in transition zones and southern regions of the United States.

Washing soil from infected plants and observing roots for symptoms often provides no useful clues. Infected roots may have light brown or fawn-colored lesions, a generalized light tan to brownish discoloration and a watersoaked appearance, which can only be seen with a microscope. Infected roots may be brown, rotted, shortened, and have twisted root tips, but in many cases there is a lack of any pronounced discoloration or lesion development on bentgrass roots, including roots that contain large numbers of *Pythium* oospores (Figure 4.5). Diagnosticians stain and search roots for the presence of the distinctive thick-walled oospores produced by *Pythium* spp. The presence of large numbers of oospores is the key diagnostic sign for PRD. Unfortunately, oospores may not be present even though roots are infected. Other diagnostic techniques involve attempting to isolate *Pythium* spp. from roots. In many cases the turf has been treated with multiple fungicide applications, complicating diagnosis. Regardless of method, diagnosis of PRD sometimes boils

FIGURE 4.5
Bronzing and scalping in the cleanup pass are symptoms of Pythium-induced root dysfunction (PRD). Diagnosis is often based on frequency and number of oospores observed in roots.

down to an "educated guess" based on signs, symptoms, and age of the turf. Molecular diagnostic methods for detecting and identifying *Pythium* species from turfgrass roots are needed to better study and understand PRD.

Symptoms of PRD-Pv

Symptoms of PRD-Pv appear as distinct, reddish-brown, or bronze-colored circular patches and primarily have been observed in Penn A-series creeping bentgrass cultivars in the southeastern United States (Figure 4.6). Penn A's were the dominant cultivars planted at the time this disease was being described, but other bentgrass cultivars also are susceptible. The following description of PRD-Pv is summarized from Kerns and Tredway (2010). The disease appears in summer when creeping bentgrass is subjected to heat and drought stress, but it also can develop in the spring, autumn, or winter during warm and dry weather. Initially, symptoms appear as circular areas of wilt followed by chlorosis or drought stress and range from 1.5 to 6.3 inches (4 to 16 cm) in diameter. The patches can increase in size from 6 to 20 inches (16 to 50 cm) in diameter and may become irregular in shape. Affected patches may appear yellow or orange and can mimic take-all patch. PRD-Pv is associated with a shortened or truncated root system. Individual roots are light tan, lack root hairs, and root tips may be bulbous. *P. volutum* is most pathogenic when soil temperatures are cool and in the range of 54 to 75°F (12 to 24°C), but symptoms of the disease do not appear until periods of heat and drought stress (Kerns and Tredway, 2008b). *P. volutum* is not actively growing in the summer, and this is why oospores and hyphae of *P. volutum* are rarely found in affected root tissues. Symptoms of PRD-Pv have not

FIGURE 4.6
Symptoms of Pythium-induced root dysfunction (PRD) incited by *Pythium volutum* include reddish-brown circular patches and a truncated root system. (Photos courtesy of L. Tredway.)

been widely reported, but similar circular patches were observed in spring in young greens in the northern United States.

Management

Pythium-induced root dysfunction must be aggressively managed by a combination of cultural and chemical means. *Pythium* grows and reproduces rapidly in wet soils; therefore, drying the soil is essential. This can be difficult because of frequent rain or because truncated root systems of infected plants may be unable to absorb and translocate sufficient amounts of water and nutrients. When soils are very wet, use of the automatic overhead irrigation should be replaced by frequent hand syringing on an as-needed basis until the rootzone dries. Affected greens only should be mowed when soils and the thatch-mat layer are relatively dry (i.e., no casual water present) using a walk-behind mower. As previously discussed, never mow affected greens on rainy days or when the surface is excessively wet. Mowing height should be increased (i.e., ≥0.150 inch; ≥4 mm) and frequency of mowing reduced to four or five times weekly. Remove grooved rollers from mowers and replace them with solid rollers. Avoid other potential mechanical stresses by delaying sand topdressing, brushing, vertical cutting, and core and water injection aeration until the disease has been controlled and turf shows signs of recovery.

Most PRD pathogens are more actively invading roots during cool and moist periods in spring and autumn in new constructions. For new constructions it would be prudent to treat greens with a preventive fungicide application in spring when soil temperatures are in the range of 54 to 68°F (12 to 20°C). Unfortunately, there are few published studies that investigated the performance of fungicides against PRD. Drenches of pyraclostrobin were shown to control PRD caused by *P. volutum*, but most conventional *Pythium* fungicides were ineffective (Kerns et al., 2009). Conventional fungicides that target *Pythium* spp. include chloroneb, cyazofamid, ethazole, fosetyl-aluminum, mefenoxam, and propamocarb. Since there are numerous *Pythium* spp. that can cause PRD, it is unlikely that all species would be sensitive to the same fungicide. Hence it is prudent to tank-mix fungicides (thereby providing different modes of action) targeting PRD curatively. For example, if symptoms mimic PRD and PRD-Pv, then a tank-mix of cyazofamid (or other conventional *Pythium* fungicide) and pyraclostrobin should be considered. If this approach fails to provide satisfactory control, a more comprehensive battery of fungicides treatments may be recommended. This may involve an application of fosetyl-aluminum + mancozeb, followed in 5 to 7 days with an etradiazol or chloroneb drench. Thereafter, alternating mefenoxam, cyazofamid, propamocarb, or a pyraclostrobin drench on a 10- to 14-day interval may be suggested. Some fungicides control or relieve the condition and some do not. Of the aforementioned, only fosetyl-aluminum + mancozeb is not watered-in. The others should be watered-in to a depth of 0.5 to 1 inch (1.3 to 2.5 cm). A syringe cycle from the irrigation system usually provides sufficient water to move fungicides into the target zone immediately follow-ing a fungicide application. Among the aforementioned fungicides, most superintendents find one or two products that perform best. In some cases, a fungicide application may be required on a 7- to 10-day interval throughout much of the growing season, particularly in wet years.

Applications of biostimulants, micronutrients, and spoonfeeding may be beneficial along with fungicides, but their overall impact on PRD is unknown. The disease in young bentgrass greens can recur for 2 to 3 years following establishment, but there are cases in greens that are much older. Except where there is chronic excessive soil wetness, mature creeping bentgrass greens gen-erally are less susceptible to injury. Conversely, annual bluegrass grown on greens in more northern climates may be more chronically affected.

Summary of Key Points

- *Pythium* species are common inhabitants of turfgrass roots.
- Pythium-induced root dysfunction (PRD) is primarily a problem in young bentgrass golf greens in new constructions, but older greens that are excessively wet may develop the disease.
- PRD in creeping bentgrass develops mostly during cool and moist periods in spring or autumn, but symptoms may not appear until the advent of higher temperatures in late spring or summer.
- PRD can be a chronic problem in annual bluegrass, particularly in more northern regions of the United States and Canada.
- Symptoms of PRD are diverse. Infected plants may appear chlorotic, reddish, or bronze. There may be proneness to wilt and the atten-dant bluish-gray or purplish color of drought stress. Symptoms first appear at the periphery of greens and eventually turf thins out or dies in irregularly shaped areas.
- Circular patches are associated with PRD incited by *P. volutum* (PRD-Pv). PRD-Pv initially produces distinctively circular and wilted patches, but infected plants eventually turn yellow, bronze, or orange in color before dying. PRD-Pv patches range from 6 to 20 inches (15 to 50 cm) in diameter and may become irregular in shape.
- PRD is diagnosed by a microscopic examination of roots for the presence of oospores. Root symptoms in infected plants range from white and healthy to light tan or brown; water-soaked bulbous or twisted root tips; lack of root hairs; or a truncated root mass.
- Pythium root diseases are best managed by a combination of cul-tural and chemical methods including the following:
  - PRD-Pv is controlled by pyraclostrobin, but in the case of other PRD incitants it may be necessary to tank-mix or rotate different *Pythium*-targeted fungicides to identify effective compounds.

- Replace automatic overhead irrigation with frequent hand syringing to promote drying of excessively wet soils.
- Increase mowing height, use a lightweight walk-behind greens mower equipped with solid rollers, and do not mow when greens are excessively wet and soft.
- Avoid mechanical stress and delay sand topdressing, grooming, and other abrasive practices until turf has recovered.

Yellow Tuft

Yellow tuft is caused by the fungus *Sclerophthora macrospora*. *Sclerophthora macrospora* attacks virtually all turfgrasses as well as several major grass crops, including rice, sorghum, and corn. A severe infection mars the appearance and playability of creeping bentgrass and annual bluegrass golf greens. Infected plants generally are not killed, and the disease primarily is a problem on golf greens. The etiology of yellow tuft disease was described by Dernoeden and Jackson (1980a, 1980b).

Symptoms

On greens and tees, the disease appears as yellow spots, 0.25 to 0.5 inch (6.2 to 12.5 mm) in diameter (Figure 4.7). A sudden appearance of numerous

FIGURE 4.7
Yellow tuft appears as small yellow spots and often is mistaken for annual bluegrass. Plants are easily detached and have many tillers and a small bunchy root system.

yellow spots often is mistaken for annual bluegrass encroachment in creeping bentgrass in the autumn. In Kentucky bluegrass, and other wider-bladed grasses, the yellow spots are 1 to 3 inches (25 to 75 mm) in diameter. Each spot consists of one or two plants that have 20 to 30 or more tillers, giving plants a tufted appearance. The tufting, or abnormal tiller production, is induced by *S. macrospora*, which causes a shift in the production of a hormone (possibly indoleacetic acid) that regulates tillering. Roots of infected plants are short and bunchy, and tufts are easily detached from the turf (Figure 4.7). During cool and moist periods in late spring and autumn, plants develop a yellow color at which time infected plants are "yellow tufted." The yellowing is the indirect result of heavy fruiting body (sporangia) and subsequent spore production by the fungus. These spores (zoospores) swim, and this accounts for why the disease is more severe in low-lying areas where water puddles. Zoospores are produced in lemon-shaped structures called *sporangia*. Sporangia develop on leaf surfaces from substomatal cavities below the leaf epidermis. *Pythium* species also produce sporangia and zoospores, but they are not as sophisticated as *S. macrospora*. During early morning hours, when leaves are wet, the pearly white sporangia can be seen with a hand lens on the upper and to a lesser extent the lower leaf surfaces of infected plants. During most summer months infected plants appear green and healthy.

Plants infected with *S. macrospora* can persist in excess of 2 years. The pathogen is a very well-adapted obligate parasite. Obligate parasites can grow and reproduce only in living tissues. As a result, most obligate parasites have evolved to the point where they generally do not directly kill plants. Instead, they debilitate or weaken plants, predisposing them to possible injury or death from other stress factors. Other diseases caused by obligate parasites include rusts, smuts, and powdery mildew.

Sclerophthora macrospora gains entry through meristematic tissue (i.e., stems, buds, or the mesocotyl region of germinating seeds), and except for roots, it spreads systemically throughout the plant. Once infected, mycelium grows upwards between cells (i.e., intercellular spaces) from buds or stem bases into sheaths and then leaves. If plants are allowed to produce seedheads, the fungus can grow upward into the culm, invade the inflorescence, and eventually infect seed. Hence, yellow tuft can be a seedborne disease. When leaves senesce, mycelium may differentiate into large spores called oospores. The oospores are thick-walled survival structures that can persist for years in dead leaf tissue. They germinate in low numbers in the presence of water and suitable temperature to produce either a germ tube or an oosporangium. Each oosporangium may contain 50 or more zoospores, and they have the same shape and appearance of asexual sporangia and zoospores. While oospores help to disperse the pathogen, the production of zoospores in asexual sporangia is the primary mechanism by which large numbers of plants become infected. Crabgrass (*Digitaria* spp.) is very susceptible to *S. macrospora*, and this weed serves as a major harborage site for the production of zoospores and oospores.

Over time, new shoots escape systemic invasion by the fungus, and eventually *S. macrospora*–free tillers replace the original plants. Escape of tillers explains the ephemeral nature of yellow tuft symptoms in older stands. Seedlings are most susceptible to infection by *S. macrospora*, which accounts for why the disease is commonly observed in the spring following autumn seeding. The disease can recur in spring or autumn in older turfs during years marked by excessively wet weather.

Management

Vertical cutting of golf greens in the spring and autumn will physically detach many tufted plants. As noted previously, in wet environments the swimming zoospores are able to move easily to uninfected plants. Thus, improving drainage helps to alleviate yellow tuft because the disease is most severe in low-lying areas where water collects. Soil compaction also contributes to water puddling, and coring and installing drains should help improve water drainage in these sites.

Yellow tuft is best controlled with mefenoxam (McDonald et al., 2009). Mefenoxam performs better when tank-mixed with either fludioxonil or chlorothalonil. Two or three mefenoxam applications may be required to eradicate the fungus in infected plants. After the application of a fungicide, however, plants can retain their tufted appearance for many weeks. It is only until new tillers replace older infected shoots that plants regain their normal appearance and growth habit. Mefenoxam works best when applied prior to rainy weather, particularly where there are seedlings or immature plants. For unknown reasons, there may be little response from a curative mefenoxam application.

Summary of Key Points

- The yellow tuft fungus (*Sclerophthora macrospora*) is a sophisticated obligate parasite that attacks nearly all turfgrasses.
- Yellow tuft is commonly found in annual bluegrass and creeping bentgrass on tees and fairways but is most problematic on golf greens where it can adversely affect aesthetics and create a bumpy surface.
- Infected plants exhibit excessive tillering and root systems are short and bunchy.
- The pathogen is spread in water by swimming zoospores that can infect only meristematic (i.e., actively cell dividing) tissues found in stems, axillary buds, and developing seedlings.
- Yellow tuft is most severe in young stands, but older turfs also are affected, especially when weather conditions are rainy for long periods or where water puddles.
- Disease-free tillers eventually replace infected shoots, which accounts for the ephemeral nature of the disease.

- Yellow tufted plants can be physically removed by vertical cutting in the spring and autumn.
- For curative yellow tuft control, multiple applications of mefenoxam tank-mixed with chlorothalonil or fludioxonil are recommended.

Anthracnose Basal Rot and Foliar Blight

Anthracnose is caused by the fungus *Colletotrichum cereale* (synonym: *C. graminicola*). *C. cereale* is a common saprophyte (i.e., living on dying or dead tissue) found colonizing thatch or dying plant tissues. This fungus often attacks weakened or senescent leaves, and its behavior as a "senectophyte" was described in Chapter 3. Under stressful conditions, *C. cereale* behaves as a pathogen. Anthracnose is primarily a serious problem of annual bluegrass and a lesser problem in creeping bentgrass. The fungus may cause either a foliar blight or a basal rot. Foliar blighting generally occurs during periods of high temperature and drought stress. Foliar blight is a distinct type or phase of anthracnose, which may or may not progress into basal rot. Plants affected by anthracnose also may be invaded by other pathogens.

Anthracnose foliar blight occurs in annual bluegrass and creeping bentgrass during high-temperature stress periods in summer, and it causes a yellowing or a reddish-brown discoloration and eventually a loss of shoot density. The distinctive fruiting bodies (i.e., acervuli) with protruding black hairs (i.e., setae) can be observed on green and discolored leaf or sheath tissue with a hand lens. Because *C. cereale* can infect green juvenile or mature leaves, it is considered a primary pathogen rather than a senectophyte in this situation. The presence of large numbers of acervuli on dead tissue in thatch does not indicate that healthy plants also are infected. Therefore, it is important to carefully look on green or discolored tissue of living plants for the distinctive fruiting bodies. The foliar blighting phase is easier to control and is less common than basal rot in most regions.

Anthracnose basal rot can be extremely destructive to golf greens and is much more common in annual bluegrass than creeping bentgrass. Anthracnose basal rot is uncommon in creeping bentgrass tees and fairways (Figure 4.8). Creeping bentgrass is most likely to be infected as a result of winter desiccation or mechanical injury (e.g., close, double cutting during stressful periods). The increased occurrence of anthracnose basal rot on golf greens can be attributed to the common practice of mowing extremely low, maintaining low nitrogen fertility, and imposing abrasive cultural practices (such as frequent sand topdressing, brushing, vertical grooming, etc.) in summer to increase green speed. Increased play on older, smaller greens also can be a factor.

FIGURE 4.8
Anthracnose is uncommon in creeping bentgrass, but basal rot can occur if turf was injured in winter or mechanically damaged. (Photos courtesy of D. Settle.)

Anthracnose basal rot occurs at varying times of the year and may produce different symptoms on different hosts. Basal rot in annual bluegrass occurs during cool to cold periods between autumn and spring, and during warm to hot periods of summer. Anthracnose basal rot can remain active in annual bluegrass throughout mild winters. Basal rot generally does not appear in bentgrass before the advent of higher summer temperatures. The pathogen is much more invasive if it enters stem tissue through wounds. The disease most often is associated with low mowing, soil compaction, traffic, wilt, and low nitrogen fertility. Hot and dry weather followed by an overcast or rainy period can trigger or intensify the disease. Shaded and wet sites are particularly vulnerable.

Symptoms

Between late autumn and spring, infected annual bluegrass plants may appear as orange or yellow speckled spots 0.25 to 0.5 inch in diameter (6.2 to 12.5 mm). This symptomatology is similar to bacterial wilt in annual bluegrass. Individual plants may have both green, healthy appearing tillers, and yellow-orange infected tillers. The central or youngest leaf is last to show the yellow-orange color change. Removal of all sheath tissue to expose the stem base reveals a water-soaked, black rot of crown tissues. By mid-to-late spring, the speckled-spot symptom is less common, and infected annual bluegrass plants coalesce into large, nonuniformly affected areas that appear yellow or reddish-brown. Large areas may thin or die out completely. The disease at this point may be confused with Bipolaris or Drechslera melting-out or red leaf spot (see Chapter 3). Anthracnose basal rot is difficult to suppress where it becomes a chronic, spring or early summer problem on annual bluegrass greens.

In summer, affected turf initially develops a reddish-brown, orange, or yellow color and thins out in irregularly shaped patterns a few inches (3 to 5 cm) to several feet (>60 cm) or more in size (Figure 4.9). Rarely, circular gray-brown patches can appear on golf greens. Annual bluegrass plants often develop a brilliant yellow color before dying, and this symptom can be confused with summer patch (Figure 4.10). When discoloration or thinning

FIGURE 4.9
Orange-yellow annual bluegrass with basal rot anthracnose contrasted with healthy slate-green creeping bentgrass.

FIGURE 4.10
Annual bluegrass develops a brilliant yellow-orange color before dying from basal rot anthracnose. (Photo courtesy of B. Corwin.)

FIGURE 4.11
Small, black fly speck infection cushions increase in size to produce black mycelial aggregates of infecting cells. (Photo courtesy of B. Corwin.)

is first observed, managers are advised to carefully look on stem bases for the infection mats, which during the early stages of the disease appear as small (pinhead sized), black "fly specks" (Figure 4.11). This will involve removing the leaf sheaths to expose the whitish inner sheath tissues or stem areas. There will be no telltale signs of the pathogen on leaf or sheath tissue during the early stages of basal rot. In advanced stages, black aggregates of fungal mycelium can be found on infected stolons or stem bases of creeping bentgrass and annual bluegrass. The spore bearing acervuli with short, black hairs and the black mycelial aggregates can be seen on stem bases and stolons with a hand lens. Once acervuli develop on sheath or leaf tissue, the basal rot phase is advanced and plants die (Figure 4.12). Note that dead tissue in the thatch-mat layer may be inundated with acervuli, but it is on the living tissue that you must look for these bodies. If acervuli only are seen in the thatch-mat, the fungus is behaving like a saprophyte. For mysterious reasons, the disease seldom attacks both annual bluegrass and creeping bentgrass on the same green or even on the same golf course.

Management

Research by Inguagiato et al. (2009) revealed that anthracnose basal rot severity in annual bluegrass was reduced dramatically by applying 0.1 lb N/1000 ft^2 (5 kg N/ha) every 7 days. They also demonstrated that increasing mowing height from 0.110 to 0.141 inches (2.75 to 3.25 mm) reduced disease injury significantly. Plant growth regulators (i.e., ethephon, mefluidide, and trinexapac-ethyl) were shown to have little impact on anthracnose although there were

FIGURE 4.12
Acervuli on a yellow annual bluegrass sheath and a bentgrass stolon.

times that some appeared to enhance the disease. Spring applications of two preemergence herbicides (i.e., bensulide and dithiopyr) intensified anthracnose, but light vertical cutting and sand topdressing had no negative impact.

Anthracnose basal rot is very difficult to control once annual bluegrass turf shows signs of thinning. This is especially true when annual bluegrass develops the disease in spring. To alleviate anthracnose basal rot, use walk-behind greens mowers and increase height of cut immediately. Divert traffic away from affected areas by rotating hole locations frequently. While irrigation often is withheld in summer to provide for firm and fast playing conditions, it is best to avoid wilt. Research has shown that daily irrigation of annual bluegrass golf greens at 80% evapotranspiration provides sufficient water to maintain growth of annual bluegrass, while helping to minimize disease severity (Roberts et al., 2011). Conversely, excessive soil wetness can exacerbate anthracnose. When the disease is active and turf is stressed by heat or wilt, avoid sand topdressing, double cutting, core aeration, brushing, vertical cutting, and other potentially abrasive practices. In autumn, after symptoms have dissipated, core aerify and overseed.

For curative fungicide control programs, include a high label rate of either chlorothalonil or fludioxonil in the mixture. The latter two fungicides are contact protectants. Chlorothalonil or fludioxonil tank-mixed with a penetrant fungicide such as a QoI or strobilorin (azoxystrobin, fluoxastrobin, pyraclostrobin, or trifloxystrobin); DMI/SI (e.g., metconazole, propiconazole, tebuconazole, triticonazole, others); fosetyl-aluminum, polyoxin D, or thiophanate-methyl should help reduce or slow the disease but will not always completely control anthracnose. Two or more applications on a 7- to 10-day interval may be required to arrest the disease. Fungicides should be applied in 88 to 100 gallons water per acre (823 to 935 L/ha) to improve stem base coverage and thus control. A modest amount of nitrogen (0.1 to 0.2 lb N/1000 ft², 5 to 10 kg N/ha) should be combined with a fungicide. It is important to note that *C. cereale* biotypes resistant to azoxystrobin and thiophanate have developed. Biotypes of *C. cereale* also may be showing reduced sensitivity to DMI/SI fungicides. Hence, the aforementioned fungicides should be used only occasionally and should be avoided if no positive response was observed.

Where basal rot is a chronic problem, fungicides should be used preventively in combination with an improved nitrogen fertility program. Moderate nitrogen levels (≥3 lb N/1000 ft²/yr or ≥150 kg N/ha/yr) are associated with less anthracnose. Greens should be spoonfed with a water-soluble nitrogen source such as urea or ammonium nitrate (i.e., 0.1 to 0.2 lb N/1000 ft² or 5 to 10 kg N/ha) every 7 to 14 days throughout summer. Weekly spoonfeeding is especially effective in encouraging the growth and spread of healthy creeping bentgrass, which will fill voids when annual bluegrass dies from the disease. Preventive applications of the aforementioned fungicides should begin 2 to 3 weeks prior to the time the disease generally appears. Superintendents observed weak responses or failures with all of the aforementioned fungicides. This is particularly true for annual bluegrass greens on older golf courses. In extreme cases, greens that are chronically infected and consist mostly of annual bluegrass may have to be re-grassed.

It is important to again note that there usually is no blighting of leaves or the appearance of acervuli on leaves or sheaths during early stages of anthracnose basal rot development. Acervuli are commonly found in dead organic matter in the thatch-mat layer and may be behaving as a saprophyte. Stems and stolons on living or discolored plants should be inspected for black infection cushions and a black, watersoaked basal rot. Anthracnose foliar blight is a phase of the disease in which acervuli are produced on green leaves and sheaths, and it can occur with or without progression to basal rot. Hence, there are actually two distinct types of anthracnose. Bentgrass greens affected with anthracnose often respond favorably to chlorothalonil tank-mixed with one of the aforementioned penetrants. Chronically infected annual bluegrass greens, however, may not respond favorably to any fungicide program. Multiple applications of flutolanil without rotation with other fungicides can promote anthracnose.

Summary of Key Points

- Anthracnose foliar blight and basal rot are different types or phases of anthracnose incited by *Colletotrichum cereale*. Anthracnose basal rot is the most damaging and difficult phase to control.
- Annual bluegrass can be chronically affected by anthracnose, but the disease is less common in creeping bentgrass.
- Anthracnose basal rot can occur in annual bluegrass from winter to autumn, but it generally develops in creeping bentgrass in the summer.

- Wilt stress, low nitrogen fertility, low mowing heights, and soil compaction intensify anthracnose.
- In summer, affected annual bluegrass on golf greens develops a golden yellow or orange color (a burned appearance to some) and thins out in irregularly shaped areas. Infected creeping bentgrass golf greens develop a reddish-brown appearance and thin out in irregular patterns.
- Anthracnose basal rot is diagnosed by inspecting living or discolored stem tissues for black infection mats or black hairs (acervuli), and/or a brownish watersoaking or a black rot. Infected annual bluegrass plants are easily detached from the surface.
- Anthracnose basal rot management strategies include:
  - Use preventive applications of fungicides (normally chlorothalonil or fludioxonil plus a penetrant) where the disease is chronic.
  - Anthracnose can be difficult to control in annual bluegrass golf greens, especially when the disease begins in spring.
  - Do not mow excessively (spongy) wet greens.
  - Increase mowing height and reduce mowing frequency.
  - Mow affected greens with a walk-behind mower equipped with solid rollers.
  - Avoid excessive irrigation and rely on syringing to prevent wilt stress in summer.
  - Spoonfeed nitrogen with a water soluble nitrogen source such as urea or ammonium nitrate (0.125 lb N/1000ft^2 or 6.0 kg N/ha) weekly throughout the summer months, especially when anthracnose is active.
  - Alleviate soil compaction and improve surface and internal water drainage.
  - Improve air circulation with fans and by pruning or removing trees in shaded environments.

Take-All Patch

Take-all patch is incited by *Gaeumannomyces graminis* var. *avenae* and is primarily a disease of bentgrass (*Agrostis* spp.) turfs. Take-all, however, has been observed in annual bluegrass (*Poa annua*) in the United Kingdom. Take-all was first reported in Holland in 1937 on a bentgrass golf green, but its occurrence in the United States was not documented until 1961 in western Washington. It was not until the 1970s that the disease was reported in the eastern United States. Take-all patch is now known to occur just about everywhere bentgrass is grown. The fungal pathogen attacks roots and stems, and there are no distinctive leaf spot or sheath lesions.

Take-all patch is most common on newly constructed golf courses, particularly those carved out of woodlands, peat bogs, or other areas that have not supported crops or grasses for decades. This disease can be especially damaging to rebuilt greens or tees on old golf courses or where a fumigant was used for renovation. Take-all also can be imported on infected bentgrass sod. The disease tends to spread more rapidly and occurs with greater severity in sandy soils and in soils that have been heavily limed. Take-all may appear immediately following the installation of infected sod, but in seeded stands it generally does not appear until the second year following seeding. Take-all seldom appears within a year of establishment but should be scouted beginning the first summer following seeding.

Symptoms

The pathogen actively attacks roots during cool and wet periods, but symptoms may not appear until the advent of warmer and drier conditions. Symptoms of take-all patch are most conspicuous from late spring, throughout the summer, and may recur in autumn. Take-all can appear in early spring in response to drought, low mowing, aggressive grooming, and the use of plant growth regulators. Bentgrass affected by take-all in the spring may recover by summer; however, if irrigation is withheld those areas affected in spring are the first to wilt and die from drought stress.

A reliable diagnostic indicator of take-all is to scout greens early in the morning for distinctive circular patches that are dew free. This indicates that the root system is being compromised by the pathogen. Soon thereafter patches will exhibit a color change. Initially, circular patches of take-all affected creeping bentgrass are only a few inches (5 to 8 cm) in diameter and are reddish-brown or orange-bronze in color. Turf in affected patches may first develop the blue-gray color associated with wilt. This is due to the impaired ability of infected roots to take up sufficient amounts of water. Patches may increase to 2 feet (60 cm) or more in diameter, particularly on chronically affected creeping bentgrass sites. Most patches range from 3 to 18 inches (7.5 to 45 cm) in diameter, but some patches may exceed 3 feet (90 cm) in diameter (Figures 4.13 and 4.14). Patches may develop in tight clusters that give the appearance of a larger diameter patch. Patches also may coalesce resulting in large, irregular areas of discolored or dead turf. When the disease is active, the perimeter of the patch usually assumes a bronzed or reddish-brown appearance, and the turf eventually turns a bleached or tan color. When the disease is relatively inactive, patches may appear yellow or white (Figure 4.15). Frog-eyes or rings of dead turf with living plants in the center, and small horseshoe-shaped crescents also are associated with take-all and sometimes are mistaken for fairy rings. Small, circular patches

FIGURE 4.13
Classic bronze-colored take-all patches 12 to 18 inches in diameter (30 to 45 cm).

FIGURE 4.14
Take-all patches (3 ft diameter; 1 m) on a golf green 10 years following fumigation. Inset shows annual bluegrass invading a take-all patch on a tee.

increase in size over a number of years, and dead bentgrass in the center of the patch may be colonized by annual bluegrass, dandelion (*Taraxacum officinale*), and other weeds. Because the fungus attacks the root system, turf in affected areas is detached easily and is reminiscent of the type of damage caused by root feeding insect pests (e.g., white grubs). Key diagnostic root signs include the presence of darkly pigmented runner hyphae and

FIGURE 4.15
Whitish-colored take-all patches. (Photo courtesy of J. Kaminski.)

FIGURE 4.16
Roots and stolon infected with the take-all fungus. Inset shows darkly pigmented runner hyphae and discolored stele of an infected bentgrass root.

discoloration of the inner portion of the root containing the xylem known as the stele (Figure 4.16).

In most cases, take-all patch will naturally decline over time, presumably due to a buildup of antibiotic-producing bacteria that antagonize or in some other way prevent *G. graminis* var. *avenae* from damaging roots. This explains why take-all is normally not a problem on older golf courses. There are, however, exceptions. Golf courses with a past problem with take-all may see the disease redevelop following a heavy application of lime or other alkaline reacting materials. In the United Kingdom, take-all redevelops on some older golf courses, especially during exceptionally wet years. It is possible in calcareous soils that calcium ions enter the soil solution when soils are wet for long periods and thus increase soil pH, which in turn can cause the disease to recur. Take-all also can persist indefinitely in highly buffered or alkaline soils or where the irrigation water has a high pH. In rare cases, it can recur on greens for 10 or more years, despite every effort to control the disease. The decline phenomenon generally occurs within 2 to 5 years from the time that the first disease symptoms were observed. During the decline phase, plants in affected patches appear wilted, reddish-brown, or yellow, and turf may or may not thin out.

Management

Take-all management involves alleviating stress and promoting root growth. It is important to increase mowing height, reduce mowing frequency, replace grooved rollers with solid rollers, and mow with lightweight mowers. Do not mow greens when they are excessively (spongy) wet due to the likelihood of mechanical damage. Avoid other practices that promote green speed such as sand topdressing, vertical cutting, brushing, and the use of plant growth regulators. Furthermore, core aeration and high-pressure water injection aeration should not be attempted when the disease is active.

There is an interesting relationship between soil pH and take-all. This disease can occur over a wide range of soil pHs, including acidic soils, but it is most severe and persistent where soil pH is above 6. Acidification of soil with ammonium sulfate (or ammonium chloride) is the primary cultural approach in managing take-all. Even in situations where soil pH is as low as 5.5, ammonium sulfate may still be recommended. Nitrate forms of nitrogen, such as potassium nitrate, calcium nitrate, and sodium nitrate, however, may intensify take-all by raising soil pH.

Early studies used to establish acidification as a method of take-all suppression involved excessively high levels of nitrogen (i.e., 8 to 12 lb N/1000 ft^2 per year or 400 to 600 kg N/ha/yr). It is now known that as little as 3 lb N/1000 ft^2 per year (150 kg N/ha/yr) from either ammonium sulfate or ammonium chloride significantly reduces but does not eliminate take-all (Dernoeden, 1987). Ammonium chloride is a fine graded material and can be safely and effectively applied only by diluting it in water and applying it through a sprayer. Ammonium sulfate or ammonium chloride should be used as the primary N-source for at least 2 years and perhaps longer where the disease is more persistent. The use of the aforementioned ammonium-based fertilizers will provide very good winter turf color, but they also encourage foliar growth. This causes slower putting green speeds and requires frequent mowing into early winter. Elemental sulfur has been used to suppress this disease in the Pacific Northwest, but sulfur can be phytotoxic to bentgrass grown in most other regions of the United States. Ammonium-based N fertilizers, like ammonium sulfate, remain the best and safest choice for take-all management in most regions. The sulfur in ammonium sulfate (i.e., SO_4 anion) is oxidized and leaches, causing no damage to bentgrass. Ammonium sulfate and ammonium chloride have high burn potentials and should be watered-in immediately after application. When using acidifying fertilizers, check soil pH often and avoid acidification if soil pH falls below 5.2. It is most important to acidify and monitor soil pH in the upper 1 to 2 inches (2.5 to 5 cm) of the profile, where stems and a large portion of the root system reside.

Acidification of the rhizosphere (i.e., the root surface and root microenvironment) in the upper 1 inch (2.5 cm) of soil is believed to be the primary factor responsible for alleviating take-all. It was suggested that acidification of soil water either directly reduces growth of the take-all fungus or favors growth of other microorganisms that compete with or in some other way antagonize *G. graminis* var. *avenae*. Manganese oxidation by microbes in soil was linked to increased take-all in wheat. Manganese sulfate applied at 0.02 to 0.04 lb elemental Mn/1000 ft² (1 to 2 kg Mn/ha) for a total of 1.8 to 3.6 lb Mn/1000 ft²/ year (88 to 176 kg Mn/ha/yr) can reduce the severity of take-all when used in conjunction with ammonium sulfate (Heckman et al., 2003). Acidification also reduces the ability of microbes to oxidize manganese, and the resulting increase in manganese availability for root uptake assists plants in their defense against the take-all fungus. Hence, manganese sulfate should be applied in an acidification program involving ammonium sulfate. Manganese sulfate is best applied on a monthly interval between early spring and late autumn (i.e., March to November in North America and Europe) in water using a sprayer and then watered-in. Uptake of manganese by plants may be improved by using a wetting agent. Even though manganese sulfate appears to be safe on fairway turf, it should be tested locally for safety on golf greens, because burning is possible with summer applications.

Applications of phosphorus (P) and potassium (K) were linked to reducing take-all severity. Phosphorus should be applied even when soil tests indicate moderate or high P levels. For best results, ammonium sulfate or ammonium chloride should be applied in an N:P:K ratio of 3:1:2. The use of lime or a sand topdressing mix with a pH above 6 should be avoided. Do not use sand topdressings that are augmented with lime. If limestone is required, use slower-acting coarse rather than fine grades in areas with a past history of take-all. Thatch-mat and soil compaction should be controlled through core aeration and vertical cutting. Core aeration should be performed when symptoms are not evident and should be delayed if it causes lifting of bentgrass turf. Hopefully, aggressive core aeration will promote root growth in the spring. A deep root system is the best defense against take-all. Check the pH of your irrigation water. In extreme cases, acid injection of irrigation water may be recommended. Frequent syringing and hand watering of affected areas often is required in the summer to prevent death of plants whose root system has been significantly damaged by the pathogen. Finally, use of preemergence herbicides, ethofumesate, bispyribac-sodium, and plant growth regulators should be avoided where take-all is a problem because they could negatively impact severity and recovery from the disease.

The fungicides azoxystrobin, fenarimol, propiconazole, pyraclostrobin, tebuconazole, or triadimefon applied preventively twice in late autumn (October and November in the United States) and twice in spring (April and May in the United States) should provide some protection against take-all patch. For additional suppression, these fungicides may have to be applied several times on a 14-day interval at the onset of symptoms. Note that high label use rates of DMI/SI fungicides (i.e., fenarimol, propiconazole, tebuconazole, and triadimefon) can discolor or injure turf, particularly on close-cut golf greens. Multiple applications of the aforementioned DMI/SI fungicides can widen leaf blades giving greens a "leafy" appearance. It is best to apply a fungicide targeting take-all in a high water volume (≥100 gallons water/acre; ≥935 L/ha) using flat fan nozzles and high pressure (>35 psi; >262 kPa) to move fungicide down to stems. Mixing a fungicide with a wetting agent and watering-in to a 1 to 2 inch (2.5 to 5 cm) depth may boost the performance of azoxystrobin and other fungicides.

Summary of Key Points

- Take-all patch primarily is a creeping bentgrass disease on new golf courses or greens renovated by fumigation. The pathogen attacks roots during cool and moist periods, but symptoms may not appear until warmer or dry weather.

- Take-all symptoms generally appear as circular patches of wilted, reddish-brown, or bronzed colored turf a few inches (5 to 8 cm) to 2 or more feet (>60 cm) in diameter.

- Take-all usually declines naturally over a 2- to 5-year period, but on some courses the disease can recur annually for many years due to high soil or irrigation water pH, and other unknown factors.

- Approaches to take-all management begin with an acidification program using ammonium sulfate in combination with other inputs as follows:
  - Use multiple and preventive applications of effective fungicides in the autumn and spring.
  - Apply phosphorus and potassium using an N:P:K ratio of 3:1:2.
  - Apply manganese sulfate monthly during the growing season (March to November in North America, United Kingdom, and Europe).
  - Avoid core aeration and water injection aeration when the disease is active.
  - Control thatch-mat and soil compaction when the disease is inactive.

- Frequently syringe and hand water during dry periods.
- Avoid using lime and nitrate forms of nitrogen.
- Inject acid (i.e., sulfuric acid) into irrigation water if recommended.
- Avoid as much as possible the use of herbicides and plant growth regulators until the disease has declined.

Dollar Spot

Dollar spot is among the most common diseases of turfgrasses worldwide. The disease was first recognized on golf turfs about 1915, but remarkably little is known about the biology of the pathogen. The causal agent initially was thought to be a species of *Rhizoctonia*, and the disease was known as small brown patch until about 1932 (Montieth and Dahl, 1932). It was not until 1937 that the pathogen was formally described and named *Sclerotinia homoeocarpa*. Through molecular techniques, it has since been learned that the dollar spot fungus is not *S. homoeocarpa*, and at some point in time it will be reclassified and assigned a new Latin binomial. Dollar spot is known to attack most turfgrass species.

Symptoms

The symptomatic pattern of dollar spot varies with turfgrass species and cultural practices. On closely mown creeping bentgrass and annual bluegrass grown on greens, tees, and fairways, the disease appears as small, circular, and white to straw-colored spots of blighted turfgrass. Initially the spots are small and about 0.25 to 0.75 inches in diameter (6 to 19 mm) and increase to about the size of a silver dollar (3.8 cm diameter) (Figure 4.17). During early spring when there are chilly nights, initial infection centers in creeping bentgrass may have a reddish rather than white or straw color. If the disease is allowed to go unchecked, surfaces can become severely pitted (Figure 4.18). With coarser-textured grasses that are suited to higher mowing practices, such as Kentucky bluegrass or perennial ryegrass, the blighted areas are considerably larger and straw-colored patches range from 3 to 6 inches (7.5 to 15 cm) in diameter. Affected patches frequently coalesce and involve large areas of turf. A fine, white or grayish-white, cobwebby mycelium may cover the diseased patches during early morning hours when the fungus is active and leaf surfaces are wet.

Grass blades often die back from the tip and have straw-colored or bleached-white lesions shaped like an hourglass. The hourglass banding on leaves often is made more obvious by a narrow brown, purple, or black band that borders the bleached sections of the lesion from the remaining

FIGURE 4.17
In close-cut turf, dollar spot infection centers are about the size of a silver dollar (1.5 in or 3.8 cm diameter), and foliar mycelium may be present.

FIGURE 4.18
When dollar spot goes unchecked surfaces may become pitted.

green portions. Hourglass bands are difficult to find on close-cut bentgrass or annual bluegrass on golf greens. On close-cut greens, lesions are oblong or oval shaped, but there is a brown-colored band of tissue where the tan or white lesion and green tissue meet (Figure 4.19). Tip die back of leaves is common, and blighted tips appear tan to white in color and also have a brown band bordering dead and green leaf tissue.

FIGURE 4.19
On close-cut bentgrass leaves dieback from the tip and lesions may be oblong or less commonly hourglass shaped.

Environmental Influences

Dollar spot is said to be favored by warm and humid weather, and when night temperatures are cool long enough to permit early and heavy dew formation. In fact, the disease develops under imperfectly understood environmental conditions. In some U.S. regions, dollar spot can be active nearly year round. In other regions, dollar spot can be active in July and August, and serious outbreaks may occur in April and as late as early December in North America. In most regions of the United States there are two dollar spot epidemics between May and October and possibly a third epidemic in late autumn. The second, late summer epidemic is of longest duration and of greatest severity in northern and transition zone United States. The reasons for the decline or stasis of dollar spot between the first and second epidemics are unknown. Dollar spot can appear at almost any time of year in the southern United States.

A growing degree day (GDD) model developed in Maryland (a mid-Atlantic state) was accurate in predicting the onset of the first dollar spot epidemic in May (Ryan et al., 2012). This model used a start date of 1 April and a base temperature of 59°F (15°C). Highly susceptible cultivars (i.e., 'Crenshaw' and 'Backspin') developed dollar spot between 60 and 70 GDD and moderately susceptible cultivars (i.e., 'Penncross', 'Providence', 'L-93' and '007') between 105 and 115 GDD. Growing degree day models for predicting plant diseases typically are weak since they are strictly temperature driven and do not take into consideration precipitation, relative humidity, leaf wetness duration, and other important variables. Most golf courses have diverse microclimates as a result of variations in topography and exposures, the presence or absence of trees and brush, and other natural features. Thus GDD are not accurate for use in diverse environments, but can be used as a tool to monitor the potential for dollar spot to appear in spring. Finally, this model needs fine-tuning since the start date would vary among regions.

Rain and long leaf wetness durations are important for dollar spot outbreaks, but average night temperature and average daily relative humidity are among the most important environmental triggers for this disease. Dollar spot is limited by average night temperatures below 54°F (12°C) and above 77°F (25°C) (Ryan, 2011). Typically, the disease activates in late spring when average night temperatures increase consistently from 54 to 59°F (12 to 15°C). Two or more days of average relative humidity >85% are needed to promote severe outbreaks of dollar spot. The disease becomes more severe when soil moisture approaches the wilting point in late summer. Very dry soil conditions earlier in the season, however, do not impact dollar spot severity.

The appearance of dollar spot during different seasons and times of year can be influenced by the genetics of both host and pathogen. Dollar spot attacks susceptible cultivars earlier and more severely than resistant biotypes. For example, dollar spot develops 7 to 14 days earlier in a highly susceptible cultivar like Crenshaw versus a moderately susceptible cultivar like Penncross (Ryan et al., 2012). There are numerous biotypes of *S. homoeocarpa*. Some biotypes are more adapted to cooler, northern regions and others to warmer, southern regions. The biotypes found in golf greens can be different than those found in fairways and roughs. Hence, a mixing of biotypes adapted to various climatic conditions would explain the occurrence of the disease across seasons. Although not clearly understood, some other contributing factors may include (1) lower inputs of nitrogen; (2) lower mowing heights, more frequent mowing, and the removal of clippings; (3) time of day mowing begins; (3) dry soil conditions in late summer; (5) traffic and wear, especially where mowers turn in fairways or approaches; (6) a lack of effective thatch-mat and soil compaction control programs; (7) the planting of very susceptible cultivars; and (8) use of improper water dilutions or nozzles when spraying fungicides.

Cultural Management

The keystone of all disease management programs is planting regionally adapted, disease-resistant cultivars. For some golf courses, the greatest contributing factor to dollar spot problems has been the seeding of creeping bentgrass cultivars highly susceptible to the disease, such as Backspin, Crenshaw, Century, Imperial, SR 1020, and Southshore. Blending cultivars may be recommended where a highly desired yet highly susceptible cultivar is utilized. For example, blending a resistant cultivar (e.g., L 93) with a susceptible cultivar (e.g., Crenshaw) results in less dollar spot than a susceptible cultivar grown in monoculture (Abernathy et al., 2001).

Nitrogen fertility and dew displacement are the key management practices for suppressing dollar spot severity. Dollar spot is said to be a disease of poorly nourished turf. It is more severe in situations of low nitrogen (N) fertility, such as in golf course roughs. Dollar spot, however, also is troublesome in properly fertilized turf. There are numerous research reports on dollar spot management with various N sources. During the 1990s, there were several studies involving organic N versus inorganic and synthetic organic N sources (Liu et al., 1995; Landschoot and McNitt, 1997). Some organic N sources were reported to provide an improved level of dollar spot control by encouraging soil microbial activity. The theory is that if microbial populations are stimulated by the organic component of the fertilizer, they will antagonize, compete with, or in some way reduce the capacity of a pathogen to cause disease. Other studies found that improved N availability, as measured by N tissue content and not necessarily enhanced microbial activity, is the mechanism of improved dollar spot suppression with a particular N source. In the most extensive study of its kind in creeping bentgrass, several important findings were elucidated (Davis and Dernoeden, 2002). In that study, several N sources (including urea, sulfur-coated urea, Ringer Lawn Restore, Milorganite, Sustane, composted sewage sludge, others) were applied primarily in the autumn, with a small amount applied in spring (i.e., May) over a 7-year period. They found that while some N sources reduced dollar spot early in the season under low to moderate pressure, no N source reduced dollar spot severity over the course of a season. The composted sewage sludge used in that study actually increased dollar spot severity. The importance of this study is that it substantiated that N only can be expected to consistently reduce dollar spot in an agronomically significant way when it is being applied continuously during an epidemic. Hence, microbial activity probably is not an important factor in dollar spot suppression with organic N sources. This is not to say that there is anything wrong with using organic N sources. Many are excellent N sources and their slow release properties are of great importance in routine fertility programs.

The preponderance of research data support the view that N is most effective in reducing dollar spot when applied routinely at the time dollar spot is active. For golf course greens, tees, and fairways, spoonfeeding with a water-soluble N source such as urea, ammonium nitrate, or ammonium sulfate is a good approach to managing dollar spot early in its development when disease pressure is low. The objective is to stimulate shoot growth so that plants can replace blighted tissue at an even or better pace than the pathogen can inflict injury. Low levels of N (0.125 to 0.2 lb N/1000 ft^2; 6.2 to 10 kg N/ha) should be sprayed on a 7- to 14-day interval when dollar spot is active. Once dollar spot becomes severe, spoonfeeding will not have an impact on the disease. Regardless, spoonfeeding may help delay the time of the first fungicide application while promoting some turf recovery. Some superintendents wisely spoonfeed with N in summer each time a fungicide is applied.

For less intensively managed roughs and lawns, use of organic or other slow release forms of N in autumn fertility programs is a good agronomic approach for any disease management program. While there is no evidence that phosphorous (P) or potassium (K) impact dollar spot, they can help to improve the overall health of turf and thus help to reduce damage and shorten recovery time. These nutrients are best applied in a 3:1:2/N:P:K or similar ratio according to soil tests in the autumn as part of a complete fertilizer program. There also is no evidence that micronutrients, molasses, biostimulants, or other similar products have any effect on dollar spot.

Raising the mowing height is among the most effective cultural approaches to promoting turf recovery from disease; however, dew removal via mowing is a more powerful approach to suppressing dollar spot. Mowing early in the morning will speed surface drying and has been linked to a significant reduction in dollar spot and improves the effectiveness of fungicides targeting this disease (Williams et al., 1996; Pigati et al., 2010b) (Figure 4.20). The earlier you can mow and get mowers out in front of sprayers, the better the dollar spot suppression. The removal of dew and leaf-surface exudates by poling, dragging, or whipping suppresses dollar spot, but not as greatly as mowing. Similarly, the use of wetting agents, which reduce leaf wetness periods, can help to reduce dollar spot severity. Rolling also has been shown to reduce dollar spot (Nikolai et al., 2002), whereas wear from turning mowers can increase dollar spot. Try to minimize wear damage by skipping perimeter mowing one or two days a week.

Thatch-mat layers and soil compaction long have been recognized to promote dollar spot and other diseases. Hence, core aeration, sand topdressing, vertical cutting, and other practices that alleviate soil compaction and control thatch-mat promote plant health and indirectly assist in reducing dollar

FIGURE 4.20
Dollar spot is less severe when turf is mowed early in the morning to displace leaf surface fluids and thus speed drying of the canopy.

spot severity. These practices are best performed during disease-free periods when turf is actively growing.

Dry soil conditions can intensify dollar spot. Turf grown on golf courses generally should be kept on the dry side throughout summer. During severe outbreaks of dollar spot in late summer and early autumn, however, soil moisture levels should be maintained above the wilting point. For example, in a bentgrass fairway study it was shown that dollar spot was less severe where soil moisture was maintained above the wilting point of the slit-loam soil (McDonald et al., 2006). Spring and early summer outbreaks of dollar spot do not appear to be influenced by soil moisture level. Hence, a balanced approach to summer irrigation is needed to maintain good soil moisture as well as playability. Contrary to early belief, night irrigation does not promote dollar spot. More importantly, low soil moisture content most significantly impacts dollar spot during late summer epidemics.

Chemical Management

Despite planting cultivars with good resistance, spoonfeeding, mowing early, and imposing all of the recommended management practices dollar spot often becomes extremely severe and needs to be controlled with fungicides. The fungicides commonly used for dollar spot control are shown in Table 4.1. Except for chlorothalonil and mancozeb, all are penetrants representing four chemical classes with different modes of action: (a) dimethylene inhibitor/sterol inhibitor (DMI/SI); (b) benzimidazole; (c) dicarboximide, and (d) carboximide. Tank mixing a fungicide with 0.1 to 0.125 lb nitrogen per 1000 ft^2 (5 to 6 kg N/ha) from urea is associated with improved recovery from dollar spot damage.

TABLE 4.1

Classes of Fungicides Targeted for Dollar Spot Control

Demethylation Inhibitor/Sterol Inhibitor (DMI/SI)	Carboximide Boscalid (Emerald®) Penthiopyrad (Velista®) Dicarboximide
Fenarimol (Rubigan®)	Iprodione (Chipco 26GT®)
Metconazole (Tourney®)	Vinclozolin (Curalan®/Touche®)
Myclobutanil (Eagle®)	
Propiconazole (Banner MAXX®)	**Benzimidazole**
Tebuconazole (Torque®)	Thiophanate-methyl (3336 Plus®, others)
Triadimefon (Bayleton®)	
Triticonazole (Trinity®, Triton®)	**Other**
	*Chlorothalonil (Daconil®, others)
	*Mancozeb (Fore Rainshield®, others

* Chlorothalonil and mancozeb are contact fungicides; whereas all other listed fungicides are penetrants.

Resistance is an important issue when planning a fungicide program for dollar spot. Resistance to thiophanate-methyl, iprodione, and DMI/SI fungicides was reported in most areas of the United States. Resistance is most commonly observed with thiophanate-methyl and least commonly observed with iprodione. In the case of DMI/SI fungicides, reduced sensitivity and not true resistance is common. Reduced sensitivity is a prelude to resistance, but it takes 8 or more years of usage to develop. Basically, in the case of DMI/SI fungicides, their longevity of control becomes less over time; however, complete failure (i.e., true resistance) is not common. The nature of fungicides and resistance is discussed in detail in Chapter 6.

The key to delaying resistance or reduced sensitivity is to rotate fungicides from different chemical classes (Table 4.1). For example, there is no advantage to rotating different DMI/SI fungicides (e.g., rotating propiconazole, triadimefon, myclobutanil, triticonazole, tebuconazole, etc., will speed reduced sensitivity). Rotating fungicides with different modes of action helps to delay the possible selection of resistant biotypes of the dollar spot pathogen (see Chapter 6 for more information). Because there has never been a report of an *S. homoeocarpa*–resistant biotype to chlorothalonil, it is especially important to use chlorothalonil as a tank-mix partner with one of the penetrants or to just rotate it into the spray program often. Mancozeb is another contact fungicide that has no known resistance problems. Mancozeb can be used as a substitute for chlorothalonil in resistance management programs but is not as effective for dollar spot control. Where dollar spot is a predictable and chronic problem, fungicides should be applied preventively. Some studies indicate that applying a dollar spot–targeted fungicide just after the second mowing in spring can reduce inoculum and delay the appearance of the disease. These early spring preventive applications were reported to work well only in colder regions of the Great Lakes and northeastern United States. In the transition zone and southern U.S. regions, however, warmer spring temperatures result in a rapid degradation of fungicides, and little or no improved or delayed control is observed following an early spring preventive treatment.

Loss of good residual dollar spot control with any particular fungicide does not necessarily mean resistance is developing. The overuse of materials within the same class or family can result in enhanced microbial degradation of the chemical. That is, continuous use of fungicides from within any of the classes listed could give rise to a buildup of bacteria and other microorganisms that use fungicides in that class as an energy source. This results in a more rapid degradation of fungicides, thus reducing their residual effectiveness. Another possible reason for less residual control is poor droplet distribution during spraying. Ideally, fungicides should be sprayed through nozzles that atomize the droplets. There is a trend to water-in fungicides, but this should only be done when targeting root diseases. Chlorothalonil and penetrants targeting dollar spot will perform well when applied to dry turf in 40 to 50 gallons of water per acre (374 to 467 L/ha). Reduced residual effectiveness, proper nozzle selection, and water dilutions are discussed in detail in Chapter 6.

Fungicides like flutolanil and azoxystrobin have little or no activity on *S. homoeocarpa* and studies have shown at times that they can encourage dollar spot. Tank-mixing azoxystrobin or flutolanil with other penetrants listed in Table 4.1, however, does help to ensure that dollar spot is not encouraged. There is no established link between dollar spot outbreaks and the common use of plant growth regulators like trinexapac-ethyl, paclobutrazol, or flurprimidol. In fact, trinexapac-ethyl has been shown to enhance the residual effectiveness of some fungicides, and paclobutrazol and flurprimidol have significant fungicidal activity that reduces dollar spot severity. The herbicides bispyrabac-sodium and ethofumesate also suppress dollar spot.

Alternative products for dollar spot control include mineral oil and wetting agents. Civitas® is a mineral oil product developed by Petro-Canada (Mississauga, Ontario) for use on turf to manage diseases. Civitas can yellow bentgrass and it thus is mixed with a green colorant to mask any potential discoloration elicited by this mineral oil. DewCure® is a wetting agent marketed by Mitchell Products (Millville, New Jersey), and it also can cause yellowing in bentgrass. Both products provide high levels of dollar spot control. It may be best to use the aforementioned mineral oil and wetting agent products during early summer epidemics when there is less probability for heat stress. Because they do not provide complete control under high disease pressure, both products can be used in rotation with fungicides for best results. Both of the aforementioned products have been linked to phytotoxicity problems when tank-mixed with certain fungicides. Therefore, strict adherence to label instructions is important. Mineral oil products also can slow green speed.

There are several biological agents that target dollar spot. These include mostly bacteria, such as *Pseudomonas aureofaciens* Tx-1 and numerous *Baccillus* species. Biological agents can suppress dollar spot, but not at levels considered commercially acceptable for golf course standards. These biological agents are discussed in detail in Chapter 5.

Ultimately, effective dollar spot suppression is going to involve combining those cultural practices that are known to suppress dollar spot into a fungicide program. In particular, nitrogen should be added to the spray tank (i.e., 0.125 to 0.2 lb N/1000 ft²; 6.2 to 10 kg/ha) from a water-soluble N-source like urea each time a dollar spot–targeted fungicide is applied to help stimulate growth and recovery of the turf. It should be noted that urea generally is safe to apply to golf greens without having to water it in. However, during hot and humid periods urea can cause a burn, especially when mixed with other chemicals. Hence it is wise to apply any water soluble nitrogen source separately (i.e., not tank-mixed with fungicide) and water it in immediately in the summer, especially when spoonfeeding golf greens. It is important to mow chronically affected greens, tees, and fairways as early as possible in the morning to speed-dry the canopy. Fungicide-treated turf, however, should not be mowed for at least 12 and preferably 24 hours after spraying a fungicide. Obviously, removal of plant tissues containing fungicide dilutes the total concentration of the active ingredient. This is why using plant growth regulators to reduce mowing frequency can sometimes help to extend the residual effectiveness of fungicides. Returning clippings is helpful, if they do not interfere with play, because they help to recycle nitrogen and other nutrients.

Summary of Key Points

- Dollar spot is ubiquitous and probably is the most important economic disease of turfgrasses.
- On creeping bentgrass greens, tees, and fairways, the disease appears as small, white to straw-colored spots that initially are small (<0.75 inch; <19 mm diameter) and increase to about the size of a silver dollar (1.5 inch; 3.8 cm diameter).
- Tip dieback, oblong, or hourglass-shaped lesions bordered by a brown band are common leaf symptoms.
- In northern and transition zones there can be as many as three annual dollar spot epidemics with the most severe occurring in late summer. Dollar spot can occur at almost any time of year in the southern United States.
- Nitrogen reduces dollar spot severity, but only when applied at times when the disease is active.
- In summer, spoonfeeding water-soluble nitrogen (0.125 to 0.2 lb N/1000 ft^2; 6 to 10 kg N/ha) on a 7- to 14-day interval helps to reduce damage and recovery time.
- Mowing early in the day or dragging to remove leaf surface exudates and rolling reduce dollar spot severity. Wetting agents also can reduce dollar spot severity.
- Maintain soil moisture at a level that promotes growth (i.e., above the wilting point) in late summer when dollar spot pressure becomes severe.
- Avoid planting highly susceptible cultivars.
- Chlorothalonil is an important fungicide for use in *S. homoeocarpa* resistance management programs.
- Rotate or tank-mix different chemistries each time a fungicide is applied when targeting dollar spot.

Rhizoctonia Diseases

Many fungi are assigned two Latin binomials and *Rhizoctonia*'s are classic. The two binomial systems include a taxonomic designation based on the imperfect or asexual state of a fungus and the perfect or sexual state of the fungus. Sometimes only the imperfect or perfect stage is found in nature and vice versa. Most often, plant pathogens are first named based on the imperfect state because asexual spores or mycelia characteristics are the first to be recognized. Later the perfect state may be found or more likely it is induced in a laboratory. The two Latin binomial system is confusing, but in the case of *Rhizoctonia*'s found attacking turf, only the asexual states are commonly observed in nature.

There are several *Rhizoctonia* species that attack turfgrasses. As previously discussed, the names of the pathogens in this group are confusing because some are based on the imperfect (i.e., asexual *Rhizoctonia*) or perfect (sexual *Waitea, Ceratobasidium, Thanatephorus*) states of these fungi. There are several synonyms for these fungi that represent old and new terminology. Regardless, all of the diseases in this section are caused by closely related fungi that have mycelia characteristics of *Rhizoctonia*. The proposed names for diseases incited by *Waitea* spp. are based on Japanese descriptions and generally are not characteristic of symptoms in cool-season grasses in the United States. Also, the perfect state of *Waitea* spp. has not been observed in the United States.

Brown patch, also known as Rhizoctonia blight, is caused by *Rhizoctonia solani* and is a common, summertime disease of cool-season turfgrasses. There are several biotypes of *R. solani*, which are assigned anastomosis groupings. *Rhizoctonia zeae* (synonyms: *Waitea circinata* var. *zeae*; *Chrysorhiza zeae*) causes leaf and sheath spot, but this name is not often used because it does not accurately describe symptoms in the United States; the disease it incites is simply referred to as *R. zeae*. *Rhizoctonia cerealis* (synonym: *Ceratorhiza cerealis*) incites yellow patch. A fourth species, *Waitea circinata* var. *circinata* (synonyms: *Chrysorhiza circinata*) incites brown ring patch or Waitea patch. (*Waitea patch* is preferred by some because *brown ring patch*—the Japanese name for this disease—is not descriptive.)

Brown Patch

Environmental conditions that favor brown patch development are day temperatures above 85°F (28°C) and high relative humidity (Fidanza and Dernoeden, 1997). A night temperature above 68°F (20°C) and periods of leaf surface wetness periods exceeding 10 hours are the most critical environmental requirements for disease development in cool-season grasses. This disease becomes extremely severe in cool-season grasses during prolonged, overcast wet periods in summer as long as average daily temperatures remain above 63°F (17°C). *R. solani* and its various biotypes known as anastomosis groups, however, can be quite active at lower temperatures. This explains its ability to damage warm-season grasses in the autumn and spring.

All cool-season turfgrasses are attacked, but the most susceptible species include creeping and colonial bentgrasses, perennial ryegrass, and tall fescue. Colonial bentgrass is perhaps the most susceptible species, which is one factor that limits its use to only a few regions in the United States. The newer high-density creeping bentgrass cultivars appear to be more susceptible to brown patch than older conventional cultivars. The symptoms of brown patch vary according to host species. On closely mown creeping bentgrass and annual bluegrass turf, affected patches are roughly circular and range from 3 inches to 3 feet (7.5 to 90 cm) or greater in diameter (Figure 4.21). In creeping bentgrass grown on greens, tees, and fairways, the disease initially may appear as very pale brown patches, but patches become distinctively brown in time. In cool-season grasses, *R. solani* primarily blights foliage. If unchecked during hot and humid weather, however, *R. solani* can severely thin out creeping bentgrass. During early morning hours, when the disease is active, a cobweb-like mycelium or tufts of hyphae can develop in sparse to large amounts on leaves laden with dew. The outer edge of the patch may develop a 0.25 to 0.5 inch (6.2 to 12.5 mm) wide smoke ring (Figure 4.22). The smoke ring is gray or black in color and its appearance is caused by mycelium in the active process of infecting leaves. Close inspection of leaf blades reveals that the fungus primarily causes a blight or dieback from the tip. This gives diseased turf its brown color. On creeping bentgrass, distinct leaf lesions generally are not evident because the leaf blades are too fine textured. Bentgrass leaves generally appear blighted or shriveled and tip dieback commonly occurs. When present, leaf lesions are a light, chocolate-brown color and are bordered by narrow, dark-brown bands.

FIGURE 4.21
Brown patch rings and patches range from 6 inches to three feet in diameter (15 to 90 cm diameter).

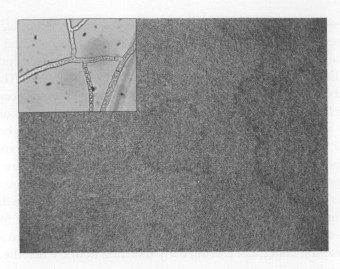

FIGURE 4.22
Brown patch smoke rings on a golf green. Inset shows classic right branching of pathogen (*Rhizoctonia solani*) hyphae.

Other Rhizoctonia Diseases

Rhizoctonia zeae is more commonly found in southern and transition zone regions of the United States, and may require higher temperatures than *R. solani* to become destructive. The disease caused by *R. zeae* was named *sheath and leaf spot*. *R. zeae* may produce yellow, bronze, or orange-brown colored rings in creeping bentgrass on golf greens when air temperatures are in a moderate to high range (Figure 4.23). This pathogen is more common in high-density creeping bentgrass cultivars (e.g., Penn A-series, 'Declaration', etc.) than traditional cultivars (e.g., L-93, Penncross, etc.). During early morning hours, a thin band (0.25 inches; 6.2 mm) of grayish-white mycelium or a grayish-black smoke ring may appear at the periphery of rings. In general, *R. zeae* blights leaves in a distinct ring pattern on golf greens.

Yellow patch is caused by *R. cereale* and is favored by extended periods of overcast, rainy, and cool to cold weather and is active primarily in spring, but autumn outbreaks can occur. Overly saturated soil conditions at snow melt are especially conducive to yellow patch. The primary hosts are creeping bentgrass and annual bluegrass grown on greens and sometimes fairways and tees. Symptoms include yellow or reddish-brown rings (Figure 4.24). Reddish ring symptoms are only associated with creeping bentgrass, whereas yellow rings and patches are associated with both annual bluegrass and creeping bentgrass. Rings may be solitary, concentric, or several rings may comingle. Large numbers of rings can appear almost overnight. *R. cerealis* is primarily a foliar blighter and symptoms can disappear abruptly with the advent of sunny, warm, and dry weather.

FIGURE 4.23
Rings, patches, and smoke rings can be associated with sheath and leaf blight caused by *Rhizoctonia zeae*.

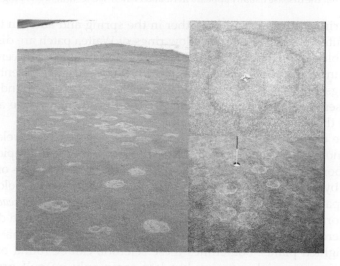

FIGURE 4.24
Yellow patch can appear as rings or patches that are yellow or reddish-brown in color.

Brown ring patch is the name given to the disease in creeping bentgrass in Japan where *W. circinata* var. *circinata* is the pathogen (Toda et al., 2005). In the United States, brown ring patch is often referred to as Waitea patch. Symptoms of Waitea patch in annual bluegrass and creeping bentgrass putting greens are similar to the yellow rings associated with *R. zeae*, but the disease appears in the spring rather than summer (Figure 4.25). When disease dissipates, the rings may appear brown and in some cases there is thinning of turf. Little is known about Waitea patch, but the disease was reported to

FIGURE 4.25
Waitea patch in annual bluegrass appears as clusters of yellow rings. (Courtesy of J. Kaminski.)
Inset shows that the disease usually appears in ribbons where it encounters creeping bentgrass.

appear during periods of mild weather in the spring and autumn throughout the United States. While the yellow rings of Waitea patch are distinctive
in annual bluegrass, they normally break up into ribbons or crescents when
they encounter creeping bentgrass in mixed stands. Turf in the center of the
yellow rings sometimes appears much darker green than surrounding turf.
In rare cases the degradation of thatch by the fungus may cause a sunken
ring effect that negatively impacts ball roll.

All three of the aforementioned species appear to be closely related, but
their taxonomy, host range, associated symptomatologies, and epidemiolo-
gies are poorly understood. Sclerotia (aka bulbils) are firm masses of hyphae
produced by some fungi and serve as resting bodies. Sclerotial color can be
used to separate some species (Smiley et al., 2005). Sclerotia of *R. zeae* are distinctively orange, round, and 0.03 to 0.06 inches (0.8 to 1.6 mm) in diameter.
R. zeae sclerotia sometimes are produced in large numbers in thatch at the
base of blighted plants. Sometimes these sclerotia can be found on leaves,
especially in perennial ryegrass, but less commonly on golf green turf.
R. solani sclerotia, however, are inconspicuous and difficult to find and usually are dark-brown and appressed to dead tissue. Sclerotia of other species
generally are not found in nature but are produced in pure culture on agar
media in the laboratory. *Waitea circanata* var. *circinata* sclerotia are orange to
dark-brown, and *R. cerealis* sclerotia are light to dark-brown in color.

Management

Proper cultural management strategies help to minimize brown patch severity. Nitrogen (N) fertilizer source and timing of application can impact

brown patch severity (Fidanza and Dernoeden, 1997). In particular, autumn applications of a slow release N source to cool-season grasses results in less brown patch the following summer, when compared to spring applications of water-soluble N. Use a complete fertilizer (i.e., N + P + K) according to soil test recommendations in the autumn. Applications of high rates of N in spring or summer can intensify brown patch. Spoonfeeding with low N rates (0.1 to 0.2 lb N/1000 ft²; 5 to 10 kg N/ha) intermittently throughout summer may slightly enhance brown patch. However, because spoonfeeding promotes turf recovery from summer stresses and injury, the practice should be continued even if brown patch is active.

Irrigation timing also impacts brown patch severity. Irrigating at dusk intensifies brown patch, whereas irrigation during early morning hours reduces disease severity. Evening irrigation intensifies brown patch by providing for longer leaf wetness duration at night. Conversely, early morning irrigation does not extend the leaf wetness period and physically knocks *R. solani* foliar mycelium off leaves (Fidanza and Dernoeden, 1997; Settle et al., 2001). Use of wetting agents as well as dragging or poling speeds leaf drying and may help to reduce disease activity. Frequent irrigation that results in saturated soil conditions favors brown patch, particularly in shaded sites with poor air circulation. Hence, soils should be kept as dry as possible in summer, especially when brown patch is active.

Brown patch develops more rapidly and intensely in dense, high-cut turf, when compared to lower mowing in more open stands. Generally, mowing high within the recommended range for the species helps turf to better tolerate summer stresses, diseases, and insect pests and helps to reduce weed invasion. For creeping bentgrass maintained on greens, tees, and fairways, adjusting mowing height for brown patch is not likely to have a significant beneficial effect. Improving drainage and air circulation, reducing thatch-mat, and alleviating soil compaction have a large impact on reducing brown patch severity in intensively managed creeping bentgrass. Use of the herbicide bispyribac-sodium (Velocity) can enhance brown patch (Kaminski and Putman, 2009).

Brown patch is more effectively controlled when fungicides are applied prior to the onset of blighting (i.e., applied preventively). Preventive applications of a strobilurin (i.e., azoxystrobin, fluoxastrobin, pyraclostrobin, trifloxystrobin), chlorothalonil, iprodione, flutolanil, fludioxonil, mancozeb, Penthiopyrad, polyoxin D, and thiophanate-methyl (*R. solani* only) effectively control brown patch. Demethylation inhibitors (i.e., DMI/SI) such as myclobutanil and triticonazole are effective when applied preventively. Other DMI/SI fungicides (e.g., metconazole, propiconazole, tebuconazole, and triadimefon) perform best when tank-mixed with a contact fungicide. For curative control, it is best to tank-mix a contact fungicide (e.g., chlorothalonil, fludioxonil, or mancozeb) with one of the aforementioned penetrants. Strobilurins, flutolanil, and polyoxin D provide the longest residual effectiveness against brown patch. Sometimes active brown patch symptoms and even some foliar mycelium may be observed the morning following a fungicide treatment. Superintendents should wait another day to assess control and avoid a second panic and possibly wasteful spray.

Diseases incited by *R. zeae* and *W. circinata* var. are disfiguring on creeping bentgrass greens. Under prolonged periods of disease activity, there may be thinning of turf. Spoonfeeding along with a fungicide may be recommended to promote rapid recovery. During grow-ins when large amounts of nitrogen fertilizer applied in spring stimulate succulent growth, however, *R. zeae* can cause rapid and serious damage to golf greens in summer. Thiophanate-methyl provides little or no protection against *R. cerealis*, *R. zeae*, or *W. circinata* var. *circinata*.

Summary of Key Points

- *Rhizoctonia cerealis* incites yellow patch. Rings and patches are yellow or reddish-brown, and the disease occurs in association with water-saturated soils during cool to cold, overcast, and rainy weather in late winter and early spring and less commonly in autumn.

- *Waitea circinata* var. *circinata* causes brown ring patch or Waitea patch. It is most often observed producing yellow-colored rings or scallops in annual bluegrass on golf greens during mild weather in spring, but less commonly may occur in creeping bentgrass in the United States.

- *Rhizoctonia zeae* causes leaf and sheath blight. Symptoms include yellow or orange-colored rings in creeping bentgrass during periods of high humidity and high temperature stress in summer. It is more common in high-density creeping bentgrass cultivars grown on golf greens.

- Brown patch (*R. solani*) is the most common *Rhizoctonia* disease. Brown patch is most severe during hot and humid periods, especially when night temperatures remain above 68°F (20°C) and there is very high humidity or rain.

- Brown patch symptoms include circular, brown-colored patches a few inches to 3 feet (7.5 to 90 cm) or greater in diameter; grayish-black smoke rings at the periphery of patches; foliar mycelium; and brown, shriveled leaves.

- Brown patch often appears overnight, and the pathogen primarily blights leaves of cool-season grasses. When disease pressure is severe, significant loss of turf density can occur.

- Brown patch is culturally managed by irrigating early in the morning rather than at night and by improving air circulation and surface water drainage.

- Chemically, brown patch is best controlled by preventive fungicide treatments in areas where the disease is chronically severe.

Stobilurins (i.e., QoI fungicides), polyoxin D, and flutolanil provide the longest residual effectiveness.

- Yellow patch and Waitea patch are mostly disfiguring but can reduce turf density and are managed by spoonfeeding in combination with an appropriate fungicide.

- Sheath and leaf blight (*R. zeae*) can severely damage high-density cultivars on golf greens overstimulated by nitrogen in summer.

Pythium Blight

Pythium aphanidermatum was isolated as early as 1926 from diseased creeping bentgrass in Virginia by John Monteith and is the most common incitant of Pythium blight. *P. ultimum* and a few other *Pythium* spp. were reported as causal agents of this disease. Pythium blight is most likely to attack mature creeping bentgrass and annual bluegrass grown under intensive management (i.e., frequent night irrigation, low mowing, and high nitrogen fertility) conditions commonly found on golf courses. *Pythium* spp. cause damping off of all seedling species. Pythium blight develops rapidly during nighttime and is among the most feared and destructive turfgrass diseases. During periods of high relative humidity, night temperatures above 70°F (21°C), and abundant surface moisture, the disease progresses with remarkable speed. The pathogen rapidly blights leaves and infects stems, thus killing plants. Huge areas of turf can be destroyed within 24 hours, particularly if there is rainfall at night. Pythium blight usually is first observed in tree-lined, shaded, poorly drained, and low-lying areas with poor air circulation.

Symptoms

On closely mown creeping bentgrass and annual bluegrass, Pythium blight appears in circular spots, patches, rings, or streaks that follow the surface water drainage pattern. Initial symptoms in creeping bentgrass appear as reddish-brown or orange-bronze colored spots 1 to 2 inches (2.5 to 5 cm) in diameter (Figures 4.26 and 4.27). There may be a gray smoke ring or grayish-white mycelium on the periphery of affected spots and patches. As the disease intensifies, patches increase in size and "frog-eye" (i.e., rings with living grass in the center) symptoms may develop. Pythium blight can occur in low and high areas, but generally the disease is most severe in low spots, swales, surface water drainage patterns, and shaded or pocketed sites (Figure 4.28). Blighting can appear as streaks on slopes or in swales as a result of draining water carrying infectious mycelia and/or spores (i.e., propagules). Sometimes, streaks develop as a result of mowers dispersing propagules and the pattern may

FIGURE 4.26
Pythium blight often appears as small reddish or bronze-colored spots in creeping bentgrass. Sometimes smoke rings and foliar mycelium are present.

FIGURE 4.27
Pythium blight inactivated by a fungicide.

appear as small, orange or reddish-brown spots in a straight line (Figure 4.29). In low-lying areas where water collects, large nonuniform, reddish-brown areas collapse. In the latter situation, most if not all plants are killed and leaves are matted. Matted leaves usually have a greasy feel.

When Pythium blight is active, a cottony web of grayish mycelium may be seen on or in the canopy during early morning hours when leaves are wet. *Pythium* spp. are capable of producing an abundance of mycelium in just a few hours. Mycelia bridge leaf blades and are responsible for the cottony

FIGURE 4.28
In low areas where water collects, large areas of turf are severely blighted and killed by *Pythium* spp.

FIGURE 4.29
Pythium streaking in tire tracks that disperse propagules of the pathogen.

appearance seen on affected turf. The fungus primarily spreads through the turf canopy by rapid mycelia growth or by movement of mycelia fragments and motile spores (zoospores) in rain or irrigation water.

Management

Water management greatly influences Pythium blight severity. It is helpful to irrigate early in the day to avoid increased moistening of the canopy and thatch-mat layer at night. Avoid using a daily irrigation schedule, especially during nighttime hours. Improving surface water drainage and air circulation by clearing brush and trees, and by installing drains and fans will help reduce disease development, but these cultural measures often are expensive and difficult to achieve. Avoiding the use of lime in alkaline soils and avoiding the application of nitrogen fertilizers at rates exceeding 0.5 lb N/1000 ft^2 (25 kg N/ha) during summer stress periods may help to reduce disease incidence and severity. Spoonfeeding nitrogen (i.e., 0.1 to 0.2 lb N/1000 ft^2; 5 to 10 kg N/ha) intermittently in the summer generally does not predispose turf to Pythium blight. An autumn fertilization program using a complete (i.e., N-P-K) fertilizer based on soil testing improves turf vigor and density. Cultural approaches to Pythium blight management, however, likely will have only minimal beneficial effects on disease suppression during high disease pressure periods (i.e., thunderstorms, and high night temperatures and humidity).

Fungicides that target Pythium blight are considered a necessity on golf courses in many regions of the United States. Before the advent of improved fungicides in the early 1980s, the disease was combated with short residual chemicals such as chloroneb and ethazole. Chloroneb and ethazole continue to be widely used and are the preferred fungicides for a knockdown of active Pythium blight. In general, a curative effect from any fungicide used to target Pythium blight may not be observed for a 1- or 2-day period. For this reason, and because Pythium blight often appears rapidly and overnight, preventive treatment is a common practice on golf courses during hot and humid periods of summer in many regions. The widespread reliance and continuous usage of metalaxyl or mefenoxam (an isomer of metalaxyl) on golf courses, however, has led to reduced residual effectiveness and in some cases the selection of *Pythium* spp. biotypes resistant to this fungicide. Reduced residual effectiveness is attributed to a buildup of microorganisms that degrade the active ingredient of a chemical. Cyazofamid, fosetyl-aluminum, and propamocarb are other fungicides that provide good, residual Pythium blight control. Phosphites are as effective as fosetyl-aluminum, when used preventively, and are less expensive than other materials (Cook et al., 2009; Ervin et al., 2009). Azoxystrobin is labeled for control of Pythium blight, but its effects are short lived, and this chemical was shown to intensify the disease in tall fescue (Settle et al., 2001).

Some observed control failures with fosetyl-aluminum and other phosphite products targeting Pythium blight, which has been linked to one species (i.e., *Pythium catenulatum*) that is tolerant of this fungicide (P. Harmon, personal communication). Symptoms of Pythium blight caused by *P. catenulatum* in creeping bentgrass can be unlike those described above. Creeping bentgrass stands infected by *P. catenulatum* may develop a lime-green canopy color before collapsing.

To avoid buildup of fungicide-resistant biotypes, and reduction of residual effectiveness of compounds, Pythium-targeted fungicides should be rotated or applied in tank-mix combinations whenever economically feasible. Resistance problems can be delayed or avoided by tank-mixing reduced rates of mefenoxam + propamocarb + fosetyl-aluminum (see Chapter 6 for more information). Consult the label for directions when using lower than normal use rates of any fungicide for tank-mix combinations. Tank-mixing cyazofamid, fosetyl-aluminum, mefenoxam, phosphite, or propamocarb with mancozeb also reduces the potential for resistant *Pythium* biotypes to develop. Fosetyl-aluminum may cause a tip burn if applied in hot weather and is incompatible with copper hydroxide and flowable formulations of mancozeb and chlorothalonil. Pythium blight–targeted fungicides can yellow putting green turf when applied during the summer, but chloroneb and ethazole are most likely to elicit a phytotoxic response when applied during warm and humid weather.

Summary of Key Points

- Several species of *Pythium* cause Pythium blight, but *P. aphanidermatum* and *P. ultimum* are most common.
- Pythium blight is most severe during hot and humid weather marked by night temperatures above 70°F (21°C) and thunderstorm activity.
- Pythium blight invariably develops overnight, and initially symptoms appear as reddish-brown or orange-bronze colored spots 1 to 2 inches (2.5 to 5 cm) in diameter. Spots, patches, rings, and streaking as well as smoke rings and foliar mycelium are associated with this disease.
- Injury is most severe in shade, water drainage patterns, swales, and low spots where affected turf collapses and leaves mat in large irregular patterns.
- Improving surface water drainage and air circulation, reducing shade, irrigating early in the day, and avoiding a daily irrigation schedule are cultural approaches to minimizing Pythium blight.
- Fungicides are best applied preventively in areas where Pythium blight commonly occurs. Mefenoxam-resistant biotypes of *Pythium* are fairly common on older golf courses in some regions of the United States.

Summer Patch

Summer patch is incited by the fungus *Magnaporthe poae* and is mostly a disease of annual bluegrass and Kentucky bluegrass. This pathogen attacks roots and causes a patch symptom. Summer patch was documented in Penn A-1 and A-4 golf greens in North Carolina (Tredway, 2005) and was found attacking creeping bentgrass greens in Southern California and Florida. Patches are 1 to 3 feet (30 to 90 cm) in diameter, initially appear wilted, and foliage then turns yellow or orange. Outbreaks of the disease in creeping bentgrass are associated with very high temperature stress and high soil pH in the range of 7 to 8 from calcareous sands, high pH irrigation water, or poor water quality.

Symptoms

Summer patch in annual bluegrass maintained under golf green conditions appears as reddish-brown, bronze, or yellow colored patches a few inches (7.6 to 15.2 cm) to about 1 foot (30 cm) in diameter (Figures 4.30 and 4.31). It

FIGURE 4.30
Summer patch in annual bluegrass. (Photo courtesy of K. Happ.)

FIGURE 4.31
Summer patch pathogen specifically attacking annual bluegrass, but creeping bentgrass is unaffected.

is common in compacted soils and in the peripheral cleanup pass on greens. In mixed annual bluegrass-creeping bentgrass greens the disease often has a nonuniform (i.e., no distinctive patches) appearance. Circular patches of creeping bentgrass may be seen as a patchwork quilt in the autumn as bentgrass fills the voids created by dead annual bluegrass (Figure 4.32). In mixed stands symptoms can mimic anthracnose and bacterial wilt (Figure 4.33). In subsequent summers, the disease often appears as yellowing annual bluegrass ribbons at the interface of bentgrass patches. Similar yellow rings,

FIGURE 4.32
Creeping bentgrass eventually fills voids, giving greens a patchwork quilt appearance.

FIGURE 4.33
Summer patch in annual bluegrass can mimic anthracnose. Infected stem bases appear water soaked and brown.

bands, or arcs are observed with Waitea patch (i.e., brown ring patch), but this is a spring and not a summer disease. Darkly pigmented runner hyphae are usually apparent on roots, and infected stem bases are water soaked and light to dark brown in color (Figure 4.31).

Management

Low mowing and frequent irrigation are the major cultural practices that exacerbate summer patch. The height of cut on golf greens should be increased to the maximum acceptable level, preferably 0.15 inches (3.8 mm). Acidification of the soil rootzone with ammonium sulfate or sulfur-coated urea also reduces disease severity over time. Conversely, nitrate forms (i.e., calcium, potassium, or sodium nitrate) of nitrogen and limestone applications should be avoided as they can intensify summer patch. The use of limestone to raise soil pH in concert with the use of ammonium sulfate, however, may not intensify summer patch on golf greens (Hill et al., 2001). Most of the annual usage of nitrogen fertilizer should be confined to autumn months, but spoonfeeding in summer with ammonium sulfate can be helpful. Core aeration alleviates damage in compacted soils, but aeration should be performed in the spring or autumn when the disease is not active. Irrigating during sunny periods will elevate soil temperature because water in thatch-mat efficiently absorbs and conducts heat (see Chapter 1). Hence, syringing golf greens to avoid wetting of soil and the thatch-mat layer is preferred to irrigation on hot and sunny days.

Summer patch, once the bane of annual bluegrass golf greens, has become less common due to the advent and regular use of QoI or strobilurin fungicides (e.g., azoxystrobin, others). Preventive applications of azoxystrobin, propiconazole, pyraclostrobin, or triadimefon are the most efficacious way of managing summer patch in annual bluegrass dominated greens. Curative applications of azoxystrobin, propiconazole, and thiophanate-methyl or drenches may provide a satisfactory level of control if applied as soon as the earliest symptoms are discerned. To control the disease in chronically affected golf greens, fungicides should be applied preventively on 21- to 28-day intervals from mid-spring (mid-May in transition and northern U.S. regions; or about 30 days after crabgrass [*Digitaria* spp.] seed germinates in your area) through summer. Most fungicides generally provide improved control when watered-in immediately after application to a depth of 1 inch (2.5 cm) or they may be applied in 100 gallons of water per acre (935 L/ha). Fungicides are rendered ineffective if turf is allowed to enter drought-induced dormancy.

Summary of Key Points

- Summer patch is primarily a disease of annual bluegrass, but creeping bentgrass greens grown in high-temperature climates can develop the disease.

- Summer patch is promoted by low mowing and wet soils as a result of rain or frequent irrigation in summer.

- In mixed annual bluegrass-creeping bentgrass greens the disease develops in yellow colored ribbons of annual bluegrass at the interface with bentgrass patches. Annual bluegrass develops a yellow, reddish-brown, or bronze color before dying.

- The disease is culturally managed by increasing mowing height, reducing soil wetness and paying careful attention to syringing, and using acidifying fertilizers.

- In chronically affected greens, fungicides are used preventively in high water volumes beginning in mid-spring.

- Since the advent of QoI fungicides, summer patch has been effectively controlled on golf greens in most regions.

Dead Spot

The causal agent of dead spot is *Ophiosphaerella agrostis*, but the disease originally was named *bentgrass dead spot* in 1999. Later, the disease was confirmed in hybrid bermudagrass (*Cynodon dactylon* × *C. transvaalensis*) golf greens in the United States. Because there are at least two hosts for the pathogen, the name of the disease was changed to *dead spot*. The etiology and management of dead spot was described by Kaminski and coworkers (2002a, 2002b, 2005a, 2005b, 2005c, 2006) and is summarized below.

Symptoms

Dead spot develops as early as May and June during warm and dry periods in creeping bentgrass grown in sand-based rootzones in transition zone regions of the United States, and can remain active throughout summer. Average daily temperatures above 77°F (25°C) are optimum for dead spot development. These daily averages generally are accompanied by sustained daytime temperatures in excess of 86°F (>30°C). The disease is less common on tees and collars, and thus far has not been found in turf grown on native soil. Disease symptoms appear initially as small, reddish-brown or copper colored spots in turf that are 0.5 to 1 inch (1.2 to 2.5 cm) in diameter (Figure 4.34). Spots enlarge to only about 3 inches (7.5 cm) in

FIGURE 4.34
Creeping bentgrass greens pitted with copper-colored dead spots.

diameter and have tan tissues in the center and reddish-brown leaves on the outer periphery of larger, active patches. The symptoms at times are similar to those associated with copper spot (*Gloeocercospora sorghi*), dollar spot (*Sclerotinia homoeocarpa*), Microdochium patch (*Microdochium nivale*), black cutworm (*Agrotis ipsilon*) damage, and ball mark injury. The disease often is confused with ball mark injury or copper spot. Should dead spot recur in subsequent years, it is possible that symptoms would include small frog-eyes (i.e., a doughnut-like pattern with living turf in the center of a dead ring). No foliar mycelium is observed on turf in the field; however, foliar mycelium will develop in a laboratory humidity chamber. Foliar mycelium is pale-pinkish-white and may take from 3 to 5 days to develop on diseased plants maintained under high humidity.

Dead spot can remain active until hard frosts occur. The spots or patches caused by *O. agrostis* sometimes coalesce, but this only occurs during severe epidemics. Sometimes depressed spots or "crater pits" develop. Darkly pigmented runner hyphae, typical of other *Ophiosphaerella* species that attack turf, are not commonly observed on roots. Runner hyphae, however, can be found on nodes of stolons and stem bases and sometimes on roots forming at nodes. The pathogen, however, can be isolated from leaves, stems, and roots of diseased plants. The pathogen appears to attack leaves first and then moves into stem bases and eventually stolons. The pathogen overwinters in or on infected tissues (especially stolons), and the disease can reactivate from infected plants adjacent to old dead spots the following year. Numerous black, flask-shaped fruiting bodies called *pseudothecia* may be found embedded in necrotic leaf and sheath tissues. The fruiting bodies contain large numbers of needle-shaped spores (i.e., ascospores). When mature, ascospores exude or are projected through a pore in the top of the neck of pseudothecia. These spores can be found in large numbers on diseased as well as nearby healthy leaves. Fruiting bodies often are produced in abundance in late summer and can be found embedded in dead tissues throughout winter months. Hence, the pathogen also overwinters as pseudothecia. Pseudothecia may be produced during early stages of disease development when the copper-colored spots first appear. No pseudothecia, however, may be found where fungicides are used frequently.

Dead spot is not specific to any single cultivar, but L-93 is among the most susceptible creeping bentgrass cultivars. The pathogen also attacks colonial (*Agrostis capillaris*) and velvet bentgrass (*Agrostis canina*). The disease has only been found in turf grown on high sand-based rootzones. To date, it has not been found in fairways or other sites where turf is grown on native soil. Normally, dead spot is most severe 1 to 4 years after seeding, but greens as old as 6 years have developed dead spot. Dead spot, however, normally declines in severity within 2 years of seeding. Warm to hot and dry weather are the environmental conditions that trigger rapid dead spot development in bentgrass. Most injury has been associated with greens in open or exposed locations, and slopes that are more prone to heat stress and wilt. The disease may develop on pocketed greens and on tees and collars.

Management

Dead spot can be controlled with boscalid, chlorothalonil, fludioxonil, iprodione, mancozeb, pyraclostrobin, propiconazole, and thiophanate-methyl. These fungicides may suppress dead spot during high pressure periods for about 7 to 10 days, after which time active symptoms can recur. Fungicides, however, may provide little curative benefit after many dead spots have developed in late summer. Hence, it is very important to apply a fungicide as soon as symptoms appear. Water-soluble nitrogen fertilizers (0.125 lb N/1000 ft^2; 6.2 kg N/ha) should be applied along with fungicides to stimulate growth of surrounding, healthy creeping bentgrass plants. Ammonium sulfate suppresses dead spot in bentgrass and bermudagrass, whereas calcium nitrate and potassium nitrate can enhance the disease. Ammonium sulfate has a very high burn potential and should be watered-in immediately. Stolon growth into the dead spots or patches is restrained or inhibited. When the disease is checked some recovery occurs as a result of tillering of adjacent healthy plants, but most dead spots present in late summer or autumn may not fully recover prior to winter. During winter, inactive spots or patches appear whitish-tan. Diseased spots often are void of living tissue, and bare soil may be evident in the center of dead areas during winter. In spring, tillers or stolons from adjacent plants cover the old dead spots, but often there is no rooting from these tillers or stolons.

Summary of Key Points

- Dead spot is a disease of young creeping bentgrass golf greens (i.e., <6 years old) grown on sand-based rootzones or where methyl bromide was used to renovate older greens.
- Dead spot appears as reddish-brown spots 0.5 to 3.0 inches (1.2–7.5 cm) in diameter.
- Symptoms can be confused with copper spot, dollar spot, and Microdochium patch, as well as cutworm damage and ball mark injury.
- In bentgrass, the disease develops during warm to hot and dry periods from late spring to autumn.
- Most bentgrass cultivars are susceptible, but thus far the disease has been restricted to turfs grown on sand-based rootzones.
- Boscalid, chlorothalonil, fludioxonil, iprodione, mancozeb, propiconazole, pyraclostrobin, and thiophanate and other fungicides arrest the disease, but only for 7 to 10 days under conditions of high disease pressure.
- Ammonium sulfate limits dead spot more than other water-soluble nitrogen sources because it rapidly acidifies soil.
- Spoonfeed weekly in summer to stimulate bentgrass regrowth into dead spots.
- Apply sufficient amounts of water-soluble nitrogen in the autumn to stimulate tillering and regrowth of plants into affected dead spots before winter.

Fairy Ring

Turf diseases known as fairy ring are caused by a plethora (>60) of fungal species; the exact number is unknown. Much of what we know about the nature of fairy rings was published by Shantz and Piemeisel in 1917. These fungi can cause the formation of rings or arcs of dead or unthrifty turf, or rings of dark-green, luxuriantly growing grass. Fairy ring fungi belong to a taxonomic group known as the basidiomycetes or "mushroom fungi." These fungi primarily colonize thatch or organic matter in soil and generally do not directly attack turfgrass plants, however, some are weakly parasitic. Fairy rings can be evident nearly year round, but they are most pronounced in appearance and most destructive in summer.

Fairy rings are classified into three types according to their effects on turf:

Type 1: Those that kill grass or badly damage it in the shape of rings or arcs.

Type 2: Those that stimulate grass by forming rings or arcs of dark-green turf.

Type 3: Those that do not stimulate grass and cause no damage, but produce mushrooms or puffballs in rings or arcs.

Obviously, the most destructive rings are of the Type 1 variety (Figure 4.35). Type 1 rings are very common, especially in lawns, roughs, and golf course fairways that previously were pasture land or where tree stumps or lumber had been buried. They also commonly appear on new, sand-based golf greens and tees, often within 1 or 2 years after seeding. The most common fungi associated with Type 1 fairy rings in general turf areas include *Marasmius oreades*, and various species of *Agaricus, Agrocybe, Chlorophyllum, Coprinus, Lepiota*, and *Lycoperdon* (known as puffballs). On U.S. golf greens, it appears that most of the fairy ring fungi involved are members of the puffball family (i.e., the *Lycoperdaceae*), which include *Arachnion album, Bovista dermoxantha*, and *Vascellum curtisii* (Miller et al., 2011).

Symptoms

Type 1 fairy rings appear initially as circles or arcs of dark-green, fast-growing grass. Hence, they are technically Type 2 rings at first, which may or may not develop into Type 1 rings. On higher-cut roughs or lawns, Type 1 rings often develop into three distinct zones: an inner lush zone where the grass is noticeably stimulated and grows luxuriantly (i.e., faster growing and

FIGURE 4.35
Type 1 killer rings are most troublesome in the summer.

darker green), a middle zone where the grass may be thinning or dead, and an outer zone in which the grass is slightly stimulated. The distance from the inside of the inner zone to the outside of the outer zone may range from a few inches (3 to 6 cm) to 4 feet (>120 cm) wide. The green stimulated zones are the result of the breakdown of organic matter, which releases nitrogen and causes more rapid foliar growth. The outer green zone is caused by the breakdown of thatch by the fairy ring fungus, which liberates nitrogen. The inner green zone develops in response to the release of nitrogen as bacteria degrade aging or dead mycelium of the fairy ring fungus produced in previous years. The formation of the three zones can be noticeable from spring to winter in higher-cut turf such as that growing in roughs, but not normally in close-cut creeping bentgrass or annual bluegrass. Mushrooms or puffballs of the fungi causing a Type 1 ring are produced in the bare zone or at the junction of the bare and outer green zone. Sometimes puffballs develop inside the center of rings on golf greens, where turf may eventually wilt and die. These fruiting bodies normally appear in abundance from late spring to late summer following heavy rainfall. Rings, however, may not produce mushrooms or puffballs, especially on closely cut turf. The underside of the mushroom cap is composed of gills, upon which spores are produced. Spores of *Lycoperdaceae* puffballs are produced inside a white-fleshy, round to egg-shaped fruiting body. Puffballs have a short stubby stipe (stem) and the outer layer may be covered with warts or fleshy spines. As puffballs mature in higher-cut turf they develop a brown or lilac color, rupture, and release their spores. On close-cut greens, tees, and fairways, puffballs either do not develop or they are damaged by mowing. The importance of spores in the spread of fairy ring fungi is not well understood.

On lower-cut bentgrass greens, tees, and fairways, fairy rings generally appear initially as Type 2 green rings. Scalping can occur due to the lush-green, stimulated growth of grass in Type 2 rings, which can interfere with playability. During extended hot and dry weather, however, they can develop into Type 1 rings at which time the turf turns blue-gray, wilts, and collapses in distinct rings or arcs (Figure 4.36). At the interface between the green stimulated zone and wilt, bentgrass may develop a yellow or orange color. On golf greens, solid circular and somewhat depressed or sunken patches (aka thatch collapse) of dark-green turf also may develop (Figure 4.37). These patches may have a yellow periphery. Thatch collapses because these basidiomycetes are wood-decaying fungi by nature and prefer the higher lignin content in the thatch layer. If not properly managed, all or most of the plants in these affected patches will wilt, turn blue-gray, and may die in a few hours or days during hot and dry weather. Fairy rings may disappear during cool and wet periods in summer but can recur with the advent of high temperature and drought stress. Fairy rings tend to decline in number and intensity over time, but they may mysteriously reappear in some years on even very old golf greens. Mushrooms and puffballs most often appear in summer and early autumn following overcast and rainy weather.

FIGURE 4.36
Type 2 fairy ring may develop into a Type 1 ring. The thatch-mat layer will have a light-brown color and a mushroom odor.

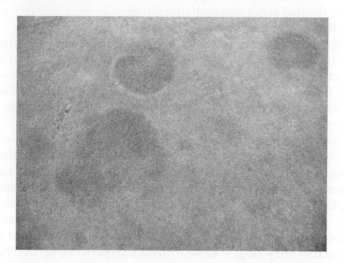

FIGURE 4.37
Thatch collapse in a creeping bentgrass golf green. (Photo courtesy of J. Kaminski.)

A ring is broken when its mycelium encounters an obstacle such as a rock, pathway, or unfavorable soil condition. The ring also may disappear for no apparent reason. In general, two fairy rings will not cross one another (i.e., at the point of intersection the growth of each ring stops). This obliteration at the point of contact is believed to be caused by the production of self-inhibitory metabolites that also will antagonize other members of the same or different species of fairy ring fungi. On slopes, the bottom of the ring is usu-ally open, giving the appearance of an arc rather than a ring. This may be due to the downward movement of self-inhibitory metabolites in soil or thatch, which prevents fungal development in turf on the lower side of the ring.

Rings vary in size from about a foot (30 cm) to 200 feet (60 m) or more in diameter, and become larger each year. The annual radial growth ranges from 3 inches (7.5 cm) to as much as 20 inches (50 cm). The rate of outward movement, as well as overall diameter of rings, is determined by imperfectly understood soil and weather conditions. In new sand-based greens, fairy rings can appear in large numbers and grow to over 12 inches (30 cm) in diameter within 1 year of seeding. They appear soon after new construction, particularly in sand-based greens and tees, because air and water are usually plentiful, and there is little microbial activity to suppress or antagonize fairy ring fungi. Growth of a fairy ring begins with the transport of fragments of fungal mycelium and probably spores. The fungus initiates growth at a central point and continues outward in all directions (i.e., radially) at an equal rate. The actual development of a fairy ring may first be observed as a cluster of mushrooms. Rings fade in the autumn or winter, but large dead zones remain visible. Loss of visibility, other than unrecovered dead zones, is due to the general brownish appearance of dormant turf during winter and because the turf is not metabolizing nitrogen in large enough quantities to produce the lush green zones.

Fairy rings have been observed in areas where soil pH has ranged from 5.1 to 7.9. It is likely that fairy rings will occur under any soil condition that will support turfgrass growth. All cultivated turfgrasses are known to be affected by fairy ring fungi, and they are found internationally. Sandy soil links courses in the United Kingdom are particularly susceptible to fairy ring problems.

Type 1 fairy ring fungi kill turf primarily by rendering infested soil hydrophobic (i.e., water repellent). The dead zone is due to mycelium of the fungus that accumulates in such large amounts that it prevents entry of rain or irrigation water or in some other way causes soils to become hydrophobic, and thus kills plants via drought stress. Probing the dead zone will reveal the soil to be much drier than adjacent soil, which is similar to that described below for localized dry spots. Sometimes, when a plug of soil is removed from the edge of an active fairy ring, a white, thread-like network of mycelium may be seen clinging to soil and to roots of grass plants (Figure 4.35). When environmental conditions are optimum for fungal growth, the white mycelium may be seen growing in the thatch layer. Fairy ring–infested soils have a mushroom odor, even if visible fungal mycelium is not evident. The mycelium usually is not observed during hot and dry periods, but the mushroom odor of soil (although faint) should be evident. Thatch color is normally buff, fawn, or light brown rather than dark brown in color where fairy ring fungi are active (Figure 4.36). Although fairy ring fungi are known to parasitize roots and produce compounds toxic to roots, it is likely that most damage to turf can be attributed to the fungal mycelium rendering soil impermeable to water. It also is noted that very high concentrations of ammonium (i.e., NH_4^+)

were measured in rootzones of Type 1 rings, which is reported to be toxic to roots (Fidanza et al., 2007).

Management

Control of fairy rings is made difficult by the impermeable nature of infested soil. Chemical control sometimes is difficult to achieve because these fungi can grow deeply into soil and lethal concentrations of fungicide do not come into contact with the entire fungal body. There are three approaches to combating fairy rings: suppression, antagonism, and eradication.

Suppression is the most practical approach to combating fairy rings. The suppression approach is based upon the premise that fairy rings are less conspicuous and less numerous where turf is well watered, properly aerated, and properly fertilized with nitrogen. Golf course superintendents report having observed effective control of some fairy rings on golf greens using fungicides (noted below) applied conventionally without drenching. The approach is preventive and fungicides generally are applied twice on a 30-day interval in spring. Some lightly water the fungicides off leaves using one or two turns on the irrigation heads.

In situations where fairy rings are chronically severe, managers need to experiment with different methods of introducing fungicide into soil, where most of these fungi reside. A common fairy ring control approach involves a combination of core aeration or spiking, deep watering, and the application of soil wetting agents (aka soil surfactants) and fungicide drenches. Core aeration and use of soil wetting agents are most beneficial because they aid in the rapid entry of water into dry soil as well as assisting in infiltration and percolation of water and fungicide. The entire area occupied by the ring, to include a 2 foot (60 cm) periphery beyond the ring, should be plugged to remove soil cores on 2- to 4-inch (5 to 10 cm) centers. If core aeration is not practical, spike or use a pitchfork to create channels through the thatch-mat and soil prior to treatment. The area should then be preirrigated to a depth of 4 to 6 inches (10 to 15 cm). A fungicide should be tank-mixed with a soil wetting agent and immediately watered-in to a depth of at least 2 to 4 inches (5 to 10 cm). On sand-based rootzones where soil wetting agents are used routinely, superintendents are cautioned not to overwater. Effective suppression of fairy rings was observed using the fungicides azoxystrobin, flutolanil, polyoxin D, and several DMI/SI fungicides including metconazole, propiconazole, tebuconazole, triticonazole, triadimefon, and possibly others (Fidanza, 1999; Settle et al., 2011). Tank-mixes of two fungicides are commonly used. The ring area should be retreated in a similar fashion on a 4-week interval or at the earliest indication of drought stress; that is, repeat the process whenever the grass turns blue-gray and begins to wilt. Frequent hand watering or water injection may be required between applications of soil wetting agents and a fungicide to alleviate wilt.

For small numbers of fairy rings, a deep root feeder or a water fork with garden hose attachment is useful to force water into dry soil. Injection of water may have to be performed many times during a hot and dry summer. High rates of a soil wetting agent applied before rings appear in late spring also are beneficial. Apply recommended amounts of nitrogen at the appropriate time of year, and spoonfeed with water-soluble nitrogen (e.g., urea) in summer or apply chelated iron to help mask Type 2 fairy rings. Fairy rings, however, can be stimulated by excessive nitrogen or organic matter.

The antagonism approach is based upon the observation that rings exhibit mutual antagonism (i.e., elimination when they come into contact with one another). This method involves removal of sod by stripping or killing turf with a nonselective herbicide. Soil must be rototilled repeatedly in several directions until the mycelium-infested soil has been thoroughly mixed. Soil then is prepared in the usual manner for seeding or sodding. This method has been shown to be promising, has had limited testing. Fairy rings, however, are known to recur in sod fields within 2 to 3 years after harvest and establishment of a new crop. This may be due to poor or inadequate mixing of soil prior to seeding. There are two methods of eradication: fumigation and excavation. Both methods are laborious, costly, impractical, and not always successful.

Superficial Fairy Ring

In addition to fairy rings, several mostly unidentified basidiomycetous (mushroom) fungi cause two other diseases of turf known as superficial fairy ring (SFR) and yellow ring. Yellow ring (*Trechispora alnicola*) is known only to occur in Kentucky bluegrass (*Poa pratensis*). Causal agents of SFR include *Trechispora cohaerens* and *Coprinus kubickae*.

Symptoms

Superficial fairy rings are sometimes referred to as "white patch" because they appear primarily in golf greens as white, circular patches that range from 3 inches (7.5 cm) to 3 feet (90 cm) in diameter (Figure 4.38). At the edge of these well-defined, circular patches are 1- to 2-inch (2.5 to 5 cm) fringes of dense white mycelium. This disease appears during warm and rainy periods in summer and is most commonly observed under conditions of low nitrogen fertility. In the United Kingdom and some regions of the United States, SFR can remain evident in the autumn and winter. Most SFR, however, go unnoticed unless golf greens are treated with a nonselective herbicide for regrassing purposes. The absence of a dense green turf canopy reveals the presence of SFR on the surface of thatch (Figure 4.39).

FIGURE 4.38
Superficial fairy ring in a bentgrass golf green.

FIGURE 4.39
Superficial fairy rings were not apparent until this green was renovated with the nonselective herbicide glyphosate.

Although SFRs are unsightly, these fungi do not appear to infect plants. Upon close inspection, one can observe that older leaves in the whitish fringe die prematurely and have a bleached-white appearance. Large amounts of SFR mycelium grow over older leaves and sheaths and block incoming sunlight. This results in the breakdown of chlorophyll in these tissues, and they senesce and die. Once leaves and sheaths die, the fungus uses the necrotic tissues as a source of nutrition. Turf density in affected patches declines, but most plants survive. Thatch degradation by the fungus may cause shrinkage, which disfigures or interferes with the trueness of a putting surface. These fungi develop principally in thatch-mat and do not penetrate more than 0.5 to 1 inch (2.5 to 5 cm) into underlying soil.

Management

Superficial fairy rings are more common in poorly nourished golf greens. Spoonfeeding with 0.125 lb N/1000 ft^2 (6 kg N/ha) on a weekly basis along with mechanical disruption of thatch-mat layer and fungal mycelium by vertical cutting, spiking, or core aeration help to mask or minimize the adverse effects of these fungi. Azoxystrobin, flutolanil, polyoxin D, and DMI/SI fungicides may suppress and in some instances control SFRs.

Summary of Key Points

- Type 1 fairy rings kill turf indirectly by rendering soil hydrophobic. Water is unable to penetrate the mycelium-infested soil and plants die from drought stress.
- Fairy rings are most noticeable during summer and sometimes mushrooms or puffballs will appear during warm and rainy periods.
- Type 1 fairy rings are most destructive during periods of heat and drought stress.
- Fairy ring–infested soil has a mushroom odor, but white mycelium may not be visible.
- Approaches to managing fairy rings include the following:
  - Golf greens with a history of fairy ring problems should be treated preventively in the spring with an appropriate fungicide.
  - Some fairy rings on golf greens can be controlled without having to drench-in fungicides.
  - Some fungicides reported to suppress or control fairy ring include azoxystrobin, flutolanil, polyoxin D, as well as triadimefon, propiconazole, and other DMI/SI fungicides.
  - For chronically severe fairy ring problems, combining core aeration or spiking with the application of a soil wetting agent plus a fungicide is the most effective approach to their management.
  - Syringing and water injection with water forks are useful for managing small numbers of fairy rings on greens or tees.
  - Partial masking Type 2 fairy rings may be achieved by applying chelated iron or by applying light rates of nitrogen fertilizer (i.e., spoonfeeding) in the summer.
- Superficial fairy ring mycelium develops in the thatch-mat layer and will grow onto lower leaves and sheaths causing tissue senescence

and sometimes depressions in the putting surface. They do not cause a hydrophobic soil condition.

- Superficial fairy rings are managed by applying nitrogen. They may be suppressed by azoxystrobin and other QoI fungicides, polyoxin D, flutolanil, and possibly DMI/SI fungicides.

Localized Dry Spot or Dry Patch

Localized dry spots are common on sand-based golf greens and native soil (i.e., push-up) greens that were aggressively topdressed for several years with a mix containing more than 80% sand by volume. Similarly, localized dry spots can develop within a few years after initiating a sand topdressing program on fairways (Figure 4.40). Some topdressing sands are hydrophobic and mixes should be tested for hydrophobicity.

Symptoms

Localized dry spots normally develop in new golf greens within 6 months to 3 years of seeding. They tend to decline in number and severity over time, but some dry spots can be a persistent problem for indefinite periods. These dry spots develop with the advent of heat and drought stress from late spring to autumn. Dry spots often are associated with areas adjacent to bunkers, where sand is routinely pitched and builds up. They often disappear during

FIGURE 4.40
Localized dry spots can develop within a few years of initiating a sand topdressing program on bentgrass fairways.

extended overcast and rainy periods. Localized dry spots appear as solid patches of wilted or dried-out turf. Their appearance is sometimes preceded by fairy ring development or simply by the presence of numerous mushrooms (Figure 4.41). Patches can be circular and range from a few inches (6 to 8 cm) to several feet (45 to 60 cm) in diameter, or they may appear as large serpentine or irregularly shaped areas of wilted or dead turf (Figures 4.42 and 4.43). Soil within patches remains bone dry, despite frequent irrigation. Water will penetrate thatch but not the thatch–soil interface, and it will run off dry spot areas. The most severe levels of hydrophobicity or water repellency associated with localized dry spots are found near the surface and decline in severity with increasing depth (Karnok and Tucker, 2008). The depth of water repellency rarely exceeds 3 inches (7.6 cm) (Karnok and Tucker, 2008). Plants within affected areas develop a blue or purplish color that is indicative of wilt, and if not addressed, turf eventually will die as a result of drought.

The cause of localized dry spots has been attributed to the decomposition activities of unidentified basidiomycetous (mushroom) fungi in the same group that cause fairy rings and other microorganisms in soil. These microorganisms cannot be isolated from hydrophobic soil and thus remain unknown. Water repellency is caused by the breakdown of older fungal mycelium and other sources of organic matter, which releases hydrophobic organic substances (e.g., fulvic acid) that coat individual sand particles. This organic coating may also result from the breakdown of plant tissues such as roots, shoots, and stems or organic soil amendments such as manures, poor-quality peats, or composts. Individually coated sand particles pack loosely together, rendering soil impervious to water infiltration. The water-repellent or hydrophobic condition is typically restricted to the upper few inches (3 to 6 cm) of soil. Removal of thatch-mat alone will not significantly improve water infiltration.

FIGURE 4.41
Localized dry spots and fairy rings on a bentgrass golf green.

FIGURE 4.42
Wilted circular areas due to localized dry spots (LDSs). Inset shows that soil in the LDS is bone dry compared to the moist adjacent soil.

FIGURE 4.43
Serpentine-shaped localized dry spots.

Another condition not unlike localized dry spots can occur in creeping bentgrass fairways that are not topdressed with sand. They differ from localized dry spots in that they do not render underlying native soil hydrophobic. Instead these basidiomycetes render the thatch layer hydrophobic. Their presence usually is preceded by the development of numerous mushrooms following rainy periods in late spring (Figure 4.44). With the advent of warm and dry conditions, affected areas develop a wilted and unthrifty appearance.

FIGURE 4.44
Mushroom fungi also cause a dry patch condition in the thatch layer of bentgrass fairways. Contrasting colors in this photo denote areas where there is dew or lack of dew as a result of a hydrophobic condition.

Management

Core aeration, water injection aeration, or hand pitchforking in combination with frequent applications of a soil wetting agent will help to alleviate this condition (Figure 4.45). Water injection aeration is least destructive and quite effective; however, it does not cure the condition. Keeping turf alive in localized dry spots may require numerous daily syringes or hand watering and treatments with soil wetting agents and water injection aeration during dry summer periods. A deep irrigation or rain that moves water through the hydrophobic zone is most helpful. Soil wetting agents reduce water repellency in sandy soils. Nonionic (contain neither a positive or negative charge) wetting agents persist longer in soil than anionic (negatively charged) or cationic (positively charged) agents and are less potentially phytotoxic. Wetting agents are organic compounds degraded by soil microorganisms and thus may require several applications in a season. Where localized dry spot is chronic, the application of a soil wetting agent should begin in spring and several weeks in advance of the time they normally appear. Without routine use of soil wetting agents, it becomes extremely difficult to wet water-repellent sand-based rootzones in the summer. Routine use of soil wetting agents allows the manager to maintain relatively dry soil conditions with confidence that soils can be rewetted quickly when needed (Karnok and Tucker, 2008). Pretreatment with soil wetting agents in conjunction with some form of aeration is the preferred preventive control strategy for this problem. Use of wetting agent tablets

FIGURE 4.45
Water injection methods wet soil and alleviate localized dry spot.

when handwatering and injecting wetting agents through the irrigation system are also helpful methods. Isolated dry spots can be individually treated by frequent probing with a water-fork or a tree deep-root feeder that injects water. Fungicides have no known impact on the incidence, severity, or control of localized dry spots.

Summary of Key Points

- Localized dry spot is a hydrophobic soil condition attributed to an organic coating of sand particles brought about by the decomposition of organic matter by microbes.
- Localized dry spots appear during warm and dry periods and are more numerous and more commonly associated with sand-based greens and tees.
- Sand topdressing fairways promotes localized dry spot.
- Affected patches may be circular, serpentine, or nonuniform in pattern.
- Localized dry spots are managed by various forms of core aeration or water injection aeration in combination with soil wetting agents and judicious syringing and hand watering.

Plant-Parasitic Nematodes

Plant-parasitic nematodes are known to cause extensive injury to turfgrasses in warm, temperate, and subtropical areas of the United States. This has been attributed in part to the long and warm growing season in the south, and thus a much longer period for parasitic nematodes to feed on roots. Parasitic nematodes can also be troublesome in the transition zone and northern regions in both creeping bentgrass and annual bluegrass grown on golf greens. The overall importance of parasitic nematodes in higher-cut (e.g., fairways) turf is unknown. There are seasonal fluctuations in parasitic nematode populations in golf greens (Jordan and Mitkowski, 2006). Populations may peak in summer for some species that prefer warmer soils, while others may peak in the autumn in cooler soils. In general, parasitic nematodes most actively feed on turfgrass roots during environmental periods favorable for growth of the root system. This most active feeding would occur in spring and autumn, but parasitic nematode populations are normally very low in the spring. Parasitic nematode populations generally begin to increase in late spring and early summer in response to warming soil temperatures. These populations peak by mid-summer and remain high throughout summer months. In the autumn, nematode populations may drop precipitously as a result of diminished food supply (i.e., loss of root systems), but in most situations the decline in populations is gradual. Thus, populations are highest during summer, and this factor combined with heat and drought stress is why they become most problematic at that time of year (Todd and Tisserat, 1993). Creeping bentgrass is more tolerant of the feeding activity of parasitic nematodes compared to annual bluegrass.

The Anguina seed and gall nematode (*Anguina pacificae*) is unlike other nematodes of turfgrasses because it attacks above-ground plant parts. This nematode produces galls at the base of annual bluegrass stems and is considered the most important parasitic nematode on Northern California golf courses (Westerdahl et al., 2005). There are other foliar damaging nematodes, including *Anguina agrositis*, which attacks floral parts of bentgrasses in seed production fields in the Pacific Northwest (Couch, 1995). Only the most common, root-feeding nematodes will be considered below.

Most parasitic nematodes are microscopic round worms that reproduce by eggs. The emerging juveniles typically molt four times before reaching adult size. Each female is capable of producing large numbers of eggs, and the entire life cycle for most species is completed in 5 to 6 weeks under suitable conditions. Because nematodes that attack plants are obligate parasites, they must feed on living tissues in order to grow and reproduce. Most parasitic nematodes are capable of attacking a wide range of plant species and can survive on weeds in the absence of turfgrasses. Most nematodes store large quantities of food, which enables them to survive long periods in soil in the absence of suitable plants. Many survive in frozen soils and may overwinter in living roots or in dead plant tissues.

Literally millions of nematodes can inhabit a few square feet (1 m²) of soil, but most are nonpathogenic. Most soil-dwelling nematodes actually perform a beneficial service in soil by helping to degrade organic matter and cycle nutrients. Although there are numerous species, only about 15 genera are

TABLE 4.2

Some Plant Parasitic Nematodes Commonly Associated with Creeping Bentgrass and Annual Bluegrass Golf Greens in the United States

Common Name[a]	Genus	Feeding Behavior On/In Roots
Lance	*Hoplolaimus*	**Endo-** and Ectoparasitic
Lesion	*Pratylenchus*	**Endo**parasitic
Ring	*Criconemella*	Ectoparasitic
Root knot	*Meloidogyne*	**Endo**parasitic
Spiral	*Helicotylenchus*	Ectoparasitic
Sting[b]	*Belonolaimus*	Ectoparasitic
Stubby root	*Trichodorus*	Ectoparasitic
Stunt	*Tylenchorhynchus*	Ectoparasitic

[a] Action thresholds vary greatly among laboratories, and you should consult a pathologist or nematologist in your region for the best possible recommendation.

[b] Encountered primarily in the southeastern United States.

known to be common turfgrass parasites. All plant-parasitic nematodes bear a hollow, spear-like structure called a stylet. The stylet is similar to a hypodermic syringe and is used to inject enzymes into plant cells. Simultaneously, partially digested food is withdrawn.

Plant-parasitic nematodes are commonly grouped according to their feeding habit on roots (Table 4.2). Endoparasitic nematodes (i.e., root knot, lesion, and lance) partially or totally burrow into plant tissues and feed primarily from within, whereas ectoparasitic nematodes (i.e., dagger, needle, pin, ring, sheath, spiral, sting, stubby root, and stunt) feed from the plant surface, although a small portion of the body may be embedded in the root. Ectoparasitic nematode species are more numerous and therefore are more commonly found than endoparasitic nematodes. Ring, stubby root, and stunt nematodes are common ectoparasitic species found in association with golf greens in many regions. Sting nematode is most commonly found in greens in the southeastern United States and is regarded as being the most destructive of all species. Endoparasitic root knot, lesion, and lance nematodes are the most common and injurious parasitic nematodes associated with both creeping bentgrass and annual bluegrass golf greens in transition and northern regions of the United States. Lance nematodes behave as both endoparasitic and ectoparasitic, with juveniles mostly feeding in roots, while adults primarily are found feeding from outside bentgrass roots (Settle et al., 2007). Nematode activity is favored by warm and moist soil conditions. Nematode populations generally peak in mid-summer, but populations normally remain high in irrigated golf greens into early autumn. Their activity is enhanced in sandy and light-textured soils and reduced in compacted or heavy soils where aeration becomes restricted. Hence, sandy soils of golf greens provide an ideal environment for parasitic nematodes. Nematodes are unable to move more than a few millimeters in soil, but they can be transported over longer distances by moving water, soil, sod, and equipment. Coring is probably the main way parasitic nematodes are dispersed any significant distances on golf courses.

Symptoms

Symptoms of parasitic nematode injury include yellowing, stunting, wilting, or early signs of drought stress, and thinning of the stand. On golf greens symptoms may mimic localized dry spot. These symptoms are related to the injury that parasitic nematodes inflict upon root systems. Therefore, symptoms of injury usually do not become noticeable until water becomes limiting or during periods of high temperature stress. Due to the similarities between environmental stress symptoms and nematode injury, the source of the problem is difficult to diagnose. Like many "weak" or secondary fungal pathogens, plant-parasitic nematodes may not cause much of a problem until environmental extremes, poor growing conditions, mechanical injury, or a fungal disease reduces turf vigor. There often is no pattern to nematode injury in creeping bentgrass or annual bluegrass. Parasitic nematodes usually clump or cluster in high population densities in streaks or small, oval-shaped areas called "hot spots." Severe infestations may result in a dramatic reduction in the vigor, color, and density of putting greens. Inspection of roots may or may not reveal some indication of nematode feeding. Roots may exhibit one or more of the following symptoms: knot-like or cyst swellings, red or brown lesions, excessive root branching, necrotic root tips, and root rot.

The search for undiagnosed problems of golf greens between spring and autumn by superintendents and diagnosticians almost always ends in assaying soil for parasitic nematodes. One problem with assay interpretation is that parasitic nematodes are invariably found in high populations during the golfing season. Proper soil sampling is required to get an accurate picture of parasitic nematode population densities. Sending one or two cup cutter plugs of soil to a laboratory provides incomplete or unreliable information. As noted previously, parasitic nematodes normally cluster in high populations in small areas, which likely is related to their breeding behavior. Therefore, soil should be collected from numerous areas at the edge or interface between healthy and wilted, unthrifty, or injured turf. A 1-inch diameter (2.5 cm) soil probe is the best sampling tool, and 15 to 20 probes (at least 0.5 pint or 250 cc of soil needed) should be taken. Some laboratories prefer a 500 gram (1.1 lb) soil sample. Soil samples should be collected from the rootzone region where most nematodes will be found; normally the upper 3 to 6 inches (7.5 to 15 cm) of soil. Samples should be kept cool and moist and shipped overnight express to a diagnostic laboratory. Healthy areas on the same green and at least one other healthy green should be sampled for comparative purposes. This is done because it is not unusual to find similar or even higher parasitic nematodes counts from a healthy appearing green, when compared to a weak green. High parasitic nematode levels from both healthy and stressed greens is an indicator that other factors, such as mechanical injury, low mowing, low nitrogen fertility, shade, or other poor growing conditions are contributing to turf decline. Sampling from severely thinned areas will yield unreliable results, because these obligate parasites will not survive in large populations in the absence of living plants. Various methods for nematode extraction are used in the laboratory.

Unfortunately, there are no reliable data correlating parasitic nematode number per sample and expected degree of turf injury in the field. Thresholds (i.e., the number of parasitic nematodes that are viewed as damaging) have been established, but their interpretation varies greatly among laboratories. Thresholds are of limited value because growing conditions greatly influence the potential importance of parasitic nematode population densities. For example, low mowing, wear or mechanical damage, other diseases, and environmental stress weaken turf, thus predisposing greens to more potential damage by fewer parasitic nematodes. Furthermore, the threshold for individual parasitic nematode species may be less important than the total number of parasitic nematodes found. Single population thresholds are especially questionable given other potential stress interactions and thus are an unreliable indicator (Settle et al., 2007). Although there is no composite threshold consensus among pathologists or nematologists, a combined count of over 1000 lance, lesion, and root knot; 50 to 100 sting; and over 2000 spiral, stubby root, or stunt per 250 cc of soil (note some labs assay 100 and 500 cc samples) from creeping bentgrass greens is cause for concern during periods of high temperature stress. For annual bluegrass, however, the combined threshold would be considerably lower in summer, especially if another disease (e.g., anthracnose or summer patch) is involved.

Parasitic nematode assays have shown that very high population densities in excess of every published threshold can be found in greens in summer that are not showing symptoms (Jordon and Mitkowski, 2006). Thus, listing of thresholds has been eliminated from this edition. Thresholds mislead and confuse golf course superintendents. Turf managers should get a proper soil assay and consult with a plant pathologist or nematologist on the proper action to implement in the event very high counts are obtained. Where parasitic nematodes are an annual concern, greens should be sampled monthly between spring and autumn to provide valuable information on the natural variations in the rise and fall of population densities. The nematologist or pathologist will ultimately make management recommendations based on the species and total number of parasitic nematodes present in a sample.

Management

There are times when golf course superintendents become obsessed with the possibility that problems with their greens are due to plant-parasitic nematodes. Given the lower environmental summer stress tolerance of annual bluegrass, it is generally accepted that parasitic nematodes are more damaging to this species versus creeping bentgrass. Furthermore, annual bluegrass growing on golf greens usually has a smaller root system in summer than creeping bentgrass, which also would increase vulnerability (Lyons et al., 2011). In many cases the activity of parasitic nematodes is that of a disease-abiotic stress complex. When combining heat and other stresses, a poor growing environment, and possible mechanical injury, parasitic nematodes may become the proverbial "straw that breaks the camel's back." In most situations involving creeping bentgrass, an absolute determination of a parasitic nematode problem from visual symptoms and even soil analysis is difficult if not impossible. Even when an assay shows population densities exceeding a threshold, it does not mean that the application of a nematicide is necessary or warranted. Creeping bentgrass golf greens containing parasitic nematode populations in excess of 10 times most published thresholds can appear perfectly healthy during summer stress periods. The best and perhaps only indication of a parasitic nematode problem is a positive response from a nematicide. Unfortunately, there currently are few nematicides available to make this determination, and none are available nationally.

Traditional nematicides were highly toxic, organophosphate (OP) derivatives and are no longer available. The soil fumigant 1,3-dichloropropene (Corfew®) has been labeled for use on greens in some Southern states. This fumigant is applied directly into soil using a knife injection system. It kills foliage on contact and the knife injection causes mechanical damage to greens. This material is extremely toxic, is injected only once annually, and can only be handled by custom applicators. Special local need labels have been approved for the insecticide abamectin (i.e., Avid®). Currently, these labels only are available in some southern states for ring and sting nematode control. Abamectin should be sprayed in a high water volume with a soil surfactant and further incorporated by overhead irrigation. This method of application is non-destructive and the material can be applied several times in a season to achieve suppression of sting nematode populations.

While there currently are few nematicides available in most of the United States, there may be someday. It therefore would be prudent not to forget what was learned in the days of fenamiphos and other OP nematicides. They had to be handled with extreme caution and with the aid of a soil assay recommendation from a nematologist. The OP nematicides could be phytotoxic and only provided short-term or poor control of endoparasitic nematodes. Annual usage or extensive reliance on OP nematicides usually resulted in an intractable problem in which parasitic nematode populations rebounded following chemical treatment. That is, the frequent use of an OP nematicide often resulted in higher parasitic nematode population densities in the long term. This occurred because OP nematicides killed beneficial nematodes and other invertebrates in soil which kept parasitic nematode populations in check by competition or antagonism. Furthermore, there can be a buildup of microbes that rapidly degrade and reduce the residual effectiveness of a pesticide. There also is the potential for some parasitic nematodes to be resistant to a nematicide or escape the effects of a nematicide by virtue of being embedded in roots. This could allow one or more species to proliferate in the absence of competition from other nematodes. Should a nematicide be available, greens should be sampled about 4 weeks after its application to determine the level of control and the species not effectively controlled.

In some situations, such as extremely high and persistent parasitic nematode population densities in chronically injured greens dominated by annual bluegrass, fumigation or treatment with a nonselective herbicide and regrassing with creeping bentgrass may ultimately become the best solution. Fumigation is preferred because it also reduces, but does not eliminate, the annual bluegrass seed-bank in soil. Deeply buried annual bluegrass seed will escape fumigation, and viable seed eventually will be brought to the surface via core aeration. Unfortunately, parasitic nematodes will eventually re-invade fumigated sites.

Because plant-parasitic nematodes only may be injurious during summer stress periods, imposing cultural practices that alleviate stress are the wisest approach to minimizing injury. Such practices would include increasing the mowing height, spoonfeeding (0.1 to 0.2 lb N/1000 ft^2; 5 to 10 kg N/ha), using a complete (N + P + K) fertilizer if soil phosphorous (P) and potassium (K) levels are low, and avoiding mechanical injury. Applications of biostimulants and micronutrients and other materials that promote rooting, as well as alleviating soil compaction, are recommended. A healthy root system will support higher parasitic nematode populations, but ultimately a well-developed root system is the best defense against these parasites. Irrigation practices (i.e., daily versus deep and infrequent) commonly used on golf courses are unlikely to impact parasitic nematode populations (Settle et al., 2006). Drying soil may be beneficial because it could force nematodes to move deeper into the soil profile in search of moisture and thereby limit their ability to survive and reproduce. Obviously, this would require vigilant syringing during summer stress periods because of the diminished capacity of root systems to find and take-up water. Settle et al. (2006), however, found in a greenhouse study that lance nematode populations were unaffected by drying the soil.

There are some research data suggesting that use of organic forms of nitrogen fertilizer (e.g., sewage sludges) can discourage development of high populations of some parasitic nematodes, when compared to the use of inorganic forms. Conversely, one long-term study showed that some composts and dehydrated manures can promote parasitic nematodes (Davis and Dernoeden, 2002). It is unlikely that any of the commonly used turf fertilizers will impact parasitic nematode levels significantly. Tests with beneficial species of nematodes or fungi as biological agents have provided mixed results. The bacterium *Bacillus firmus* is being developed to manage parasitic nematode problems in turfgrasses. The mechanism of protection is that *B. firmus* applied to turfgrasses will colonize roots and its presence discourages feeding activity by parasitic nematodes. Currently, *B. firmus* appears promising when targeting sting nematode and some other nematode species in warm-season grasses. Granular chitin-protein materials do not have any activity against plant parasitic nematodes found in turfgrasses. Mustard seed extract has shown promise for reducing populations of sting and possibly other parasitic nematodes in naturally infested soils (Cox et al., 2007). Most other natural products have demonstrated little if any effect on controlling parasitic nematodes found in turf.

Summary of Key Points

- Plant-parasitic nematodes are extremely common and often found in large populations in turfgrass stands where they can injure plants by feeding on roots.
- Plant-parasitic nematodes are most problematic in golf greens during summer stress periods.
- Symptoms of parasitic nematode injury on greens include yellowing, stunting, a proneness to wilt, and thinning of the stand.
- Parasitic nematodes often cluster in oval-shaped, "hot spot" areas, which may cause wilt and give the appearance of localized dry spots on golf greens
- Proper soil sampling is required to determine if plant-parasitic nematode population densities warrant concern.
- There are no reliable data correlating parasitic nematode number per sample and expected level of injury. Growing conditions and other stress factors greatly influence the relative importance of plant-parasitic nematode population densities. It is best to have your nematode assay interpreted by a turfgrass nematologist or pathologist in your state or region.
- Currently, there are few nematicides available for use on turfgrasses. The soil fumigant 1,3-dichloropropene and the insecticide abamectin are labeled for use on golf greens in some southern states for the control of nematodes. Both require special handling.
- Plant-parasitic nematodes are best managed by improving the growing environment.
- Cultural practices that help to minimize injury caused by parasitic nematodes include:
  - Increase mowing height and reduce mowing frequency.
  - Syringing frequently in summer to prevent wilt.
  - Spoonfeeding with nitrogen and applications of biostimulants may help plants to better withstand injury from parasitic nematodes.
  - Core in the spring and autumn and fertilize properly to promote a deeper, more extensive root system.

Bibliography

Abad, Z.G., H.D. Shew, and L.T. Lucas. 1994. Characterization and pathogenicity of *Pythium* species isolated from turfgrass with symptoms of root and crown rot in North Carolina. *Phytopathology* 84:913–921.

Abernathy, S.D., R.H. White, P.F. Colbaugh, M.C. Engelke, G.R. Taylor, and T.C. Hale. 2001. Dollar spot resistance among blends of creeping bentgrass cultivars. *Crop Sci.* 41:806–809.

Cook, P.J., P.J. Landschoot, and M.J. Schlossberg. 2009. Inhibition of *Pythium* spp. and suppression of Pythium blight of turfgrasses with phosphonate fungicides. *Plant Dis.* 93:809–814.

Couch, H.B. 1995. *Diseases of Turfgrasses.* Krieger, Malabar, FL.

Cox, C.L., B. McCarty, and S.B. Martin. 2007. Suppressing sting nematodes using botanical extracts. *Golf Course Management* 75(9):94–97.

Davis, J.G., and P.H. Dernoeden. 2002. Dollar spot severity, tissue nitrogen and soil microbial activity in bentgrass as influenced by nitrogen source. *Crop Sci.* 42:480–488.

Dernoeden, P.H. 1987. Management of take all patch of creeping bentgrass with nitrogen, sulfur, and phenyl mercury acetate. *Plant Dis.* 71:226–229.

Dernoeden, P.H., and J. Fu. 2008. Fungicides can mitigate injury and improve creeping bentgrass quality. *Golf Course Management* 76(4):102–106.

Dernoeden, P.H., and N. Jackson. 1980a. Infection and mycelia; colonization of gramineous hosts by *Sclerophthora macrospora. Phytopathology* 70:1009–1013.

Dernoeden, P.H., and N. Jackson. 1980b. Managing yellow tuft disease. *J. Sports Turf Research Institute* 56:9–17.

Dernoeden, P.H., N.R. O'Neill, M.P.S. Câmara, and Y. Feng. 1999. A new disease of *Agrostis palustris* incited by an undescribed species of *Ophiosphaerella. Plant Dis.* 83:397.

Elliot, M.L. 1998. Use of fungicides to control blue-green algae on bermuda grass putting-green surfaces. *Crop Protection* 17:631–637.

Endo, R.M. 1961. Turfgrass diseases in California. *Plant Dis. Reptr.* 45:869–873.

Ervin, E.H., D.S. McCall, and B.J. Horvath. 2009. Efficacy of phosphate fungicides and fertilizers for control of Pythium blight on a perennial ryegrass fairway in Virginia. *Applied Turfgrass Science* Online. doi: 10:1094/ATS-2009-1019-01-BR.

Feng, Y., and P.H. Dernoeden. 1999. *Pythium* species associated with root dysfunction of creeping bentgrass in Maryland. *Plant Dis.* 83:516–520.

Fidanza, M.A. 1999. Conquering fairy ring disease with new tools. *Golf Course Management* 67(3):68–71.

Fidanza, M.A., J.L. Cisar, S.J. Kosta, J.S. Gregos, M.J. Schlossberg, and M. Franklin. 2007. Preliminary investigations of soil chemical and physical properties associated with type-I fairy ring symptoms in turfgrass. *J. Hydrological Processes* 21:2285–2290.

Fidanza, M.A., and P.H. Dernoeden. 1996. Influence of mowing height, nitrogen source, and iprodione on brown patch severity in perennial ryegrass. *Crop Sci.* 36:1620–1630.

Fidanza, M.A., and P.H. Dernoeden. 1997. A review of brown patch forecasting, pathogen detection, and management strategies for turfgrasses. *Intern. Turfgrass Soc. Res. J.* 8:863–874.

Fidanza, M.A., P.H. Dernoeden, and A.P. Grybauskas. 1996. Development and field validation of a brown patch warning model for perennial ryegrass turf. *Phytopathology* 86:385–390.

Heckman, J.R., B.B. Clarke, and J.A. Murphy. 2003. Optimizing manganese fertilization for the suppression of take-all patch disease on creeping bentgrass. *Crop Sci.* 43:1395–1398.

Hill, W.J., J.R. Heckman, B.B. Clarke, and J.A. Murphy. 2001. Influence of liming and nitrogen on the severity of summer patch of Kentucky bluegrass. *Intern. Turfgrass Soc. Res. J.* 9:388–393.

Hodges, C.F., and L.W. Coleman. 1985. *Pythium*-induced root dysfunction of secondary roots of *Agrostis palustris. Plant Dis.* 69:336–340.

Inguagiato, J.C., J.A. Murphy, and B.B. Clarke. 2009. Anthracnose disease and annual bluegrass putting green performance affected by mowing practices and lightweight rolling. *Crop Sci.* 49:1454–1462.

Jordon, K.S., and N.A. Mitkowski. 2006. Population dynamics of plant-parasitic nematodes in golf course greens turf in Southern New England. *Plant Dis.* 90:501–505.

Kaminski, J.E., and P.H. Dernoeden. 2002a. Geographic distribution, cultivar susceptibility and field observations on bentgrass dead spot. *Plant Dis.* 86:1253–1259.

Kaminski, J.E., and P.H. Dernoeden. 2005a. Dead spot of creeping bentgrass and hybrid bermudagrass. *App. Turfgrass Sci.* Online. doi: 10.1094/ATS-2005-0419-01-DG.

Kaminski, J.E., and P.H. Dernoeden. 2005b. Nitrogen source impact on dead spot (*Ophiosphaerella agrostis*) recovery in creeping bentgrass. *Intern. Turfgrass Soc. Res. J.* 10:214–223.

Kaminski, J.E., and P.H. Dernoeden. 2006. Dead spot severity, pseudothecia development, and overwintering of *Ophiosphaerella agrostis* in creeping bentgrass. *Phytopathology* 96:248–254.

Kaminski, J.E., P.H. Dernoeden, and N.R. O'Neil. 2002b. Reactivation of bentgrass dead spot and growth, pseudothecia production, and ascospores germination of *Ophiosphaerella agrostis. Plant Dis.* 86:1290–1296.

Kaminski, J.K., P.H. Dernoeden, and N.R. O'Neil. 2005c. Environmental influences of the release of *Ophiosphaerella agrostis* ascospores under controlled and field conditions. *Phytopathology* 95:1356–1362.

Kaminski, J.E., and A.I. Putman. 2009. Safety of bispyribac-sodium on colonial bentgrass and influence on brown patch severity. *Intern. Turfgrass Soc. Res. J.* 11:219–226.

Karnok, K., and K. Tucker. 2008. Using wetting agents to improve irrigation efficiency. *Golf Course Management* 76(6):109–111.

Kennelly, M.M., and R.E. Wolf. 2009. Effect of nozzle type and water volume on dollar spot control in greens-height creeping bentgrass. *Applied Turfgrass Science* Online. doi: 10.1094/ATS-2009-0921-01-RS.

Kerns, J.P., M.D. Soika, and L.P. Tredway. 2009. Preventive control of Pythium root dysfunction in creeping bentgrass putting greens and sensitivity of *Pythium volutum* to fungicides. *Plant Dis.* 93:1275–1280.

Kerns, J.P., and L.P. Tredway. 2008a. Pathogenicity of *Pythium* species associated with Pythium root dysfunction of creeping bentgrass and their impact on root growth and survival. *Plant Dis.* 92:862–869.

Kerns, J.P., and L.P. Tredway. 2008b. Influence of temperature on pathogenicity of *Pythium volutum* towards creeping bentgrass. *Plant Dis.* 92:1669–1673.

Kerns, J.P., and L.P. Tredway. 2010. Pythium root dysfunction of creeping bentgrass. *Plant Health Progress* Online. doi: 10.1094/PHP-2010-0125-01-DG.

Landschoot, P.J., and A.S. McNitt. 1997. Effects of nitrogen fertilizers on suppression of dollar spot disease of *Agrostis stolonifera* L. *Intern. Turfgrass Soc. Res. J.* 8:905–911.

Lee, H., D. Jong, and S. Shim. 2000. Chemical control of algae on creeping bentgrass golf greens. *Korean J. Turfgrass Sci.* 14:263–272.

Liu, L., T. Hsiang, K. Carey, and J.L. Eggens. 1995. Microbial populations and suppression of dollar spot disease in creeping bentgrass with inorganic and organic amendments. *Plant Dis.* 79:144–147.

Lyons, E.M., P.J. Landschoot, and D.R. Huff. 2011. Root distribution and tiller densities of creeping bentgrass cultivars and greens-type annual bluegrass cultivars in a putting green. *HortScience* 46:1411–1417.

McDonald, S.J., P.H. Dernoeden, and C.A. Bigelow. 2006. Dollar spot and gray leaf spot severity as influenced by irrigation, paclobutrazol, chlorothalonil and a wetting agent. *Crop Sci.* 46:2675–2684.

McDonald, S., P. Dernoeden, J. Kaminski, and M. Agnew. 2009. Curative control of yellow tuft in creeping bentgrass. *Golf Course Management* 77(5):106–110.

Miller, G.L., L.F. Grand, and L.P. Tredway. 2011. Identification and distribution of fungi associated with fairy rings on golf putting greens. *Plant Dis.* 95:1131–1138.

Monteith, J., Jr., and A.S. Dahl. 1932. Turf diseases and their control. *Bulletin of the U.S. Golf Association Green Section* 12(4):85–188.

Nelson, E.B., and C.M. Craft. 1991. Identification and comparative pathogenicty of *Pythium* spp. from roots and crowns of plants exhibiting symptoms of root rot. *Phytopathology* 81:1529–1536.

Nikolai, T.A. 2002. More light on lightweight rolling. *U.S. Golf Association Green Section Record* 40:9–12.

Pigati, R.L., P.H. Dernoeden, A.P. Grybauskas, and B. Momen. 2010b. Simulated rain and mowing impact fungicide performance when targeting dollar spot in creeping bentgrass. *Plant Dis.* 94:596–603.

Putman, A.I., and J.E. Kaminski. 2011. Mowing frequency and plant growth regulator effects on dollar spot severity and duration of dollar spot control by fungicides. *Plant Dis.* 95:1433–1442.

Roberts, J.A., J.C. Inguagiato, B.B. Clarke, and J.A. Murphy. 2011. Irrigation quantity effects on anthracnose disease in annual bluegrass. *Crop Sci.* 51:1244–1252.

Ryan, C.P. 2011. Seasonal development of dollar spot epidemics in Maryland and nitrogen effects on fungicide performance in creeping bentgrass. M.S. thesis. Department of Plant Science and Landscape Architecture, University of Maryland, College Park, MD.

Ryan, C.P., P.H. Dernoeden and A.P. Grybauskas. 2012. Seasonal development of dollar spot epidemics in six creeping bentgrass cultivars in Maryland. *HortSci.* 47:422–426.

Sanders, P.L. 1984. Failure of metalaxyl to control Pythium blight on turfgrass in Pennsylvania. *Plant Dis.* 68:776–777.

Sanders, P.L., W.J. Houser, P.J. Parish, and H. Cole, Jr. 1985. Reduced-rate fungicide mixtures to delay fungicide resistance and to control selected turfgrass diseases. *Plant Dis.* 69:939–943.

Settle, D.M., J.D. Fry, G.A. Milliken, N.A. Tisserat, and T.C. Todd. 2007. Quantifying the effects of lance nematode parasitism in creeping bentgrass. *Plant Dis.* 91:1170–1179.

Settle, D.M., J.D. Fry, and N.A. Tisserat. 2001. Development of brown patch and Pythium blight in tall fescue as affected by irrigation frequency, clipping removal and fungicide application. *Plant Dis.* 85:543–546.

Settle, D.M., J.D. Fry, T.C. Todd, and N.A. Tisserat. 2006. Population dynamics of the lance nematode (*Hoplolaimus galeatus*) in creeping bentgrass. *Plant Dis.* 90:44–50.

Settle, D., T. Sibicky, and R. Rinker. 2011. Fungicides and alternatives for curative control of fairy ring on a bentgrass-*Poa annua* green in Chicago, 2010. *Plant Disease Management Reports* 5:T037.

Shantz, H.L., and R.L. Piemeisel. 1917. Fungus fairy rings in eastern Colorado and their effects on vegetation. *J. Agric. Res.* 11:191–245.

Smiley, R.W., P.H. Dernoeden, and B.B. Clarke. 2005. *Compendium of Turfgrass Diseases*, 3rd ed. APS Press, St. Paul, MN.

Smith, J.D., N. Jackson, and A.R. Woolhouse. 1989. *Fungal Diseases of Amenity Turf Grasses*. E. & F.N. Spon, London and New York.

Tani, T., and J.B. Beard. 1997. *Color Atlas of Turfgrass Diseases*. Ann Arbor Press, Chelsea, MI.

Toda, T., T. Mushika, T. Hayakawa, A. Tanaka, T. Tani, and M. Hyakumachi. 2005. Brown ring patch: A new disease on bentgrass caused by *Waitea circinata* var. *circinata*. *Plant Dis.* 89:536–542.

Todd, T.C., and N. Tisserat. 1993. Understanding nematodes and their impact. *Golf Course Management* 61(5):38, 42, 44, 48, 52.

Tredway, L.P., and J.P. Kerns. 2005. Defining the nature of creeping bentgrass root diseases. *U.S. Golf Association Green Section Record* 43(4):7–9.

Tredway, L.P. 2005. First report of summer patch of creeping bentgrass caused by *Magnaporthe poae* in North Carolina. *Plant Dis.* 89:204.

Vargas, J.M., Jr. 2005. *Management of Turfgrass Diseases*. Wiley, Hoboken, NJ.

Westerdahl, B.B., M.A. Harivandi, and L.R. Costello. 2005. Biology and management of the nematodes of turfgrass in northern California. *U.S. Golf Association Green Section Record* 43(5):7–10.

Williams, D.W., A.J. Powell, Jr., P. Vincelli, and C.T. Dougherty. 1996. Dollar spot on bentgrass influenced by displacement of leaf surface moisture, nitrogen, and clipping removal. *Crop Sci.* 36:1304–1309.

5

Biological Approaches to Turf Disease Management

Although genetically transformed grasses may ultimately be the answer to several common pest problems, alternative nonchemical methods and technologies are being studied for their potential to reduce the use of fungicides on turfgrasses. Composts and natural organic fertilizers are being used in the hope that they can boost the activity of beneficial soil microbes. It is generally believed that enhanced soil microbial activity results in more competition with or antagonism of potential pathogens, thereby resulting in fewer disease problems. Natural organic materials may be helpful, but the few field studies that directly compared natural organic to synthetic organic (i.e., manufactured) fertilizers did not provide a great deal of evidence that natural fertilizers are superior. Furthermore, when natural and synthetic organics were shown to reduce diseases, such as brown patch or dollar spot, the level of suppression generally was not commercially acceptable to superintendents or golfers throughout the entire "active disease" period. There are two general approaches to biological management of turfgrass diseases: (1) apply overtop of turf or incorporate into soil natural organic fertilizers or composts to boost microbial activity, and (2) apply microbial inoculants overtop the turf or incorporate them in soil.

Organic Fertilizers and Amendments

Golf course greens and tees are topdressed several to many times a year with straight sand or a mixture of sand and some kind of organic matter (normally a type of peat, usually sphagnum). Sphagnum peat does not provide a good medium for the growth of disease-suppressive microorganisms. Instead of amending topdressings with sphagnum peat, golf course superintendents could use organic amendments that would support microbial activity (Nelson, 1997a,b). Disease suppression from compost materials, natural organic fertilizers, and some other fertilizers was correlated with increased microbial activity in thatch and soil (Nelson et al., 1994). Another reason for using organics would be to avoid nitrate leaching associated with inorganic fertilizers applied to sandy soils during rainy periods.

When selecting a natural organic fertilizer or amendment (i.e., composts), consideration should be given to the overall particle size of the material and the mineral ash content (i.e., fine particles remaining following biological decomposition) of the product. Organic amendments high in mineral ash or the presence of significant "fines" may plug pore spaces in sand-based rootzones and impede drainage. Although this is a valid concern, a few studies demonstrated that it would likely require substantial quantities well in excess of application rates associated with normal fertility practices. Furthermore, the ash from many commercially available fertilizer products is soluble in water and is not a concern (McNitt et al., 2007; Moeller et al., 2009). In one study natural organic fertilizers were applied to three bentgrass cultivars, 'Penncross', 'L-93', and 'Penn A-4', with contrasting shoot densities and incorporated into the rootzone following seasonal coring events for 4 years. There was not a strong effect of fertilizer source on soil surface organic matter levels. Cultivar was the factor most strongly affecting increasing organic matter accumulation (i.e., Penn A-4 > L-93 or Penncross) and thus a greater potential for compromised physical properties in sand-based rootzones (Moeller et al., 2009).

Several successful field experiments used organic amendments to suppress turfgrass diseases. Nelson and Craft (1992) observed that sand topdressing amended with poultry waste organic fertilizers (e.g., Ringer Compost Plus®, Ringer Greens Restore®, and Sustane®) significantly suppressed the severity of dollar spot. They also reported that selected composts prepared from turkey litter and sewage sludge as well as noncomposted blends of plant and animal meals also consistently suppressed dollar spot. Furthermore, studies showed that amending sand-based greens with a municipal biosolids compost, a brewery sludge compost, or a reed sedge peat induced a high level of Pythium root rot (*Pythium graminicola*) disease suppression (Nelson, 1997a,b). These amendments provided control of Pythium root rot 6 months after incorporation into sand-based putting greens, and retained their suppressive properties for up to 4 years.

Melvin and Vargas (1994) evaluated five fertilizers (Ringer Lawn Restore®, Sustane®, Bio Groundskeeper® + NITRO-26®, Nitroform®, and urea) in combination with three irrigation regimes for management of necrotic ring spot (*Ophiosphaerella korrae*) in Kentucky bluegrass (*Poa pratensis*). Necrotic ring spot severity was reduced after 2 years by monthly applications (1 lb N/1000 ft^2; 49 kg N ha^{-1}) of Ringer Lawn Restore® (composed of feathermeal, bonemeal, and soybean-meal), when compared to urea with P + K (9N-4P-4K). After 3 years of applications, however, no nitrogen source had a significant effect on necrotic ring spot suppression. All fertilized plots had less severe necrotic ring spot than nonfertilized plots at the end of the fourth year.

There are reports showing that natural organic fertilizers are not generally superior to synthetic organics in suppressing turfgrass diseases. Landschoot and McNitt (1997) examined seven nitrogen sources to determine if dollar spot suppression in creeping bentgrass is greater with natural organic fertilizers, when compared to synthetic organic N-sources. They tested Ringer Commercial Greens Super®, Ringer Compost Plus®, Sustane®, Milorganite®, and Harmony®, all of which are derived mainly from natural organic sources. They compared the natural organic N-sources to ureaform and urea, both of which are synthetic organic fertilizers. Their results showed that urea provided equal or better dollar spot control, when compared to natural organic fertilizers. They also reported that on the majority of rating dates, dollar spot severity was correlated with turf color, indicating that as nitrogen availability increased, disease severity decreased.

Davis and Dernoeden (2002) evaluated the effect of nine organic fertilizers and composts on the sustainability of disease-suppressive microorganisms in fairway height creeping bentgrass. Several natural organics including Milorganite®, Ringer Lawn Restore®, Sustane®, and Com-Pro® and two synthetic organic (urea and sulfur-coated urea) fertilizers were evaluated. The N-sources were applied annually at 4 lb N/1000 ft^2 (200 kg N ha^{-1}) for 7 years and dollar spot data were collected in the last 3 years of the study. Ringer Lawn Restore® (a natural fertilizer derived from poultry waste), urea, and sulfur-coated urea were the only N-sources that reduced dollar spot from mid-to-late spring, when disease pressure was in the low to moderate range. Com-Pro® composted sewage sludge enhanced dollar spot and increased thatch. No N-source reduced dollar spot when data were averaged over the entire epidemic. Data revealed little or no correlation between soil microbial activity and dollar spot suppression. During low to moderate disease pressure periods in one year, however, leaf tissue N levels were negatively correlated (i.e., as tissue N increased, disease injury decreased) with dollar spot severity. Hence, the mechanism of dollar spot suppression was likely due to better N availability accorded by the more readily available fertilizers rather than their ability to promote disease-suppressive soil microorganisms.

Peacock and Daniel (1992) compared turfgrass growth, N-uptake, and brown patch incidence under greenhouse conditions following applications of a natural organic fertilizer. Turf Restore®, which was microbiologically amended with *Bacillus* spp., *Trichoderma viride*, and actinomycetes (filamentous bacteria found in soil), was compared to the identical, nonamended material and urea. Their results indicated that microbiologically amended organic fertilizer did not afford any advantage in disease suppression, even in sterilized soil. They concluded that the addition of soil microorganisms as biological amendments to natural organic fertilizers was not beneficial. Fidanza and Dernoeden (1996) conducted field studies to determine the influence of N-source on brown patch (*Rhizoctonia solani*) severity. They applied eight fertilizers (Sustane®, Ringer Lawn Restore®, Ringer CP®, IBDU, SCU, methylene urea, ammonium sulfate, and urea) to perennial ryegrass. None of the N-sources consistently reduced brown patch in at least 2 of the 3 years of the study.

Microbial Activity in Sand-Based Rootzones

Organic matter increases soil microbial activity, water retention, cation exchange capacity (i.e., the ability of a soil to hold nutrients), nutrient transformation, recycling and retention, pesticide degradation, and other properties. Golf greens are typically constructed with high sand content rootzone mixtures (>80% by volume) to resist compaction and maintain drainage. High sand growing media possess lower nutrient-holding capacities and are regarded as being somewhat less productive in terms of microbial activity. Plants grown in soils containing higher organic matter and an active microbial population are typically more vigorous and less pest prone than those grown in soils devoid of organic matter.

As previously noted, an approach to increasing soil organic matter in established turf is through the use of natural organic (i.e., bio-organic) fertilizers and amendments. Mancino and coworkers (1993) compared the influence of a water-soluble fertilizer and the natural organic Ringer Greens Restore® (hydrolyzed poultry meal) for their effect on soil and thatch microbial populations in a mature creeping bentgrass green (80:20 mix). They found that the sand rootzone mixture without being amended with a natural organic material can support microbial populations similar to those found in some finer-textured native soils. The thatch layer had much higher levels of most types of microbes than the underlying rootzone, particularly bacteria. Greens Restore® increased fungal counts in the soil, and plots receiving natural organic fertilizer had slightly less thatch when compared to plots treated with water-soluble nitrogen. Thatch in plots receiving Greens Restore®, however, was thicker than that observed in plots receiving no fertilizer. The magnitude of the difference in thatch development in the 2-year study between nitrogen sources was small (≤0.4 mm). They concluded that using Green Restore® to boost microbial activity to stimulate thatch decay was not very important, because thatch naturally develops very high levels of indigenous microbes. Similarly, a study conducted in bentgrass grown on a sandy loam native soil and maintained at fairway height showed that numerous bio-organic fertilizers applied over a 7-year period had little impact on promoting soil microbial activity, improving soil organic matter levels, or influencing thatch thickness (Davis and Dernoeden, 2002).

Bacteria are the most abundant microorganisms found in soil. They degrade organic matter, and improve soil quality; numerous species have disease-suppressive properties. In 2001, Elliott and Des Jardin published results of their investigation on the effects of bacterial populations in new hybrid bermudagrass golf greens grown on sand (>80%). They built several small greens, fumigated the soil with methyl bromide, dazomet, or metam-sodium, and monitored selected groups of bacterial populations. They found that even when using 10 times the recommended rate, none of the fumigants eliminated all bacterial groups. The bacterial levels in soil recovered to prefumigation numbers within 25 to 70 days of fumigation. Additionally, they found that bacterial populations on bermudagrass roots within 20 weeks of fumigation were similar to those observed in a bermudagrass green that was 7 years old. They concluded that bacterial populations stabilize very early during putting green establishment. Similarly, Bigelow et al. (2002) found that bacterial populations in newly constructed sand-based rootzone mixtures achieved levels similar to mature sand-based creeping bentgrass greens within 6 months following a late summer seeding. These results further document that soil bacterial levels stabilize in sand-based rootzones soon after establishment. In a third study, plots in a creeping bentgrass research green were amended with either hydrolyzed poultry meal (Ringer Greens Super®) or humate to determine their effects on soil microbial activity, rooting, thatch, and other properties (Kaminski et al., 2004). The green was newly constructed with a rootzone mix of heat-sterilized sand plus sphagnum peat moss (80:20 by volume). In the humate-incorporated plots, bentgrass root mass increased during the first 3 months following establishment, but no root mass differences were subsequently detected. Microbial activity generally was greater in soil of poultry meal-treated plots during the first 14 months following seeding, but not thereafter. Poultry meal-incorporated plots always were fertilized with poultry meal, whereas the humate plots were fertilized with either water-soluble nitrogen or methylene urea (i.e., non-bio-organic N). Using poultry meal as the sole source of nutrients resulted in slower establishment, generally inferior turf quality, but over time the turf developed slightly less thatch. Thatch depth differences were statistically significant, but the magnitude of the difference was small (i.e., 0.9 mm). Hence, incorporating either poultry meal or humate provided mostly short-term benefits.

In the three aforementioned studies, microbial activity had stabilized in 70 days (bermudagrass), and 6 and 14 months (creeping bentgrass) following establishment, respectively. In those studies, bacterial populations or soil microbial activity were measured in new greens over a 2-year period. In time, however, soil microbial biomass does increase. In a Nebraska study, 47 greens, 8 to 28 years old, were analyzed for microbial biomass (Kerek et al., 2002). Microbial biomass is primarily associated with the particulate organic matter fraction (i.e., organic matter produced by decaying turfgrass roots) of soil. Microbial biomass increased with age of the individual greens from about 26% in an 8-year-old green to 78% in a 28-year-old green. Hence, soil microbial activity is going to naturally increase as roots and stems are degraded creating more organic matter.

Mounting evidence indicates that using natural organic fertilizers and amendments have short-term benefits in increasing microbial activity. Soil and thatch microbial activity does not appear to be greatly influenced for long periods by using natural organic fertilizers, even in turf grown on sand-based rootzones. Turfgrass roots and stems provide the major source of organic matter in soil, and the organic matter produced by decaying turfgrass tissues is most effective in promoting soil microbial activity. Therefore, cultural practices (i.e., proper core aeration and fertility) that promote a greater root biomass are the best strategy for improving soil organic matter levels and attendant soil microbial properties. It should be noted in the studies cited that different methods were used to measure either specific groups of microbes (i.e., bacteria, fungi, and actinomycetes) or general soil microbial activity (i.e., a composite measure of all groups). It is difficult to draw definitive conclusions when comparing results involving different assay methods and microbial groups. Furthermore, traditional methods of measuring soil microbial activity are considered crude, but the development of better techniques may provide a more accurate measure of the different microbial groups found in soil and thatch.

Biological Inoculants

Use of biological inoculants to suppress disease is becoming more popular, and there are numerous products that claim disease suppression. Bio-Trek® 22G was the first biofungicide to receive registration (1997) for turfgrass disease control from the U.S. Environmental Protection Agency (EPA). Products such as Bio-Trek® are applied as a granular material or as sprayable spore/propagule suspensions overtop the turf. They reduce disease by one or more mechanisms including: parasitizing the pathogen (i.e., mycoparasitism); antagonizing by outcompeting pathogens for space and infection sites or by producing antibiotic substances that suppress the growth or kill sensitive pathogens; or triggering the plant's natural defense mechanisms.

Mycoparasitism

Fungi, including some species of *Trichoderma* and *Gliocladium*, work in part by parasitizing the vegetative body (i.e., mycelium) or propagules (spores, sclerotia, etc.) of sensitive pathogens. They may provide their beneficial effects in the turf canopy, thatch, or soil.

Antagonism and Competition

Antagonists include bacteria, actinomycetes, and fungi that produce antibiotic substances that inhibit growth or spore production, or the antibiotic may be toxic enough to kill sensitive microbes. This phenomenon also is known as *antibiosis*. Antagonists are likely to be excellent competitors that grow rapidly to displace potential pathogens as they attempt to colonize roots, leaves, and other infection sites on plant tissues. Common soil bacteria from the

genera *Bacillus, Enterobacter, Stenotrophomonas,* and *Pseudomonas* are among
the best-known antagonists.

A classic example of the benefits of natural antagonists in turf involves take-
all patch (*Gaeumannomyces graminis* var. *avenae*) disease in bentgrass. Take-all
is more common and severe in young creeping bentgrass turf. After turf has
matured over a 2- to 5-year period, take-all usually declines in severity due
to the buildup of antibiotic-producing bacteria, including *Pseudomonas* spp.
Dead spot is another disease that is most severe in young bentgrass stands,
which usually naturally declines in 1 to 3 years. The mechanism of dead spot
decline is unknown, and greens as old as 6 years have been reported to have
this disease.

Triggering Natural Defenses

Some microbes can be used to "trigger" the plant's natural defense mechanisms. Plants can produce substances (i.e., phytoalexins) in response to infection that are toxic to pathogens. Phytoalexins are phenolic compounds that
are toxic to or inhibit the growth of a pathogen. Plants also may produce an
array of physical barriers to prevent further ingress of a pathogen following
infection. Australian researchers demonstrated that the fungus *Phialophora
graminicola* can prevent take-all disease of creeping bentgrass in laboratory
studies (Wong and Siviour, 1979). When *P. graminicola* was placed on bentgrass roots it penetrated only a few cells deep and thus triggering resistance
mechanisms, and the fungus caused no further injury to plants. When the
take-all fungus (*G. graminis* var. *avenae*) was introduced after *P. graminicola,*
the take-all pathogen was unable to become highly invasive due to the triggering of plant defenses.

Stenotrophomonas maltophilia strain C-3 was shown to provide control of leaf
spot (*Bipolaris sorokiniana*) in tall fescue by both antagonism and induction of
host resistance (Yuen et al., 2001). Strain C-3 produces the enzyme chitin-
ase. Most fungi have cell walls composed of chitin. In the case of C-3, chi-
tinase inhibits cell wall production in germinating spores. The researchers
also found strong evidence that C-3 triggers the plants' own natural defense
mechanism in leaves but not roots. The application of C-3 cells to any portion of a tall fescue leaf inhibited the germination of *B. sorokiniana* spores on
the entire surface of the same leaf. Hence, placement of C-3 cells on just the
leaf tips resulted in a reduction of spore germination on lower segments of
the same leaf that was not treated with C-3. The phenomenon occurred only
on leaves treated with C-3 and therefore the effect was not systemic. The
nature of induced host resistance is imperfectly understood. Evidently, presence of C-3 cells on a leaf sends a signal to the plant that the leaf is about to
be infected by a pathogen. The signal is believed to be elicited by substances
exuded by a microbe on the plant surface. Cells in the leaf respond by producing specialized cells that inhibit infection or leaf cells may produce antifungal chemicals known as phytoalexins. Phytoalexins are toxic to or inhibit

the growth of a pathogen, which may be preformed or induced as a result of infection by specialized plant cells. Other laboratory and field studies conducted by Yuen and coworkers showed that C-3 suspensions reduced the level of blighting by *Rhizoctonia solani* (i.e., brown patch) in both tall fescue and perennial ryegrass.

Bio-Trek® 22G is an inoculant meant to be applied overtop of turf. The agent, *Trichoderma harzianum* strain 1295-22 (T-22), was shown to significantly reduce the severity of dollar spot, brown patch, and *Pythium* root rot in New York. The mechanism of disease suppression is attributed to the production of enzymes by T-22 that digest cell walls of sensitive microbes, and possibly by colonizing roots and outcompeting pathogens for root infection sites. Bio-Trek, however, failed to provide any suppression of brown patch, dollar spot, or summer patch in studies conducted in the mid-Atlantic region (Couch, 1998; Dernoeden, unpublished). These findings suggest possible quality control problems and demonstrate that the performance of a biological fungicide is likely to be dependent on the level of disease pressure, relative susceptibility of hosts, soil type, and regional environmental conditions.

There are significant barriers to successful pest control with biological agents, and perhaps the greatest problem is maintaining high populations of introduced agents in the soil and thatch, and on leaves and sheaths. Biological agents are generally common microbial species found in most soils. Many biotypes within a species evolve in place, but when introduced into different environments or soil types, they are not likely to survive, be competitive, or reproduce in high enough populations to provide the beneficial effects. For example, the introduced biological agent must be able to grow and reproduce in these new soil or thatch niches where they are likely to encounter unfamiliar types of environmental extremes (i.e., high and low temperature, wet and dry conditions, soil pH and aeration variations, etc.). They also must be able to compete with a myriad of naturally occurring antagonists.

Little is known about the survivability of most candidate biological agents or how to handle or formulate these agents for use in turf. Important questions needing answers include when should applications begin each year for each potential pathogen, and should the agents be watered-into the root zone or are they better placed in the thatch or maintained on leaves and sheaths? We do not know the thresholds (i.e., how many propagules should be applied per unit area) or how often they need to be applied. Furthermore, we know many agents (especially *Pseudomonas* spp.) are sensitive to ultraviolet (UV) light, indicating that daytime application will result in many cells dying rapidly. Furthermore, little is known about the influence of other pesticides, fertilizers, wetting agents, biostimulants, and other products routinely used and how they may impact the viability of these introduced agents. Then there are production and storage (i.e., some must be refrigerated) problems that must be overcome.

Many of these potential problems were addressed in studies involving the Bio-Ject Biological Management System®. The Bio-Ject system ferments large numbers of bacteria and injects them directly into the irrigation system for nighttime delivery. As noted previously, some microbes are sensitive to the UV light in sunshine; therefore, night is the preferred time to deliver some biological inoculants. The bacterium *Pseudomonas aureofaciens* strain Tx-1 was registered as a biofungicide (Spotless®) by the U.S. EPA in 1999. Davis and Dernoeden (2001) evaluated the BioJect system over a 4-year period using the bacterium *P. aureofaciens* strain Tx-1. Strain Tx-1 was shown to effectively suppress dollar spot (*Sclerotinia homoeocarpa*) in laboratory and field studies (Powell and Vargas, 2000). The mechanism of disease suppression is the production of phenazine-1 carboxylic acid (PCA) by Tx-1. The PCA is an inhibitor of fungal growth (i.e., antibiosis) and hence the mechanism of action of Tx-1 is antagonism. Davis and Dernoeden (2001) monitored the number of colony-forming units (cfu's) of Tx-1 fermented and delivered through the irrigation heads. The number of Tx-1 cfu's found in the foliage plus thatch and soil were quantified. Results showed that the Bioject effectively fermented and delivered high populations of Tx-1 to the foliage plus thatch (average = 40 million cfu's per gram of tissue) and soil (average = 340,000 cfu's per gram soil). The Tx-1 also was shown to survive winter in low populations in thatch and soil. Dollar spot was suppressed on average by 33% in Crenshaw creeping bentgrass. It should be noted that Crenshaw is highly susceptible to dollar spot. There were dates when Tx-1 was shown to provide dramatic reductions in dollar spot during high disease pressure periods. In one year, Tx-1 was shown to increase the residual effectiveness of the fungicides iprodione and propiconazole by 7 to 10 days. These fungicides and chlorothalonil had little impact on the levels of Tx-1 recovered from soil. While some brown patch (*R. solani*) suppression was noted under low disease pressure, Tx-1-treated plots were more severely blighted than nontreated control plots during periods of moderate to high disease pressure. Similar results were reported by Hardebeck et al. (2004).

The aforementioned studies showed that the Bioject fermentation and delivery technology work well and demonstrated that Tx-1 could significantly reduce dollar spot in a highly susceptible cultivar. The level of control, however, was not commercially acceptable. Hence, the research provided more evidence that it is unlikely that biological agents will replace the need for fungicides on golf courses. Use of Tx-1 and other biological agents may extend the residual effectiveness of some fungicides and reduce the potential for *S. homoeocarpa*–resistant biotypes from developing. This system addresses the problem of delivering high populations of microbes on a frequent basis during nighttime hours. Much needs to be learned about how to ferment high enough populations that can be delivered uniformly over a large area on a frequent basis. The ability to deliver these microbes to turf will only be as good as the distribution provided by your irrigation system. Furthermore, the practice of supplementing irrigation every night may lead to excess soil moisture, which can promote nontarget disease development and poorly conditioned surfaces (i.e., not firm and fast).

Other inoculants include various sprayable and granular formulations of *Bacillus subtilis* (Rhapsody®), *B. licheniformis* (EcoGuard®), *Pseudomonas florescens*, and *Burkholderia casidae*, which have been shown to provide some suppression of brown patch and dollar spot under low disease pressure. In most studies the level of disease suppression did not meet with commercially acceptable quality standards for golf course turfs in general and for golf greens in particular. The bacterium *Bacillus firmus* is being developed to manage parasitic nematode problems in turfgrasses. The mechanism of protection is that *B. firmus* applied to turfgrasses will colonize roots and its presence discourages feeding activity by parasitic nematodes. Currently, *B. firmus* appears promising when targeting sting nematode and some other nematode species in warm-season grasses. Search for effective biological agents that will provide an acceptable level of disease suppression continues.

It is not unusual to see inconsistencies in pest suppression using biological agents from year to year or among regions. This is due in large part to environmental conditions. Microbes applied to plants must have favorable moisture and temperature conditions in order to grow and reproduce. In many situations, the optimal environmental conditions that favor growth of a biological agent may not prevail or may not prevail long enough at the times they are applied. Furthermore, success and failure of biological agents can be influenced by pathogen biotype at the site (e.g., there are at least four biotypes of *R. solani* that attack turfgrasses), level of disease pressure (i.e., low versus high), relative susceptibility of the turf (e.g., Declaration, L-93, and 007 creeping bentgrasses have good dollar spot resistance when compared to Backspin, Crenshaw, and Southshore), formulation and shelf-life of biological agents, timing of application, and numerous other known and unknown factors. Hence, like all new products and technologies, biological agents must be tested over a wide geographic area, a range of environmental and soil conditions, and over a course of several years to learn how to attain the best possible level of disease suppression. When testing any new biocontrol agents or pesticides, always leave an untreated area so that you can objectively evaluate the effectiveness of the product.

Genetic Transformation

Advances in biotechnology have made it possible to develop turfgrasses with improved tolerance to stresses and pests. Even though genes with improved disease and stress tolerance were identified in the 1990s and early 2000s, commercial release of transgenic turfgrasses has not yet become a reality. The U.S. federal regulating agency responsible for approving transgenic plants is the Animal Plant Health Introduction Service (APHIS). Most transgenic plants have been annual field crops (e.g., corn, cotton, and soybean) rather than perennial crops like turf. The first transgenic turfgrasses contained a gene to provide nonselective herbicide (i.e., glyphosate) resistance. During the developmental stages fears arose that the trait would outcross to native grasses. Pollen transport studies later showed that the gene for glyphosate tolerance could outcross up to a mile from breeders' fields. Environmental impact studies were then required by APHIS, which has slowed the introduction of transgenic turfgrasses, including those with benign but highly desirable traits including disease resistance and drought and heat resistance. Germplasm was developed to use in transgenic breeding programs, but transformed grasses remain to be commercialized.

The transformation technique involves identifying and isolating genes from the DNA in other plants or microbes that contain a desirable trait (e.g., pest, heat, and drought resistance). These desirable genetic traits are identified using molecular markers that make it faster and easier to find desirable characteristics in even unrelated organisms. These genes are coated onto gold particles that are physically propelled using a helium-powered "gene gun" into callus tissues (i.e., masses of knot-like cells) grown from seed on media in petri dishes. The gene enters a cell to become part of its DNA. Transformed grasses must then be grown out in the greenhouse and tested in various ways to determine if the desirable gene was successfully incorporated and expressed. Transformed plants must then be grown in the field to determine if plants retain the gene and to determine if transformed grasses possess desirable turf characteristics (i.e., the same fine texture, density, good color, etc., of the original cultivar). Before these cultivars enter the commercial marketplace, several years of field testing in various geographic regions is needed. Many scientists believe that genetic transformation eventually will yield grasses that require fewer inputs of pesticides, fertilizers, and water.

Summary

- Biological control of turf diseases generally involves use of natural organic fertilizers or composts or the application of inoculants to boost microbial activity in thatch and soil.
- Various fertilizers including natural and synthetic organic and inorganic fertilizers have been demonstrated to suppress or reduce the activity of specific diseases.
- Beneficial microbes reduce disease severity by antagonizing, parasitizing or outcompeting pathogens, and by triggering natural plant defenses.

- The performance of biological inoculants often is erratic due to the influence of environmental factors on survival and growth of introduced microbes.

- Biological approaches to turfgrass pathogen suppression generally do not reduce injury to commercially acceptable aesthetic and playability standards expected on golf courses.

- Biological control methods should be used in conjunction with sound cultural practices to achieve the best combination of disease suppression and reduced inputs of synthetic fungicides.

- Advances in biotechnology and genetic transformation will likely yield disease and insect pest resistant grasses, which will provide for significant reductions in the future use of pesticides.

Bibliography

Bigelow, C.A., A.G. Wollum II, and D.C. Bowman. 2002. Characterization of microbial properties in newly constructed sand-based rootzones. *Crop Sci.* 42:1611–1614.

Couch, H.B. 1998. Efficacy of *Trichoderma harzianum* in the control of two major diseases of turfgrasses. *Agron. Abst.* 127.

Davis, J.G., and P.H. Dernoeden. 2001. Fermentation and delivery of *Pseudomonas aureofaciens* strain Tx-1 to bentgrass affected by dollar spot and brown patch. *Int'l. Turfgrass Soc. Res. J.* 9:655–664.

Davis, J.G., and P.H. Dernoeden. 2002. Dollar spot severity, tissue nitrogen and soil microbial activity in bentgrass as influenced by nitrogen source. *Crop Sci.* 42:480–488.

Elliott, M.L., and E.A. Des Jardin. 2001. Fumigation effects on bacterial populations in new golf course bermudagrass greens. *Soil Biol. Biochem.* 33:1841–1849.

Fidanza, M.A., and P.H. Dernoeden. 1996. Influence of mowing height, nitrogen source, and iprodione on brown patch severity in perennial ryegrass. *Crop Sci.* 36:1620–1630.

Hardebeck, G.A., R.F. Turco, R. Latin, and Z. Reicher. 2004. Application of *Pseudomonas aureofaciens* through irrigation for control of dollar spot and brown patch on fairway-height turf. *HortScience* 39:1750–1753.

Kaminski, J.E., C.A. Bigelow, and P.H. Dernoeden. 2004. Soil amendments and fertilizer source effects on creeping bentgrass establishment, soil microbial activity, thatch and disease. *HortScience* 39:620–626.

Kerek, M., R.A. Drijber, W.L. Powers, R.C. Sherman, R.E. Gaussoin, and A.M. Streich. 2002. Accumulation of microbial biomass with particulate organic matter of aging golf greens. *Agron. J.* 94:455–461.

Landschoot, P.J., and A.S. McNitt. 1997. Effects of nitrogen fertilizers on suppression of dollar spot disease of *Agrostis stolonifera* L. *Int'l. Turfgrass Soc. Res. J.* 8:905–911.

Liu, L., T. Hsiang, K. Carey, and J.L. Eggens. 1995. Microbial populations and suppression of dollar spot disease in creeping bentgrass with inorganic and organic amendments. *Plant Dis.* 79:144–147.

Mancino, C.F., M. Barakat, and A. Maricic. 1993. Soil and thatch microbial populations in an 80% and 20% peat creeping bentgrass putting green. *HortScience* 28:189–191.

McNitt, A.S., D. Petrunak, and T. Serensits. 2007. The effects of organic fertilizers on the saturated hydraulic conductivity of USGA rootzones. *U.S. Golf Association Turfgrass Environ. Res.* Online. 6(12):1–5. http://usgatero.msu.edu/v06/n12.pdf

Melvin, B.P., and J.M. Vargas Jr. 1994. Irrigation frequency and fertilizer type influence necrotic ring spot of Kentucky bluegrass. *HortScience* 29:1028–1030.

Moeller, A., C. Bigelow, J. Nemitz, and G. Hardebeck. 2009. Natural organic fertilizers and bentgrass cultivar effects on rootzone organic matter content. American Society of Agronomy, International Annual Meeting. Abstracts.

Nelson, E.B. 1997a. Biological control of turfgrass diseases. *Golf Course Management.* 65(7):60–69.

Nelson, E.B. 1997b. Microbial inoculants for the control of turfgrass diseases. *Int'l. Turfgrass Res. Soc. J.* 8:791–811.

Nelson, E.B., and C.M. Craft. 1991. Identification and comparative pathogenicty of *Pythium* spp. from roots and crowns of plants exhibiting symptoms of root rot. *Phytopathology* 81:1529–1536.

Nelson, E.B., and C.M. Craft. 1992. Suppression of dollar spot on creeping bentgrass and annual bluegrass turf with composted-amended topdressings. *Plant Dis.* 76:954–957.

Nelson, E.B., L.L. Burpee, and M.B. Lawton. 1994. Biological control of turfgrass diseases. In *Handbook of Integrated Pest Management for Turf and Ornamentals*, A. Leslie, Ed., pp. 409–428. CRC Press, Boca Raton, FL.

Peacock, C.H., and P.F. Daniel. 1992. A comparison of turfgrass response to biologically amended fertilizers. *HortScience* 27:883–884.

Powell, J.F., and J.M. Vargas. 2000. Management of dollar spot on creeping bentgrass with metabolites of *Pseudomonas aureofaciens* (Tx-1). *Plant Dis.* 84:19–24.

Wong, P.T.W., and T.R. Siviour. 1979. Control of Ophiobolus patch in *Agrostis* turf using avirulent fungi and take-all suppressive soils in pot experiments. *Ann. Appl. Biol.* 92:191–197.

Yuen, G.Y., and O. Lilic. 2001. Evidence of induced resistance in the control of *Bipolaris sorokiniana* in tall fescue by *Stenotrophomonas maltophilia* C3. *Int'l. Turfgrass Res. Soc. J.* 9:736–741.

6

Understanding Modern Turf Fungicides

Fungicides are integral in the maintenance of intensively managed turfgrasses and are a key component of integrated pest management (IPM) programs. One of the largest markets for fungicides in the United States is in the turfgrass industry. Fungicides are used most extensively by golf course superintendents. The game of golf is played on grass, and golf greens are the most important and intensively managed part of the golf course. As demand has mounted to provide excellent playing conditions from tee to green and oftentimes rough areas as well, the use of fungicides on golf courses has increased. Competitive pressures and pride among professional turfgrass managers to provide for superior quality also is a factor. Because disease can impair or diminish the quality of golf turfs, fungicides are used extensively in some regions of the United States to maintain high standards for both aesthetic and functional (i.e., playability) quality. In some regions of the United States, the cost of fungicides used on golf courses exceeds those of all other pesticides combined.

Factors Governing Fungicide Use Decisions

Effective fungicide programs hinge upon knowledge of past disease problems at the facility, the ability to distinguish turfgrass species and sometimes cultivars within a species, the ability to diagnose diseases at an early stage in their development, the ability to monitor and recognize weather conditions and patterns that favor disease activity, and the proper selection and application of the most appropriate product. Arriving at the decision of whether to apply a fungicide to any turf area often is based on economic considerations. Aside from cost, the primary determinants in using a fungicide are based on the prevailing environmental conditions, the susceptibility of the host species and cultivars present, the pathogen involved, and the level of quality desired. Unique factors in turfgrass pathology include the intensity and nature of turfgrass management practices, which greatly influence plant vigor and therefore the severity of diseases.

Promoting vigorous growth through sound cultural practices is the first step in minimizing disease injury. Frequently, however, environmental stresses, poor growing conditions and environments, traffic, and inappropriate management practices weaken plants, predisposing them to invasion by pathogens. When disease symptoms appear, it is imperative that a rapid and accurate diagnosis of the disorder be made. The prudent manager also attempts to determine those environmental and cultural factors that may have led to the development or contributed to the intensity of a disease. A common cause for extensive disease injury frequently can be related to poor growing environments or improper management practices. Poor growing environments include shade, poor air or water drainage, soil compaction, and the development of excessive thatch or mat layers in some species. Management practices that can exacerbate diseases include frequent and close mowing, mechanical practices (i.e., coring, grooming, vertical cutting, rolling, sand topdressing, etc.) during periods of environmental stress, light and frequent or excessive irrigation, and applications of inadequate or excessive amounts of fertilizer. Despite hard work and adherence to sound management practices, however, diseases may become a serious problem. This normally occurs when environmental conditions reduce plant growth and vigor while favoring growth of a pathogen. Once the competitive advantage tilts toward the pathogen and away from turfgrass plants, diseases become problematic and destructive. In this situation, fungicides may be recommended in conjunction with cultural practices that promote turf vigor.

Fungicides may be applied preventively (i.e., before anticipated disease symptoms appear or pre-plant infection) or curatively (i.e., when disease symptoms first become evident or post-plant infection). Preventive or prophylactic fungicide treatment generally is recommended for chronically damaging diseases in high-input management turfgrasses and for managing high-risk pathogens for resistance. Preventive applications are based on past history of turf diseases at a facility as well as weather conditions conducive to disease development. Curative fungicide applications can be effective when they are made as soon as disease symptoms are discerned. Curative applications often involve higher use rates and possibly more frequent or closer spray intervals to subdue a disease. Preventive applications generally involve lower rates and longer spray intervals until or unless disease symptoms appear. The preventive approach is employed commonly on high-quality golf course turfs such as putting greens and generally targets chronically severe diseases. Curative applications can be more economically wise for less severe damaging diseases. The key to a successful curative fungicide program is vigilant scouting and a rapid response to building disease epidemics. The performance of a preventive or curative fungicide program is improved by imposing more conservative cultural practices during stressful periods.

Fungicide Mobility in Plants

The behavior of fungicides on or inside of plants and how they physiologically affect microorganisms are not completely understood. In international literature, fungicides are subdivided into two types: contact and systemic. Systemic fungicides are loosely defined as those that enter plant tissue and may or may not move from the site of uptake. A more descriptive nomenclature was offered by Couch (1995), in which he divided fungicides into three types based on their behavior in or on plants as either contact, penetrant, or systemic. Contact fungicides remain on the surface of tissues and are not translocated into tissue (i.e., nonsystemic or nonmobile). A systemic chemical by definition is capable of moving throughout a plant in both the phloem (i.e., symplast) and xylem (i.e., apoplast) from leaves to stems to roots and vice versa. The phosponates (i.e., fosetyl-aluminum and salts of phosphorous acid) are "true" systemics. Most other systemics are better referred to as penetrants, because they either remain localized inside tissue or primarily move upwards in xylem via the transpiration stream. The translocation behavior of fungicides used on turfgrasses is summarized in Table 6.1.

Acropetal penetrants move upwards in response to transpiration in the xylem from the point of uptake. Azoxystrobin, boscalid, flutolanil, fluxostrobin, mefenoxam/metalaxyl, propamocarb, thiophanate-methyl, and dimethylene inhibitor/sterol inhibitor (DMI/SI) fungicides are acropetal penetrants. Site absorption or localized penetrants are absorbed into tissues, but they do not move significantly beyond the site of uptake. Localized penetrants are taken up by leaves, and molecules of some compounds can move across leaves from upper to lower leaf surfaces (i.e., translaminar) and vice versa. Cyazofamid, Fluopicolide, iprodione, polyoxin D, pyraclostrobin, and vinclozolin are localized penetrants. Cyazofamid is mostly a contact protectant, but has limited ability to penetrate tissues (and move translaminar), and is capable of some curative activity. Trifloxystrobin is similar to a localized penetrant, but its behavior in plants is unique. After application, trifloxystrobin can be redistributed in a vapor phase to nearby treated or untreated leaves, but most molecules become fixed within the waxy portion of the cuticle. Equilibrium exists between trifloxystrobin molecules inside the leaf and those embedded in the cuticle, which is referred to as being *mesostemic* (Knauf-Beiter et al., 2000).

Contact fungicides provide a protective coating (i.e., protectants) to contacted tissues but have no activity on fungi that have gained entry into plant tissue. Because contact fungicides are subjected to removal as leaves grow by mowing and other environmental and physical forces (i.e., wash-off, volatization, photodecomposition, microbial breakdown, etc.), they tend to provide a relatively short period of protection compared to penetrants. The number of contact fungicides available for use on turfgrasses is dwindling and currently includes chloroneb, chlorothalonil, ethazol or etridiazole, fludioxonil, maneb, mancozeb, pentachloronitrabenzene (PCNB or quintozene), and thiram. Penetrants generally provide longer protection than contacts because some active ingredient enters tissues where it is protected from environmental forces. Penetrant fungicides have activity on the pathogen after infection has occurred and also provide protection on plant surfaces.

TABLE 6.1

Common Trade Names, Common Chemical Name, Chemical Groups, Plant Mobility, Fungicide Resistance Action Committee (FRAC) Code, and Risk for Resistance of Turfgrass Fungicides Used in the United States

Common Trade Names[a]	Common Name	Group Name[b]	Mobility[c]	FRAC Code[d]	Resistance Risk
Heritage®	Azoxystrobin	QoI	AP	11	High
Emerald®	Boscalid	Carboximide	AP	7	Medium
Terramec SP®	Chloroneb	Aromatic hydrocarbon	C	14	Low
Daconil®, others	Chlorothalonil	Chloronitrile	C	M3[c]	Low
Segway®	Cyazofamid	Cyanoimidazole	LP	21	Medium to high
Koban®, Terrazole®	Ethazole/Etridiazole	Triadiazole	C	14	Low to medium
Rubigan®	Fenarimol	DMI	AP	3	Medium
Stellar (premix with propamocarb)	Fluopicolide	Benzamide	LP	43	Unknown
Medallion®	Fludioxonil	Phenylpyrrole	C	12	Low to medium
Disarm®	Fluoxastrobin	QoI	AP	11	High
Chipco Signature®	Fosetyl-aluminum	Phosphonate	S	33	Low
ProStar®	Flutolanil	Carboximide	AP	7	Medium
Chipco 26 GT®	Iprodione	Dicarboximide	LP	2	Medium to high
Pentathalon®	Maneb	Dithiocarbamate	C	M3	Low
Dithane®, Fore Rainshield®	Mancozeb	Dithiocarbamate	C	M3	Low
Subdue MAXX®	Mefenoxam	Phenylamide	AP	4	High
Subdue®, Apron® Seed treatment	Metalaxyl	Phenylamide	AP	4	High
Tourney®	Metconazole	DMI	AP	3	Medium
Eagle®	Myclobutanil	DMI	AP	3	Medium

TABLE 6.1 (continued)

Common Trade Names, Common Chemical Name, Chemical Groups, Plant Mobility, Fungicide Resistance Action Committee (FRAC) Code, and Risk for Resistance of Turfgrass Fungicides Used in the United States

Common Trade Names[a]	Common Name	Group Name[b]	Mobility[c]	FRAC Code[d]	Resistance Risk
Velista®	Penthiopyrad	Carboximide	AP	7	Medium
Alude®, Magellan®, Vital®, others	Phosphite	Phosphonate	S	33	Low
Affirm®, Endorse®	Polyoxin D	Polyoxin	LP	19	Medium
Banol®	Propamocarb	Carbamate	LP	28	Low to medium
Banner MAXX®	Propiconazole	DMI	AP	3	Moderate
Insignia®	Pyraclostrobin	QoI	LP	11	High
PCNB®, Terraclor®, others	Quintozene	Aromatic hydrocarbon	C	14	Low
Torque®	Tebuconazole	DMI	AP	3	Moderate
3336 Plus®	Thiophanate-methyl	Benzimidazole	AP	1	High
Spotrete®	Thiram	Dithiocarbamate	C	M3	Medium
Bayleton®	Triadimefon	DMI	AP	3	Medium
Trinity®, Triton®	Triticonazole	DMI	AP	3	Medium
Compass®	Trifloxystrobin	QoI	LP	11	High
Curalan®	Vinclozolin	Dicarboximide	LP	2	Medium to high

[a] No discrimination is intended against trade names or generic products not mentioned.

[b] Group name is based on chemical structure and site of action from FRAC Code List: Fungicides sorted by mode of action, 2010.

[c] Contact (C), contact protectant fungicides are active only on leaf and sheath surfaces; acropetal penetrant (AP), fungicide is absorbed by contacted tissue, then moves upward (acropetal) in the xylem and can provide activity both on the outside and inside of plant tissues; local penetrant (LP), fungicide is absorbed into leaves and may move across the lamina (translaminar) providing activity both on the outside and inside of plant tissues; systemic (S), true systemic fungicides are absorbed and translocated throughout the plant via the phloem and xylem.

[d] From Fungicide Resistance Action Committee, 2010.

Chemical Groups and Modes of Action

Most fungicides are fungistatic. That is, most fungicides only prevent growth or development of fungi (i.e., fungistatic), but they do not actually kill them (Lyr, 1995). Several contact fungicides can kill fungal spores as they germinate, but even most contacts are fungistatic. Fungicides may be grouped either by chemistry or mode of action. Fungicides can be subdivided further based on mode of action into single site and multisite. Single-site fungicides may be referred to as being site specific, and they disrupt a single biochemical or physiological process controlled by a single gene in a sensitive fungus. Single-site fungicides are more likely to encounter resistant problems. *Monogenic* is the term used to describe the development of resistance that involves mutation of only a single gene. *Polygenic* is a term used to describe fungicides that require several mutations to occur in the shifting pathogen population in order for resistance to occur. Benzimidazole, dicarboximide, demethylation inhibitors (DMI/SI or SBI), phenylamides and QoI (strobilurins) fungicides are single-site specific (Copping and Hewitt, 1998; Brent and Holloman, 2007). Boscalid, flutolanil, propamocarb, and polyoxin D also are believed to be single-site specific (Brent and Hollomon, 2007). The aforementioned fungicides are considered at higher risk for resistance as described below; however, boscalid, flutolanil, polyoxin D, and propamocarb have had no reported resistance problems for any turfgrass disease. Resistance is highly unlikely with contact and other multisite fungicides (i.e., those that disrupt multiple biochemical processes) because many spontaneous mutations in the pathogen population would be required before resistance could develop (Brent and Hollomon, 2007).

Benzimidazoles such as thiophanate-methyl and benomyl have a high affinity for tubulin proteins and prevent spindle fiber development during mitosis so cell division does not occur (Copping and Hewitt, 1998). Iprodione and vinclozolin are grouped as dicarboximides. Dicarboximides interfere with signals that trigger phosphorylation at hyphal membrane receptors, thus inhibiting further hyphal formation (Leroux et al., 2002). Metalaxyl and mefenoxam are grouped as phenylamides. Mefenoxam is an isomer (i.e., a mirror image) of the metalaxyl molecule, which has several-fold greater activity than metalaxyl but the same mode of action. Phenylamides interfere with an RNA polymerase and inhibit the production of ribosomal RNA (Copping and Hewitt, 1998; Brent and Hollomon, 2007). The QoI fungicides provide disease control by interfering with respiration at the QoI site of cytochrome bc in mitochondria in sensitive fungi (Kim et al., 2003; Brent and Hollomon, 2007). That is, QoI fungicides disrupt respiration and deplete the production of the key energy compound (ATP) needed by living cells to grow and function. The QoI fungicides also are known by their chemical strobilurin grouping. Strobilurin was derived from a mushroom fungus (*Strobilurus tenacellus*) and stabilized to prevent rapid deterioration in the environment. Benzimidazoles, dicarboximides, phenylamides, and QoI's are single site, monogenic and at high risk of developing resistance (Fungicide Resistance Action Committee [FRAC], 2010).

The demethylation or sterol-inhibiting (DMI/SI) fungicides interrupt the production of ergosterol in sensitive fungi by blocking a single demethylation reaction (Koller and Scheinpflug, 1987; Kuck et al., 1995). Ergosterol is a sterol (a type of lipid) used in minute quantities by fungi to produce membranes. In the absence of this sterol, membrane form and function are impaired and the fungus cannot grow. The DMIs are also known as sterol biosynthesis inhibitors (SBIs) or simply sterol inhibitors (SIs). Most DMI fungicides are chemically classified as triazoles (e.g., metconazole, mycobutanil, propiconazole, tebuconazole, triadimefon, tritconazole). Fenarimol is the exception and is chemically grouped as a pyrimidine. The DMI fungicides are single site, polygenic, and are considered at medium risk for developing resistance.

Boscalid, flutolanil, and penthiopyrad inhibit a respiratory enzyme, and their risk for resistance is medium. Polyoxin D is an antibiotic produced by the bacterium *Streptomyces cacaoi* var. *asoensis*. This antibiotic inhibits cell wall synthesis in sensitive fungi by interfering with chitin production. Polyoxin D is considered at medium risk for resistance (FRAC, 2010).

Cyazofamid, fluopicolide, propamocarb, and phosphonates are used primarily on turf to target diseases incited by *Pythium* species. Fosetyl-aluminium and phosphorous acid salts (i.e., phosphites) are grouped as phosphonates. Phosphonates degrade to phosphonic acid in water, soil, and plants. It is believed that phosphonic acid disrupts phospholipid metabolism, thus inhibiting growth of sensitive fungi (Copping and Hewitt, 1998). Phosphonates may trigger phytoalexins and other inherent plant defense mechanisms and are at low risk for resistance (Schwinn and Staub, 1995; FRAC, 2010). It is proposed that propamocarb interferes with lipids and cell membrane function and has low to medium risk for resistance (FRAC, 2010). Cyazofamid inhibits respiration in the cytochrome bc 1 site in mitochondrial membranes of sensitive fungi. The risk for resistance with cyazofamid is medium to high (FRAC, 2010). Fluopicolide interferes with mitosis and cell division and its risk for resistance is unknown (FRAC, 2010).

The modes of action of contact fungicides generally are imperfectly understood or unknown. Contact fungicides normally interfere with the production of several enzymes and thus several metabolic processes in susceptible fungi. Resistance is known to occur with contact fungicides in some crops, but there currently are no known resistance problems with contact fungicides that target turfgrass pathogens. Anilazine, cadmium, and mercury-based fungicides are contact fungicides with known resistance to the dollar spot fungus (*Sclerotinia homoeocarpa*), but they are no longer commercially available. Chlorothalonil, mancozeb, maneb, thiram are multisite and have a low risk for resistance (FRAC, 2010). It was proposed that chloroneb, etridiazole, and quintozene interfere with lipids and membrane function and have a low to medium risk for resistance (FRAC, 2010). Fludioxonil interrupts signal transduction, has a mode of action similar to dicarboximides, and has a low to medium risk of resistance (Leroux et al., 2002; FRAC, 2010).

Evolution in the Use of Turf Fungicides

Bordeaux mixture (a combination of copper sulfate and limestone) was the first fungicide used on turf and targeted brown patch (Monteith and Dahl, 1932). The development of turf fungicides thereafter focused on organic and inorganic forms of mercury and quintozene in the 1930s and cadmium and cycloheximide in the 1950s (Smith et al., 1989). Both Bordeaux mixture and quintozene remain available today, but restrictions are placed on the use of quintozene and Bordeaux mixture is no longer used in the U.S. turf industry. The dithiocarbamates (i.e., maneb, mancozeb, and thiram) were developed during the 1940s due to shortages in mercury and cadmium needed during World War II (Smith et al., 1989). The dithiocarbamates were used for managing Bipolaris and Drechslera leaf spots, red thread, rusts, and brown patch. Chloroneb and ethazole were introduced in the late 1960s and were the first fungicides to offer some persistent level of Pythium blight control. Chlorothalonil was registered for use on turf in 1966 and offered improved and longer periods of dollar spot, Drechslera and Bipolaris leaf spot, red thread, brown patch, and gray leaf spot control compared to the dithiocarbamates. Chlorothalonil remains one of the most commonly used turfgrass fungicides and is valued for its use in resistance management programs. Similarly, mancozeb remains widely used for brown patch and gray leaf spot control and has some applications in dollar spot resistance management programs.

Most fungicides available for use on turf prior to 1960 were contact protectants. A new era in fungicide use began with the introduction of the benzimidazoles in the late 1960s (Delp, 1995). Benzimidazoles include benomyl, thiophanate ethyl, and thiophanate methyl, and offered more effective control of dollar spot, brown patch, and snow molds at lower use rates, when compared to contact protectants available at that time. Dicarboximide and DMI fungicides became available in the late 1970s and early 1980s; however, new DMIs appeared in the market as late as 2007 (metconazole and triticonazole) and 2010 (tebuconazole). Dicarboximides target Drechslera and Bipolaris leaf spots, dollar spot, red thread, brown patch, snow molds, and others, and generally offer improved residual control of the aforementioned diseases compared to contact protectants. The DMIs are effective against dollar spot, but unlike the benzimidazoles or dicarboximides, they provide a greater level of consistency against rusts, stripe smut, and root diseases such as take-all patch, spring dead spot, necrotic ring spot, and summer patch. The DMIs suppress or control numerous other diseases including anthracnose, brown patch, gray leaf spot, snow molds, and others.

Metalaxyl was introduced in 1984 and surpassed the residual effectiveness of chloroneb and ethazole for Pythium blight control. Flutolanil also entered the market in the mid-1980s and provided longer residual brown patch control compared to contact protectants, benzimidazoles, and dicarboximides. Flutolanil was the first fungicide labeled for suppression of some types of fairy ring. Propamocarb and fosetyl-aluminum were developed in the late 1980s to target Pythium diseases and to date have not had the resistance problems associated with metalaxyl or its isomer mefenoxam. Nonaluminum phosphorous acid products made from mono- and dipotassium salts of phosphorous acid were registered for use against Pythium blight in 1997 but did not achieve widespread use until the late 2000s. The phosphorous acid fungicides and fertilizers provide equivalent Pythium blight control compared to fosetyl-aluminum and are less expensive (Cook et al., 2009; Ervin et al., 2009). Cyazofamid is the most recent (2007) fungicide to be marketed for control of Pythium blight.

The first QoI fungicide to enter the turf market was azoxystrobin in 1997, followed by trifloxystrobin (1999), pyraclostrobin (2004), and fluoxastrobin (2005). The QoI fungicides are noted for their low use rates and improved residual control of anthracnose, brown patch, gray leaf spot, red thread, snow molds, and other diseases. Azoxystrobin, fluoxastrobin, and pyraclostrobin also target several important root diseases including take-all patch and summer patch, as well as some types of fairy ring and Pythium-incited root disorders.

Fungicides of Microbial Origin

Use of natural products as fungicides is not new in the turfgrass market. Cycloheximide, sold under the trade name of Acti-dione®, was used in the 1960s and 1970s on turf to control dollar spot, leaf spot, and other diseases. Cycloheximide was a by-product in the production of the antibiotic streptomycin. Streptomycin was derived from the bacterium *Streptomyces griseus*. Cycloheximide was expensive to produce, could be phytotoxic to turf, had potential health problem considerations, and its registration was withdrawn in the United States in 1981.

Fungicides of microbial origin can be produced by fermentation (i.e., growing large quantities of a desired microbe in an aerated vat) or the antifungal properties of the microbe can be identified and synthesized in the laboratory. Regardless of how products are produced, they must undergo the same U.S. Environmental Protection Agency (U.S. EPA) registration rigor that is required for all other pesticides. The QoI (strobilurins), polyoxin D, and fludioxinil are commonly used fungicides of microbial origin.

Azoxystrobin was the first fungicide to be marketed from a new class of chemistry referred to as QoI fungicides. The origin of the first identified compound (i.e., strobilurin A) was a fungus from the mushroom family named *Strobilurus tenacellus*. The strobilurin-based compound was stabilized by adding molecules to the structure to ensure that it was not rapidly broken down by solar radiation. Because the original compound was slightly changed in the laboratory, azoxystrobin is best described as a synthesized analog of a natural substance. Other QoI fungicides include fluoxastrobin, pyraclostrobin, and trifloxystrobin.

Fludioxonil was registered in 1998 for use in controlling *Rhizoctonia* diseases in turf. Fludioxonil is an analog of a natural product called pyrollnitrin. Pyrollnitrin is produced in nature by the bacterium *Pseudomonas pyrocinia*. As was the case with azoxystrobin, pyrollnitrin had to be stabilized to protect it from rapid degradation by bright sunlight. Fludioxonil is a contact fungicide. As such, its residual activity is short lived and may need to be tank-mixed with a penetrant to extend its effectiveness. Fludioxonil is the first contact fungicide to enter the turf market in over 30 years.

The *polyoxins* are a class of antifungal compounds produced by fermentation of *Streptomyces cacaoi* var. *asoensis*. Polyoxin D was first registered for use on turf in 1997. Polyoxin D is a localized penetrant that was shown to be extremely effective in controlling brown patch. It also controls anthracnose and some fairy ring fungi. Cell walls of most fungi contain chitin. Chitinous inhibitor polyoxin D prevents sensitive fungi from producing cell walls.

Hence, turfgrass disease management strategies were expanded to include not only the direct application of biological agents but also development of microbial-based analogs of naturally occurring, antifungal compounds. These new products are generally applied at much lower rates than conventional fungicides and therefore represent a step forward in environmental stewardship.

Summary of Key Points

- The key to an effective fungicide program is knowing the disease history of the golf course and being able to rapidly and accurately diagnose diseases.
- It is best to apply fungicides preventively when targeting chronically severe diseases. Curative applications may require higher rates and shorter spray intervals to suppress a disease.
- Fungicides are divided into three groups based on mobility in plant tissues. Contact fungicides are nonmobile and remain on the outside of contacted tissue. Penetrant fungicides enter plant tissue and may be localized or move acropetally (i.e., upwards) from the point of contact. Systemics move from leaves to stems to roots and vice versa. Only phosphonates are true systemics.

- Most fungicides prevent fungal growth (i.e., fungistatic) rather than killing them.

- Fungicides can be grouped by mode of action or by chemistry.

- Single-site fungicides disrupt only a single biochemical reaction controlled by a single gene in a sensitive fungus. Single-site fungicides are most likely to encounter resistance problems. Multisite fungicides disrupt two or more biochemical reactions and resistance is unlikely.

- Modes of action of penetrants include mitotic or cell division inhibitors, respiration inhibitors, ergosterol synthesis inhibitors, and others. The exact mode of action of contact fungicides is unknown, but it is believed that they interfere with the production of several enzymes in the metabolic processes of sensitive fungi.

- Common chemical classes of fungicides include benzimidazole, carboximide, dicarboximide, stobilurin or QoI, demethylation or sterol synthesis inhibitor (DMI/SI), phosphonate, and others.

- Bordeaux mixture (copper sulfate and lime) was the first fungicide used on turf in the 1920s. Mercury fungicides were developed in the 1930s, dithiocarbamates in the 1940s, and cadmium in the 1950s. Chlorothalonil was licensed in 1966. Benzimidazoles were the first penetrants introduced in the late 1960s. The DMI/SI and dicarboximide fungicides were first used about 1980 and the first QoI's entered the market in 1997.

- Some fungicides were derived from natural compounds produced by fungi and bacteria and include strobilurin or QoI fungicides, fludioxonil, and polyoxin D.

Fungicide-Induced Resistance

The indiscriminate use or employment of numerous fungicide applications led to problems with fungicide-induced resistance. Resistance occurs when a fungicide totally fails to control the target pathogen, but reduced efficacy is a prelude to resistance for some fungicides such as DMIs. The benzimidazoles, dicarboximides, QoI's, and phenylamides developed resistance problems within a few years of commercial introduction (Brent and Hollomon, 2007). The DMI fungicides developed resistance more gradually. The primary mechanisms of fungicide-induced resistance are a decrease in the affinity or binding of a fungicide at the site of action within a fungal cell and a change in the amount of fungicide taken up, resulting in less active ingredient reaching the site of action (Georgopoulos, 1986). Mutations leading to resistance

can result in a reduction in the amount of fungicide that penetrates mycelium, an increase in the amount of fungicide actively excreted by mycelium,
or a change in an enzyme that affects binding of a fungicide to a target site
(Leroux et al., 2002).

The development of turfgrass pathogens resistant to fungicides is well documented. High-risk turfgrass diseases and pathogens for resistance include
dollar spot (*Sclerotinia homoeocarpa*), Pythium blight (*Pythium aphanidermatum*), gray leaf spot (*Pyricularia grisea*), and anthracnose (*Colletotrichum cereale*).
There are no scientific criteria that determine or predict the risk of a pathogen
to develop resistance to a fungicide. In general, pathogens that undergo many
short disease cycles and sexual recombination per season and are capable
of abundant sporulation and widespread dispersal of spores over space and
time are at higher risk (Brent and Hollomon, 2007; FRAC, 2005). Given the ability of a pathogen to produce large populations each season results in greater
genetic diversity and therefore a greater likelihood of a resistant strain.

Resistant biotypes of the dollar spot fungus first appeared in the late
1960s and early 1970s as a result of repeated usage of anilizine, cadmium,
mercury-based fungicides, and benzimidazoles (i.e., benomyl and thiophanate) (Cole et al., 1968; Nicholson et al., 1971; Smith et al., 1989, Vargas, 2005).
Biotypes of the dollar spot fungus resistant to iprodione (Detweiler et al.,
1983) and DMI fungicides (Golembiewski et al., 1995; Hsiang et al., 2007)
were first reported in the 1980s and 1990s. Formal reports of *S. homoeocarpa*
resistance toward DMI and dicarboximide fungicides, however, have been
sporadic, and reduced sensitivity rather than resistance appears to be more
commonplace with these fungicides (Jo et al., 2006; Koch et al., 2009). *Pythium
aphanidermatum* biotypes resistant to metalaxyl from turfgrasses were first
documented by Sanders in 1984. The anthracnose pathogen was reported
to develop resistance to benzimidazoles (Shane and Danneberger, 1989) and
QoI (Wong et al., 2007) fungicides. Wong and Midland (2007) also observed
reduced sensitivity of the anthracnose pathogen (*C. cereale*) isolates to DMI
fungicides, indicating that resistance may soon appear. Vincelli and Dixon
(2002) were first to document the development of resistance of *P. grisea* to QoI
fungicides. The potential for thiophanate resistance in *P. grisea* was observed
in the field but has not been formally documented (Dernoeden, unpublished). *Microdochium nivale* (aka Microdochium or Fusarium patch) biotypes
resistant to iprodione were reported in 1982 (Chastagner and Vassay, 1982).
Mann (2002) describes reduced fungicidal sensitivity to large percentages of
M. nivale isolates to carbendazin (benzimidazole), chlorothalonil, and iprodione in the United Kingdom.

There are two types of resistance: cross-resistance and multiresistance.
Cross-resistance occurs when a pathogen is not controlled by fungicides having the same mode of action. For example, *S. homoeocarpa* biotypes resistant to
one DMI fungicide theoretically will be resistant to all other DMI fungicides.
Similarly, *S. homoeocarpa* biotypes resistant to iprodione (a dicarboximide)
will exhibit resistance to other dicarboximides such as vinclozolin. Multiple

resistance is a phenomenon that occurs when a pathogen develops resistance to fungicides with different modes of action. For example, *S. homoeocarpa* isolates were shown to be resistant to benomyl, iprodione, propiconazole, and thiophanate (Detweiler et al., 1983; Bishop et al., 2008). There also are multiple resistant strains of *C. cereale* to benzimidazole, DMIs, and QoI fungicides (Vargas, 2005; Wong et al., 2007). Thiophanate-methyl resistance occurs rapidly in turf. In a study in which 192 isolates were collected from 55 Ohio golf courses, it was revealed that 62% of the golf courses surveyed had *S. homoeocarpa* resistance to thiophanate-methyl (Jo et al., 2006).

The buildup of resistant biotypes of fungi occurs in response to a selection process that eventually enables a small but naturally occurring subpopulation of resistant biotypes or mutants to predominate in the fungicide-treated turfgrass microenvironment. Spontaneous mutations continually occur and it is assumed that the mutant gene already exists prior to any fungicide use (Brent, 1995). The survival and spread of these rare resistant mutants in response to fungicides can be rapid resulting from a single gene mutation (monogenic) or gradually from multigene mutations (polygenic) (Brent, 1995). Thus, the two biological steps responsible for resistance include the existence of a naturally occurring yet very small population of resistant strains that arise from spontaneous mutation and the selection of resistant strains by continuous exposure to fungicides having the same biochemical mode of action, which diminishes the population of sensitive strains through selective pressure (Jung and Jo, 2008). The development of resistance can be rapid in single-site fungicides because mutation of only a single gene is required. For example, as little as two applications of thiophanate-methyl can generate resistant *S. homoeocarpa* mutants (Jo et al., 2008). Hence, resistance occurs when fungicides having the same modes of action are used frequently and essentially diminish populations of the sensitive wild type (i.e., original sensitive strain) in the ecosystem, while allowing the more genetically fit resistant strains to dominate. Resistance to monogenic fungicides often appears rapidly and usually within 2 to 5 years of their introduction. For example, resistance strains of *P. aphanidermatum* from turf were documented in the year that metalaxyl was introduced into the turf market (Sanders, 1984). Azoxystrobin was introduced in 1997 and resistance to *Pyricularia grisea* was observed in 2000 (Vincelli and Dixon, 2002).

In the case of DMI fungicides, the development of resistance is gradual and may take 10 or more years. The mode of action of DMIs affects a single site in sensitive fungi, and the demethylation step that it uncouples is regulated by multiple genes. This is what is meant by *polygenic*. In the polygenic process, resistance occurs when different genes affecting several mechanisms are altered, and the more genes that mutate, the greater is the level of resistance. In the case of DMIs, the gradual shift toward resistance occurs in steps (i.e., mutations of several genes taking place over time) and will first be noted as reduced sensitivity (i.e., shorter duration of control or less control) in the field. Thus, for DMIs, resistance builds slowly in steps before reduced effectiveness is observed. For example, DMI fungicides were introduced into the turfgrass market in 1979, but field failures in the control of dollar spot were not recognized until 1991 (Golembiewski et al., 1995). The repeated use of DMI fungicides resulted in a gradual evolution toward the loss of sensitivity to naturally occurring strains (i.e., wild-type or sensitive strain) of *S. homoeocarpa*. That is, there is a population shift from sensitive strains to those that are less sensitive. The less-sensitive strains reproduce faster, passing on their genes, and dominate in response to the selection pressure imposed by the continuous use of fungicides of the same mode of action. Survival fitness refers to resistant strains that compete better with sensitive strains for survival. Fit populations are more successful in passing their genes along to a greater number of individuals and eventually predominate in an ecological niche (Nelson, 1979). Reduced *S. homoeocarpa* sensitivity to DMI fungicides is more commonplace than resistance (i.e., complete failure) (Jo et al. 2006; Koch et al., 2009). Strains that have a high level of fitness are likely to persist indefinitely, such as strains of *S. homoeocarpa* resistant to benzimidazoles. Strains of *S. homoeocarpa* resistant to DMI and dicarboximide fungicides have a low level of fitness (Jo et al., 2006). Niches where these strains exist would be expected to survive only 2 to 3 years after use of all DMI fungicides was suspended (Couch, 2003). Hence, DMIs and dicarboximides could theoretically be rotated back into a dollar spot control program after several years of suspended usage but obviously should be applied only once or twice annually thereafter when targeting the once-resistant population.

Resistance Management

The general rules of thumb in resistant management programs are to minimize the use of high-risk compounds, and rotate or tank-mix fungicides of varying modes of action (Dekker, 1986; Brent, 1995). Most resistant problems in turfgrasses are associated with anthracnose (azoxystrobin and thiophanate), dollar spot (most penetrants), gray leaf spot (azoxystrobin and possibly thiophanate), and Pythium blight (metalaxyl and mefenoxam). Hence, a simple rule is to minimize the use of or, in the case of confirmed resistance, to avoid the aforementioned fungicides and their biochemical mode of action relatives at times when high-risk diseases are active. The Fungicide Resistance Action Committee (FRAC) is an international industry-based organization with a steering committee and several working groups. A code system was developed by FRAC to assist end users in choosing compounds having different chemical modes of action. Each fungicide is assigned a number, and the user simply chooses fungicides that have different FRAC codes when tank-mixing or rotating fungicides. These codes and the relative risk of each fungicide for resistance are shown in Table 6.1.

Imposing a combination of resistant strategies uniformly is necessary to achieve their full biological effect (Brent, 1995). Those strategies include tank-mix or rotate fungicides with diverse modes of action, restrict the number of applications of high-risk fungicides per season, use manufacturers' recommended dose, employ integrated pest management techniques, and apply mixtures of known synergists. By alternating or tank-mixing fungicides with different modes of action, the companion (often a contact) will inhibit the growth of resistant biotypes and will dilute the selection pressure exerted by the high-risk partner (Brent, 1995; Brent and Hollomon, 2007). The resistant population, however, may continue to build when alternating or tank-mixing because the pressure exerted by the risky fungicide remains constant. Therefore, it is essential to interrupt the selection process by using nonrisky fungicides for long periods between use of risky fungicides. A general recommendation therefore is to schedule at least two multisite fungicides (i.e., low-risk fungicides) between the use a of single-site fungicide (Dekker, 1986). The single-site fungicide used the second time should be of a different mode of action than the first single-site fungicide. Similarly, tank-mixing a contact or two penetrants with different modes of action also would be expected to delay or prevent resistance.

An important but commonly overlooked approach to delaying or avoiding resistance is to limit the use of single-site fungicides to one or a few applications annually when targeting high-risk diseases such as dollar spot, gray leaf spot, anthracnose, or Pythium blight. This approach can slow the selection of resistant biotypes, but not always. A substantial break in the time period between uses of high-risk compounds when the pathogen is multiplying can allow for the resurgence of sensitive forms (Brent, 1995). For several years golf course superintendents used reduced rates under conditions of low disease pressure, often with favorable results. Although FRAC considers the use of reduced rates as conducive to enhancing the development of resistance, the effects of using lower than label rates on resistance remain unclear. The consensus view today is that the risk of resistance increases for high-risk monogenic fungicides just as the level of control increases with dose. According to Brent and Hollomon (2007), the degree of disease control is proportional to the selection pressure on monogenic-resistant mutants. Conversely, use of very low doses would be expected to increase the risk of polygenic resistance. Thus, in high-risk dollar spot situations it would be prudent to avoid lower than label rates of DMI fungicides.

Integrated pest management (IPM) is also a major strategy for avoiding or delaying resistance. Integrated pest management approaches that help to minimize the potential for resistance are plant disease-resistant cultivars and species, impose sound cultural practices, and use biological agents and other nonchemical methods of disease control. Unfortunately, biological agents available today generally fail to provide commercially acceptable levels of consistent disease control. One of the most effective biological agents is *Pseudomonas aureofaciens* strain Tx-1, which targets dollar spot (Powell and

Vargas, 2000). Field research showed that frequent applications of Tx-1 can provide on average 33% to 37% dollar spot control (Davis and Dernoeden, 2001; Hardebeck et al., 2004). The use of Tx-1 in combination with fungicides sometimes extends the residual overall effectiveness of some fungicides and thus may be useful in IPM and resistance management programs. Chemicals that activate plant defense mechanisms such as acibenzolar-S-methyl show promise for targeting dollar spot and using in IPM programs (Lee et al., 2003). Other biological agents, mineral or horticultural oils, and silicates may be useful in IPM programs but require more field study to learn how to best exploit their usefulness.

Strict adherence to resistant management principles does not eliminate the possibility of fungicide-resistant populations developing. Vargas (2005) observed that golf course superintendents followed the aforementioned resistance management procedures and cross- and multiresistance dollar spot problems have arisen on golf courses. Dekker (1986) noted that resistance can occur when fungicides of different modes of action are mixed or rotated, but in most cases following these guidelines would be expected to delay the buildup of a resistant population. Vargas (2005) proposed that superintendents use fungicides having the same mode of action continuously until resistance appears and then switch to fungicides having a different mode of action. To be useful, however, this strategy would depend on the registration of fungicides having new modes of action that are unavailable today.

Summary of Key Points

- Resistance occurs when a fungicide totally fails to control the target disease and happens when fungicides of the same mode of action are used excessively.
- The mechanisms for resistance are a decrease in the amount of fungicide that is taken up by the pathogen or a reduction in binding of the chemical to the site of action in the fungus.
- Resistance develops in response to a selection process of naturally occurring resistance biotypes and can occur within 2 (e.g., benzimidazole and QoI) to 10 years (DMI/SI) after a fungicide is first introduced.
- Cases of resistance involve single-site fungicides and include dollar spot (benzimidazole, dicarboximide, and DMI/SI fungicides), Pythium blight (metalaxyl/mefenoxam), gray leaf spot (QoI and possibly benzimidazole fungicides), and anthracnose (benzimidazole and QoI fungicides).
- Cross-resistance means that a pathogen is not controlled by fungicides having the same mode of action. Multiresistance occurs when a pathogen is not controlled by fungicides having two or more different modes of action. Multiresistance is most common with dollar spot.

- Resistance is gradual with DMI/SI fungicides, which exhibit reduced sensitivity before resistance. Reduced sensitivity refers to a shorter period of effective control and should not be confused with resistance (i.e., total failure). Reduced sensitivity is more common than resistance to DMI/SI fungicides.

- Resistance management involves imposing a combination of strategies to include tank-mixing or rotating fungicides with diverse modes of action, restricting the number of applications of high-risk fungicides per year, using label rates, employing sound cultural practices and improving the growing environment where possible, and employing integrated pest management techniques to include biological agents and alternative chemicals when available.

- Strict adherence to the aforementioned resistance management techniques does not eliminate the possibility of resistance but can delay the probability.

Other Nontarget Effects of Fungicides

In addition to resistance, there are other potential negative nontarget effects of fungicides as follows: enhanced microbial degradation of the active ingredient, resulting in reduced residual effectiveness; fungicide-induced disease resurgence; encouragement of nontarget diseases; turf phytotoxicity; encouragement of blue-green algae (i.e., cyanobacteria); and toxicity to earthworms. Conversely, there are some nontarget beneficial responses associated with fungicides, which include turfgrass quality enhancement in the absence of disease.

A phenomenon associated with fungicides, which may be confused with resistance, is reduced residual effectiveness due to enhancement of microbial populations that degrade fungicides. Some microbial populations can build up in response to the continuous use of certain fungicides (Sigler et al., 2003). As a result of a fungicide being more rapidly degraded, the residual effectiveness becomes less and less over time. The majority of fungicides applied to turfgrasses are captured in the canopy; however, most of the chemical is degraded in the thatch or soil (Sigler et al., 2003). Mocioni et al. (2001) provided evidence that the reduced effectiveness of iprodione for dollar spot control on golf courses in Italy was related to enhanced degradation by soil microorganisms. The overall significance of microbial-based enhanced reduced effectiveness of fungicides applied to turfgrasses is unknown. As previously noted the loss of residual effectiveness may be an indicator that resistant biotypes are building in the turf. Reduced residual effectiveness can be caused by a multitude of factors including improper application of fungicides, use of nozzles and spray volumes that do not provide adequate canopy coverage, planting of highly susceptible cultivars, and other management decisions (i.e., time between application and mowing or rain) and environmental influences.

Some diseases may recur more rapidly and severely in turfs previously treated with fungicides, when compared to adjacent untreated areas (e.g., treated fairways versus untreated roughs). The phenomenon is called *fungicide-enhanced resurgence*, and dollar spot is among the most common diseases to exhibit this phenomenon (Couch, 2003). Resurgence of brown patch was also observed (Bigelow et al., 2003). The mechanism responsible for fungicide-enhanced resurgence is unknown. Disease resurgence is attributed to a fungicide reducing populations of beneficial microorganisms, which naturally antagonize and keep disease-causing fungi in abeyance (Smiley et al., 2005). It is conceivable that non-fungicide-treated turf that is blighted, yet able to recover prior to the time the disease recurs, is better prepared to resist future infections due to natural defense systems in plants having been activated by the initial infection.

Fungicides applied to control one disease may encourage other diseases. Triadimefon may enhance Drechslera leaf spot; iprodione can increase yellow tuft; azoxystrobin and flutolanil may enhance dollar spot; chlorothalonil can increase summer patch and stripe smut; propiconazole can increase injury from Microdochium patch; thiophanate-methyl can increase Waitea patch; and azoxystrobin may enhance Pythium blight (Dernoeden and Jackson, 1980; Dernoeden and McIntosh, 1991a; Reicher and Throssell, 1997; Settle et al., 2001; Smiley et al., 2005). The mechanism responsible for the encouragement of nontarget diseases is unknown but may be attributed to offsetting the balance between antagonistic and pathogenic microorganisms in the ecosystem. It is important to note that using a selected fungicide will not invariably result in an increase in a nontarget disease. These problems are sporadic and enhancement of nontarget diseases cannot occur unless environmental conditions are conducive for the disease to occur naturally.

The phytotoxicity that accompanies the usage of some fungicides generally is not severe. Most phytotoxicity problems occur when fungicides are applied to closely mown golf greens during periods of high-temperature stress. Fungicides formulated as emulsifiable concentrates are most likely to cause a foliar burn when applied during hot weather, but most manufacturers have developed improved formulations (such as microemulsion, dry flowable, soluble powders, and liquids) with little burn potential. Chloroneb, etridazole, PCNB, fosteyl-aluminium, and phosphorous acid products can burn foliage of golf green turfs when applied during periods of high temperature and humidity (Figure 6.1). Granular forms of PCNB that wash after a rainstorm and accumulate in low areas can be quite phytotoxic. Flowable formulations of PCNB tend to discolor more than granular formulations (Landschoot et al., 2001). Spring applications of PCNB for snow mold control can yellow or cause a light brown foliar discoloration

FIGURE 6.1
Chloroneb and ethazole can burn turf when applied in warm and humid weather.

if temperatures rise unexpectedly (Landschoot et al., 2001). Repeated applications of high label rates of DMI fungicides can elicit a blue-green or reddish-brown color, widen leaf blades, suppress foliar growth, and in some cases injure creeping bentgrass maintained at golf green height. Triticonazole and triadimefon (both DMI) can injure annual bluegrass. Tank-mixes of phosphate fungicides and fertilizers with the plant growth regulator trinexapac-ethyl can cause yellowing in bentgrass, especially when applied during periods of heat stress.

When used repeatedly, certain fungicides slightly increase thatch accumulation, but these increases are agronomically insignificant. Benzimidazole fungicides (benomyl and thiophanate) and sulfur-containing fungicides (mancozeb, maneb, and thiram) can cause thatch to accumulate by acidifying soil. The effect of acidifying fungicides is indirect—that is, acidification is presumed to inhibit the thatch decomposition capacity of microorganisms. The primary mechanism by which fungicides enhance thatch, however, is mostly an indirect effect of promoting stem, stolon, and rhizome survival (Dernoeden et al., 1990). Benzimidazoles and chlorothalonil can be toxic to or decrease the activity of earthworms (Potter et al., 1990; Reicher and Throssell, 1997). Earthworms help reduce thatch by mixing soil with organic matter, and they improve soil aeration and other soil properties.

Some DMI fungicides can promote the growth of blue-green algae (aka cyanobacteria) on golf greens. The mechanism for this phenomenon is unknown. Open canopies or less dense turf favor growth of blue-green algae in part by improving sunlight penetration to the thatch or soil surface. It is possible that the growth regulator effects of DMI fungicides may cause leaves to grow more upright, thus promoting sunlight penetration to the thatch layer. These fungicides also can slow turf growth and thus give blue-green algae a competitive advantage. Conversely, chlorothalonil, fosetyl-aluminum, and mancozeb suppress blue-green algae growth on golf greens.

It should be noted that the harmful side effects just described were mostly isolated events or occurred only after repeated use of one chemical class of a fungicide over the course of several years. There is little evidence that frequent applications of fungicides to turfgrasses cause serious nontarget effects (Reicher and Throssell, 1997). It should be noted that fungicides can improve the quality of turf in the absence of disease and can even mitigate mechanical damage (Dernoeden and Fu, 2008; Dernoeden and McIntosh, 1991a, 1991b) (Figures 6.2 and 6.3). Furthermore, fungicides were shown

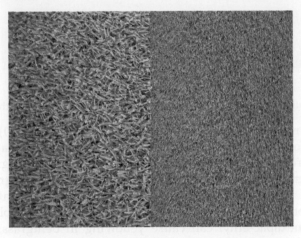

FIGURE 6.2
Fungicides can mitigate mechanical injury. Left side treated with a tank-mix of fosetyl-aluminum and mancozeb contrasted with the control.

FIGURE 6.3
Pigments in some fungicides mask chlorosis and minor mechanical injury.

to have no impact on the *Acremonium* endophyte in perennial ryegrass or population densities of plant parasitic nematodes (Dernoeden et al., 1990). Arbuscular mycorrhizal levels in roots of annual bluegrass (*Poa annua*) were shown to be unaffected by applications of chlorothalonil, fenarimol, or iprodione to golf greens (Bary et al., 2005). As a general rule, nontarget effects are sporadic and they do not occur in most situations. It is obvious that scientists do not understand the mechanisms that cause both deleterious and beneficial nontarget effects of fungicides.

Environmental Fate of Fungicides

Environmental fate information on most fungicides is available in the form of fact sheets from the U.S. Environmental Protection Agency (EPA) (2008, 2010) and from the European Food Safety Authority (2010). The environmental fate of fungicides applied to turfgrasses was reviewed by Sigler et al. (2000) and Magri and Haith (2009). The following discussion contains the main points from those reviews. Loss of a fungicide from the target site begins before the compound leaves the sprayer and ultimately is impacted by several biotic and abiotic factors. Microbial decay generally is the primary means by which pesticides are degraded in the turfgrass environment (Magri and Haith, 2009). There are exceptions such as azoxystrobin, fludioxnil, and iprodione, which are photodegraded (Sigler et al. 2000; EPA, 2008, 2010). Some amount of fungicide will adhere to plastic and rubber components in the sprayer. Upon leaving the sprayer, the compound may be lost to drift or volatization. Once the fungicide alights on the turfgrass canopy, it may undergo more volatilization, photodecomposition, or microbial degradation, or it may be washed off by rain or irrigation. The majority of fungicides applied to turfgrasses initially are intercepted in the canopy or by thatch. Fungicides translocated into tissues may be metabolized and broken down by plant cells. Sigler et al. (2003) found that metalaxyl, triadimefon, and vinclozolin applied to turf are tightly sorbed by leaf tissue and are not degraded by microbes on leaf surfaces. Therefore, most of the chemical ultimately is degraded in soil and especially in thatch. The rate at which a fungicide is degraded by microbes in soil and thatch is dependent on moisture, temperature, soil texture, soil pH, and other factors. The higher the clay and organic matter content of soils, the greater is the microbial activity and thus there is more rapid dissipation. Rate of degradation is reduced in wet soils when low oxygen levels limit the aerobic activity of microbes. Similarly, microbial degradation is slowed when temperatures are above or below the optimum for growth of microbes. Soil pH is important because many fungicides undergo chemical alterations depending on pH. For example, some compounds will bind to a greater or lesser extent in an acid versus an alkaline soil. The solubility of fungicides in water, as well as their sorption and desorption characteristics, greatly impact their ability to bind to plant surfaces, soil, and organic matter. The sorption characteristics of a fungicide determine its tendency to leach in soil or runoff. Half-life in the persistence of a pesticide is expressed as the time required for 50% of the active ingredient to degrade. Half-life of an active ingredient found in soil is measured in a laboratory under controlled conditions and should be considered an estimate and not an absolute value (Kenna and Snow, 2000). Dissipation and retention characteristics of several turfgrass fungicides are listed by Sigler et al. (2000). Dissipation to half-life of fungicides can be variable ranging from 1 to 2 months for chlorothalonil to 1 to 2 hours for fosetyl-aluminum (Sigler et al., 2002). Magri and Haith (2009) state that the half-life of pesticides is much shorter in a turfgrass environment because the thatch layer can contain 10 times greater microbial biomass compared to underlying soil (Raturi et al., 2004). For example, the half-life of chlorothalonil in soil averages 48 days; however, Magri and Haith (2009) calculate that in turf its half-life is only 4.2 days. The environmental fate of all pesticides remains a concern to the public. Turfgrass surfaces to include thatch, mat, and the rootzone, however, provide a large beneficial impact on the environment and society (Beard, 2000).

Summary of Key Points

- Aside from resistance, other negative nontarget effects of fungicides include enhanced microbial degradation and reduced residual control, resurgence, encouragement of nontarget diseases, growth of blue-green algae (i.e., cyanobacteria), phytotoxicity, and toxicity to earthworms.
- Reduced residual effectiveness of a fungicide can be confused with resistance. It occurs due to a buildup of microbes over time, which are capable of degrading the active ingredients in fungicides.
- Fungicide-enhanced resurgence occurs when fungicide-treated turfs are more severely damaged than adjacent untreated areas. This most commonly is observed with dollar spot.
- Phytotoxicity is uncommon with most fungicides. Chloroneb, etridazole, and PCNB are most likely to injure golf greens when applied in warm weather. Fosetyl-aluminum and phosphites can cause a tip burn. DMI fungicides can elicit objectionable discoloration and other growth regulator effects when used frequently.
- DMI fungicides can enhance leaf spot diseases and promote blue-green algae (aka cyanobacteria) growth on greens.
- Thiophanate-methyl and chlorothalonil can be toxic to earthworms.
- Some fungicides can improve turf quality and provide some protection against mechanical damage in the absence of disease.

- Fungicides are primarily degraded by microbes in thatch and soil, but some can be degraded by sunlight and other mechanisms.

Tank-Mixing Fungicides and Synergism

Fungicides are sometimes tank-mixed for purposes of increasing their spectrum and duration of activity, improving their curative activity, and as a measure to delay or avoid their resistance. Another potential reason to tank-mix is to achieve a synergistic effect. Synergistic tank-mixes of fungicides having different biochemical modes of action can be important in disease resistance management programs. When tank-mixing two or more fungicides having different modes of action, the resulting level of effectiveness may be synergistic, antagonistic, or additive. There are no universally accepted definitions of synergism, antagonism, or additive. In general, scientists measure synergistic and antagonist responses of tank mixtures and compare them statistically to some reference or model. A synergistic response occurs when two or more fungicides are mixed and are then able to provide a level of control greater than predicted by a reference model. A more practical definition is offered by Latin (2011), who states that synergistic mixtures provide a level of control that exceeds the sum of control afforded by its individual components. Hence, one component renders the target pathogen more sensitive to another fungicide in the mixture, thus resulting in a response greater (e.g., longer duration of control) than would be predicted. Antagonism occurs when one of the components of a mixture reduces chemical activity of the mixture and less control is achieved, when compared to the predicted level of control. An additive effect indicates that none of the components in a mixture reduce or enhance the toxicity of any single component and control does not exceed the predicted level. An additive effect may occur when the level of control provided by fractional rates of two or more fungicides is effective, but not more effective than the low label rate of any of the components applied singly.

Most scientists who study synergism limit their evaluation to fractional or sublabel rates of fungicides. Mixtures of label use rates can be synergistic as well as antagonistic or additive. There have been few studies to determine either synergism or the usefulness of using fungicide mixtures to combat resistance (Sanders et al., 1985; Couch and Smith, 1991; Burpee and Latin, 2008). In the aforementioned studies, mixtures of sublabel rates of two or three fungicides were assessed. Sanders et al. (1985) evaluated half label rates of metalaxyl mixed with the half label rates of propamocarb, mancozeb, or fosetyl-aluminium, and the study was conducted using field-grown perennial ryegrass transferred to incubation chambers. They found that all fungicide mixtures provided excellent control of Pythium blight. The researchers concluded from these results, and population cycling experiments with *P. aphanidermatum*, that these mixtures would be useful in controlling the target pathogens and would delay fungicide resistance when targeting Pythium blight. Sanders et al. (1985), however, did not compare the mixtures to the low label rate of each fungicide applied separately and therefore could not determine if synergism had occurred. Couch and Smith (1985) reported that sublabel rates of mancozeb + propamocarb, fosetyl-aluminium + metalaxyl, and fosetyl-aluminium + propamocarb were synergistic, whereas mancozeb + chloroneb was antagonistic when targeting Pythium blight. Burpee and Latin (2008) used a mathematical formula to statistically compare actual and expected values of duration of disease control to determine if synergism had occurred with tank-mixtures of sublabel rates of iprodione, propiconazole, triadimefon, and vinclozolin targeting dollar spot. Several experiments were conducted over a 3-year period on creeping bentgrass in Georgia and Indiana. There were a few rating dates when synergism was detected, but Burpee and Latin (2008) concluded that there was no consistent evidence for synergism among their treatments and that there was a low level of probability that turfgrass managers would obtain a synergistic response from the mixtures that they evaluated. Burpee and Latin (2008) found some evidence of antagonism but generally observed that the tank-mixes evaluated were additive rather than synergistic. Using tank-mixes of sublabel rates of fungicides that effectively control a disease would be expected to involve less total fungicide applied and some protection against resistance (Couch and Smith, 1991).

Fungicide Application

Most fungicides are diluted in water and sprayed onto turfgrasses. Nearly all efficacy research with fungicides involved sprayable formulations. Little effort has been devoted to comparing sprayable with granular forms. Because of lack of published research information, it is difficult to predict the performance of granular versus sprayable formulations. In general, granular forms of fungicides provide a shorter period of less effective control than their sprayable counterpart. Exceptions include some products containing azoxystrobin, mefenoxam, and propiconazole, which were demonstrated to provide high levels of disease control. Granular fungicides can be applied rapidly to small areas where unexpected disease appears. They are particularly useful in small units where diseases are localized and spraying is impractical.

Achieving complete spray coverage using appropriate nozzles and water-carrier volumes is very important in maximizing fungicide performance. Nozzles that produce small water droplets provide improved disease control because of better coverage compared to nozzles that produce large or coarse droplets, especially when targeting foliar pathogens (Couch, 1984; Shepard et al., 2006). Smaller droplets, however, are more susceptible to drift. Coarse or larger droplets do not provide as good coverage as small droplets but are less likely to drift. Nozzles that produce coarse to very coarse droplet sizes, however, would be appropriate for applying fungicides that target root pathogens, which also may require watering-in (Shepard et al., 2006). Manufacturer's label recommendations for water volume are vague and range from 30 to 200 gallons per acre (GPA) (280 to 1870 L/ha). Most golf course superintendents who apply fungicides use water volumes ranging between 44 and 100 GPA (411 and 935 L/ha). Superintendents can use higher water volumes to spray greens because they constitute relatively small areas. For logistical reasons, they prefer to use low water volumes to spray fairways. Fungicides applied to greens generally are applied in water volumes ranging from 44 (i.e., 1 gallon/1000 ft²) to 100 GPA (412 to 935 L/ha), whereas 44 to 60 GPA (411 to 561 L/ha) are used for spraying fairways. Higher water volumes are recommended for targeting root pathogens. In general, spray volumes ranging from 44 to 50 GPA (411 to 486 L/ha) and 88 to 100 GPA (814 to 935 L/ha) are used for targeting foliar and root pathogens, respectively.

Nearly all spray volume research focused on dollar spot in creeping bentgrass. Dollar spot primarily is a foliar blight, but the pathogen can attack stems. Results of the dollar spot studies revealed no compelling evidence to recommend using a higher (i.e., 88 to 100 GPA; 823 to 935 L/ha) versus a lower (i.e., 44 to 50 GPA; 411 to 468 L/ha) spray volume when targeting dollar spot in greens or fairway height creeping bentgrass (McDonald et al., 2006; Kennelly and Wolf, 2009; Pigati, 2010a). Obviously, using a lower spray volume would be advantageous because less water, time, labor, and equipment would be required to apply fungicides. Even though there are no corroborative studies to make a confident recommendation, fungicides targeting foliar pathogens other than *S. homoeocarpa* would be expected to perform well when applied in 44 to 50 GPA (411 to 468 L/ha).

Higher water volumes are more appropriate when targeting stem or root pathogens. In the case of basal rot anthracnose (a stem disease), fludioxonil was shown to provide better disease control when applied in 88 versus 44 GPA (823 versus 411 L/ha) (Fidanza et al., 2009). Again, there is little information due to lack of research on the appropriate water volume for targeting root pathogens with fungicides. Some root diseases, like Pythium-induced root dysfunction, are controlled more effectively by watering-in fungicides as noted below. The manner in which a fungicide is formulated may also influence performance in different water volumes. For example, newer formulations such as the microemulsion formulation may not require as much water as older emulsifiable concentrate or wettable power formulations to provide effective control (Shepard et al., 2006).

Fungicides targeting foliar diseases should be allowed to dry on leaves prior to rainfall or irrigation to obtain maximum effectiveness. Pigati et al.

(2010b) researched dollar spot in creeping bentgrass by assessing the performance of boscalid, chlorothalonil, iprodione, and propiconazole as influenced by simulated rain (imposed 30 minutes after application) and mowing timing (plots mowed when wet with dew in the morning versus in the afternoon when the canopy was dry). The percent reduction in dollar spot control associated with simulated-rain versus rain-free treatments ranged as follows: chlorothalonil 67% and 83%; propiconazole 42% and 79%; boscalid 48% and 70%; and iprodione 33% and 66%. Chlorothalonil, a contact fungicide, was most adversely affected by simulated rain. Boscalid and iprodione were judged to be most rain-fast, and propiconazole was intermediate. When disease pressure was low, all fungicides subjected to simulated-rain provided effective dollar spot control for 7 or more days following the initial application. When disease pressure was moderate to high, the fungicides (except chlorothalonil, which was ineffective) reduced disease severity when subjected to simulated rain, but the level of control was not great. The average percent reduction in dollar spot associated with morning mowing in all fungicide-treated plots ranged from 54% to 65%. The reduction in dollar spot severity in morning-mowed plots improved the performance of all fungicides.

There are few published studies involving the influence of watering-in fungicides. Fungicides that target root diseases are watered-in, but there are no guidelines on how much water to utilize or to what depth soil should be moistened. Thiophanate-methyl provides better summer patch control when watered-in to a soil depth of about 1 inch (2.5 cm) before it had time to dry on leaf surfaces, but the results with triadimefon were inconclusive (Clarke and Dernoeden, unpublished). Pyraclostrobin was shown to be effective when watered-in with 0.125 inch (3.1 mm) of water to target *Pythium volutum*, a casual agent of Pythium-induced root dysfunction (Tredway and Kerns, 2005; Kerns et al., 2009). Phosphonates are true systemics and technically should not require watering-in when using them to target Pythium-incited root diseases.

Fungicides should be sprayed through nozzles that atomize droplets. Flat-fan nozzles that produce fine droplet sizes generally are more efficient than nozzles that deliver a large droplet, such as flood-jet, raindrop, or hollow cone nozzles (Shepard et al., 2006). When droplets alight on a surface they spread and create an area about twice the diameter of the droplet (Ebert and Downer, 2006). Hence, nozzles delivering small water droplets provide more droplets per unit area and thus better fungicide coverage than larger water droplets. Atomized droplets delivered by flat-fan nozzles, however, are subject to drift. Air-induction and Turbo TeeJet® nozzles offer drift-reducing advantages and are as effective as standard flat-fan nozzles (Shepard et al., 2006; Fidanza et al., 2009). Low pressure produces larger droplets and can be another cause of reduced effectiveness. Pressure in the spray boom at delivery should be in the range of 30 to 60 psi (207 to 414 kPa). It is important to use enough water and pressure to force fungicide into the turf canopy so that the chemicals can wash down between leaf sheaths and contact stem bases when targeting stem and root pathogens. Proper pressure and water volume are needed to form an effective spray angle (i.e., 80° to 100°).

Little information exists regarding chemical interactions of tank-mixes. Most well-known chemical incompatibilities are noted on manufacturer labels. There are two general types of incompatibilities: chemical and physical. Chemical incompatibilities generally occur when pH of the final solution or the presence of one of the compounds reduces the efficacy or increases the phytotoxicity of a pesticide. Some examples of chemical incompatibilities are as follows: mixing lime or an alkaline reacting fertilizer with a benzimidazole or a dithiocarbamate fungicide can reduce their effectiveness, tank-mixing nonchelated iron sulfate with an emulsifiable concentrate may cause phytotoxicity, and tank-mixing a DMI or phosphite fungicide with some plant growth regulators may discolor or damage some turfgrasses. Fosetyl-aluminum or acid-reacting fertilizers (especially phosphoric acid and phosphate) can dramatically drop pH of the mixture. Hence, fosetyl-aluminum may not be compatible with some fertilizers, fungicides (especially flowable formulations), herbicides, or copper-based pesticides. The ideal pH of the final tank-mixture should be between 6.5 and 7. Additives are available for adjusting the pH of spray solutions.

Physical incompatibility normally is associated with excessive foaming or settling-out of particles. Some wettable powder and flowable fungicides may cause the formation of a precipitate. Mixing flowable formulations of chlorothalonil or mancozeb with fosetyl-aluminum also can form a precipitate. Physical incompatibility can indicate that there is an equipment problem. For example, wettable powders mixed without sufficient agitation or without a sufficient amount of water will clog screens. Prewetting and creating a slurry are helpful in getting wettable powders into suspension, especially when spraying with a small quantity of water. It is important to always keep the agitation system running, even during breaks or when in transit.

Summary of Key Points

- When two different fungicides are mixed the resulting level of effectiveness can be synergistic, antagonistic, or additive.
- Synergistic mixes provide a level of control greater than would be predicted by a model or afforded by each component of the mixture. Mixing fractional rates of two or more Pythium-targeted fungicides can be synergistic and of value in resistance management programs. There are no other known synergistic mixtures.
- Most fungicides are mixed in water and sprayed. Most fungicides are more effective when sprayed, but some azoxystrobin, mefenoxam, and propiconazole granular formulations work well.

- In general, 44 to 50 GPA (411 to 468 L/ha) is an effective dilution when targeting foliar blighting pathogens. When targeting stem or root pathogens, an 88 to 100 GPA (823 to 935 L/ha) dilution is recommended.
- There is little information on how much water to use when drenching a fungicide. In general, enough water to penetrate the upper 1- to 2-inch (2.5 to 5 cm) profile is recommended.
- Flat-fan nozzles that atomize droplets are more efficient than those delivering a large droplet size.
- Air-induction and Turbo TeeJet nozzles offer drift reducing advantages and are as effective as flat-fan nozzles.
- Fungicides should be applied at 30 to 60 psi (207 to 414 kPa).
- Fungicides should be allowed to dry on leaves before rain or irrigation. Contact fungicides are less rain-safe than penetrants. Several days of control can be achieved when disease pressure is low and it rains within 30 minutes of fungicide application. Under higher disease pressure control is not likely to be adequate.
- There are two types of incompatibilities: chemical and physical. Excessive foaming and settling-out of particles are most common.

Bibliography

Bary, F., A.C. Gange, M. Crane, and K.J. Hagley. 2005. Fungicide levels and arbuscular mycorrhizal fungi in golf putting greens. *J. Appl. Ecol.* 42:171–180.

Beard, J.B. 2000. Turfgrass benefits and the golf environment. In *Fate and Management of Turfgrass Chemicals*, J.M. Clark and M.P. Kenna, Eds., pp. 36–44. American Chemical Society, Washington, DC.

Bigelow, C.A., J.E. Kaminski, P.H. Dernoeden, and J.M. Krouse. 2003. Brown patch control and resurgence with various fungicides, 2002. *Fungicide and Nematicide Tests* 58:T044.

Bishop, P., J. Sorochan, B.H. Ownley, T.J. Samples, A.S. Windham, M.T. Windham, and R.N. Trigiano. 2008. Resistance of *Sclerotinia homoeocarpa* to iprodione, propiconazole, and thiophanate-methyl in Tennessee and Northern Mississippi. *Crop Sci.* 48:1615–1620.

Brent, K.J. 1995. Fungicide resistance in crop pathogens: How can it be managed? FRAC Monograph No. 1. Global Crop Production Federation, Brussels.

Brent, K.J., and D.W. Hollomon. 2007. Fungicide resistance: The assessment of risk. FRAC Monograph No. 2, revised. CropLife International, Brussels.

Burpee, L., and R. Latin. 2008. Reassessment of fungicide synergism for control of dollar spot. *Plant Dis.* 92:601–606.

Chastagner, G.A., and W.E. Vassey. 1982. Occurrence of iprodione-tolerant *Fusarium nivale* under field conditions. *Plant Dis.* 66:112–114.

Cole, H., Jr., B. Taylor, and J.M. Duich. 1968. Evidence of differing tolerances to fungicides among isolates of *Sclerotinia homoeocarpa*. *Phytopathology* 58:683–686.

Cook, P.J., P.J. Landschoot, and M.J. Schlossberg. 2009. Inhibition of *Pythium* spp. and suppression of Pythium blight of turfgrasses with phosphonate fungicides. *Plant Dis.* 93:809–814.

Copping, L.G., and H.G. Hewitt. 1998. Chemistry and mode of action of crop protection chemicals. The Royal Society of Chemistry, Cambridge.

Couch, H.B. 1984. Turfgrass fungicides II: Dilution, nozzle size, nozzle pressure and disease control. *Golf Course Management* 52(8):73–80.

Couch, H.B., and B.D. Smith. 1991. Synergistic and antagonistic interactions of fungicides against *Pythium aphanidermatum* on perennial ryegrass. *Crop Protection* 10:386–390.

Couch, H.B. 1995. *Diseases of Turfgrasses*. Krieger, Malabar, FL.

Couch, H.B. 2002. Better dollar spot control with less fungicide. *Golf Course Management* 70(11):89–93.

Couch, H.B. 2003. Strategies for preventing and managing fungicide resistance. *Golf Course Management* 71(5):111–115.

Davis, J.G., and P.H. Dernoeden. 2001. Fermentation and delivery of *Pseudomonas aureofaciens* strain Tx-1 to bentgrass affected by dollar spot and brown patch. *Int'l. Turfgrass Soc. Res. J.* 9:655–664.

Dekker, J. 1986. Preventing and managing fungicide resistance. In *U.S. National Research Council Committee on Strategies for the Management of Pesticide-Resistant Plant Populations, Pesticide Resistance: Strategies and Tactics for Management*, pp. 347–354. National Academy Press, Washington, DC.

Delp, C.J. 1995. Benzimidazole and related fungicides. In *Modern Selective Fungicides: Properties, Applications, Mechanisms of Action*, H. Lyr, Ed., pp. 291–303. Gustav Fischer Verlag, New York.

Dernoeden, P.H., and J. Fu. 2008. Fungicides can mitigate injury and improve creeping bentgrass quality. *Golf Course Management* 76(4):102–106.

Dernoeden, P.H., and N. Jackson. 1980. Managing yellow tuft disease. *J. Sports Turf Res. Inst.* 56:9–17.

Dernoeden, P.H., L.R. Krusberg, and S. Sardanelli. 1990. Fungicide effects on *Acremonium* endophyte, plant-parasitic nematodes, and thatch in Kentucky bluegrass. *Plant Dis.* 74:879–881.

Dernoeden, P.H., and M.S. McIntosh. 1991a. Disease enhancement and Kentucky bluegrass quality as influenced by fungicides. *Agron. J.* 83:322–326.

Dernoeden, P.H., and M.S. McIntosh. 1991b. Seasonal quality responses of perennial ryegrass as influenced by fungicides. *HortScience* 26:1181–1183.

Detweiler, A.R., J.M. Vargas, Jr., and T.K. Danneberger. 1983. Resistance of *Sclerotinia homoeocarpa* to iprodione and benomyl. *Plant Dis.* 67:627–630.

Ebert, T.A., and R.A. Downer. 2006. A different look at experiments on pesticide distribution. *Crop Protection* 25:299–309.

Ervin, E.H., D.S. McCall, and B.J. Horvath. 2009. Efficacy of phosphate fungicides and fertilizers for control of Pythium blight on a perennial ryegrass fairway in Virginia. Online. *Appl. Turfgrass Sci.* doi: 10:1094/ATS-2009-1019-01-BR.

European Food Safety Authority (EFSA). 2010. Rapporteur Member State assessment reports submitted for the EU peer review of active substances used in plant protection products. Available at http://dar.efsa.europa.eu/dar-web/provision. EFSA, Parma, Italy.

Fidanza, M.A., J.E. Kaminski, M.L. Agnew, and D. Shepard. 2009. Evaluation of water droplet size and water-carrier volume on fungicide performance for anthracnose control on annual bluegrass. *Int. Turfgrass Soc. Res. J.* 11:195–205.

Fungicide Resistance Action Committee (FRAC). 2005. Pathogen risk list. Available at http:frac.info/frac/index.htm. CropLife International, Brussels.

Fungicide Resistance Action Committee (FRAC). 2010. FRAC code list: Fungicides sorted by mode of action (including FRAC Code numbering). Available at http:frac.info/frac/index.htm. CropLife International, Brussels.

Georgopoulos, S.G. 1986. Plant pathogens. In *U.S. National Research Council Committee on Strategies for the Management of Pesticide-Resistant Plant Populations, Pesticide Resistance: Strategies and Tactics for Management*, pp. 100–111. National Academy Press, Washington, DC.

Golembiewski, R.C., J.M. Vargas, Jr., A.L. Jones, and A.R. Detweiler. 1995. Detection of demethylation inhibitor (DMI) resistance in *Sclerotinia homoeocarpa* populations. *Plant Dis.* 79:491–493.

Hardebeck, G.A., R.F. Turco, R. Latin, and Z. Reicher. 2004. Application of *Pseudomonas aureofaciens* through irrigation for control of dollar spot and brown patch on fairway-height turf. *HortScience* 39:1750–1753.

Hsiang, T., A. Liao, and D. Benedetto. 2007. Sensitivity of *Sclerotinia homoeocarpa* to demethylation-inhibiting fungicides in Ontario, Canada, after a decade of use. *Plant Pathol.* 56:500–507.

Jo, Y.-K., S.W. Chang, M. Boehm, and G. Jung. 2008. Rapid development of fungicide resistance by *Sclerotinia homoeocarpa* on turfgrass. *Phytopathology* 98:1297–1304.

Jo, Y.-K., A.L. Niver, J.W. Rimelspach, and M.J. Boehm. 2006. Fungicide sensitivity of *Sclerotinia homoeocarpa* from golf courses in Ohio. *Plant Dis.* 90:807–813.

Jung, G., and Y. Jo. 2008. New challenge to an old foe, dollar spot fungicide resistance. *Golf Course Management* 76(2):117–121.

Kenna, M.P., and J.T. Snow. 2000. The U.S. Golf Association turfgrass and environmental research program overview. In *Fate and Management of Turfgrass Chemicals*, J.M. Clark and M.P. Kenna, Eds., pp. 2–35. American Chemical Society, Washington, DC.

Kennelly, M.M., and R.E. Wolf. 2009. Effect of nozzle type and water volume on dollar spot control in greens-height creeping bentgrass. *Applied Turfgrass Science* Online. doi: 10.1094/ATS-2009-0921-01-RS.

Kerns, J.P., M.D. Soika, and L.P. Tredway. 2009. Preventive control of Pythium root dysfunction in creeping bentgrass putting greens and sensitivity of *Pythium volutum* to fungicides. *Plant Dis.* 93:1275–1280.

Kim, Y., E.W. Dixon, P. Vincelli, and M.L. Farman. 2003. Field resistance to strobilurin (QoI) fungicides in *Pyricularia grisea* caused by mutations in the mitochondrial cytochrome b gene. *Plant Dis.* 93:891–900.

Knauf-Beiter, G., W. Fisher, D. Laird, and J.P. Genay. 2000. The mesostemic activity of trifloxystrobin. Sixth ANPP International Conference on Plant Diseases. Tours, France.

Koch, P.L., C.R. Grau, Y. Jo, and G. Jung. 2009. Thiophanate-methyl and propiconazole sensitivity of *Sclerotinia homoeocarpa* isolates collected from golf course putting greens, fairways, and roughs in Wisconsin and Massachusetts. *Plant Dis.* 93:100–105.

Koller, W., and H. Scheinpflug. 1987. Fungal resistance to sterol biosynthesis inhibitors: A new challenge. *Plant Dis.* 71:1066–1074.

Kuck, K.H., H. Scheinpflug, and R. Pontzen. 1995. DMI fungicides. In *Modern Selective Fungicides: Properties, Applications, Mechanisms of Action*, H. Lyr, Ed., pp. 205–258. Gustav Fischer Verlag, New York.

Landschoot, P.J., B.S. Park, and W. Uddin. 2001. Nontarget effects of PCNB on putting green turf. *Intern. Turfgrass Soc. Res. J.* 9:679–684.

Latin, R. 2011. *A Practical Guide to Turfgrass Fungicides*. The American Phytopathological Society of America, St. Paul, MN.

Lee, J., J. Fry, and N. Tisserat. 2003. Dollar spot in four bentgrass cultivars as affected by acibenzolar-S-methyl and organic fertilizers. *Plant Health Progress* Online. doi: 10.1094/PHP-2003-0626-01-RS.

Leroux, P., R. Fritz, A. Debieu, C. Lanen, J. Bach, M. Gredt, and F. Chapeland. 2002. Mechanism of resistance to fungicides in field strains of *Botrytis cinerea*. *Pest Manage. Sci.* 58:876–888.

Lyr, H. 1995. Selectivity in modern fungicides and its basis. In *Modern Selective Fungicides: Properties, Applications, Mechanisms of Action*, H. Lyr, Ed., pp. 13–22. Gustav Fischer Verlag, New York.

Magri, A., and D.A. Haith. 2009. Pesticide decay in turf: A review of processes and experimental data. *J. Environ. Qual.* 38:4–12.

Mann, R.L. 2002. *In vitro* fungicide sensitivity of *Microdochium nivale* isolates from the UK. *J. Turfgrass Sports Surf. Sci.* 78:25–30.

McDonald, S.J., P.H. Dernoeden, and C.A. Bigelow. 2006. Dollar spot control in creeping bentgrass as influenced by spray volume and application timing. *Applied Turfgrass Science* Online. doi:10.1094/ATS-2006-0531-01-RS.

Mocioni, M., M. Gennari, and M.L. Gullino. 2001. Reduced sensitivity of *Sclerotinia homoeocarpa* to fungicides on some Italian golf courses. *Int'l. Turfgrass Soc. Res. J.* 9:701–704.

Monteith, J., Jr., and A.S. Dahl. 1932. Turf diseases and their control. *Bulletin of the U.S. Golf Association Green Section* 12(4):85–188.

Nelson, R.R. 1979. The evolution of parasitic fitness. In *Plant Disease: An Advance Treatise. Volume IV. How Pathogens Induce Disease*, J.G. Horsfall and E.B. Cowling, Eds., pp. 23–46. Academic Press, New York.

Neumann, St., and F. Jacob. 1995. Principles of uptake and systemic transport of fungicides with the plant. In *Modern Selective Fungicides: Properties, Applications, Mechanisms of Action*, H. Lyr, Ed., pp. 53–73. Gustav Fischer Verlag, New York.

Nicholson, J.F., W.A. Meyer, J.B. Sinclair, and J.D. Butler. 1971. Turf isolates of *Sclerotinia homoeocarpa* tolerant to Dyrene. *Phytopathol. Z.* 72:169–172.

Pigati, R.L., P.H. Dernoeden, and A.P. Grybauskas. 2010a. Early curative dollar spot control in creeping bentgrass as influenced by fungicide spray volume and application timing. *Applied Turfgrass Science* Online. doi: 10.1094/ATS-2010-0312-03-RS.

Pigati, R.L., P.H. Dernoeden, A.P. Grybauskas, and B. Momen. 2010b. Simulated rain and mowing impact fungicide performance when targeting dollar spot in creeping bentgrass. *Plant Dis.* 94:596–603.

Potter, D.A., M.C. Buxton, C.T. Redmond, C.G. Patterson, and A.J. Powell. 1990. Toxicity of pesticides to earthworms (*Oligochaeta:Lumbricidae*) and effects on thatch degradation in Kentucky bluegrass thatch. *J. Econ. Entomol.* 83:2362–2369.

Powell, J.F., and J.M. Vargas. 2000. Management of dollar spot on creeping bentgrass with metabolites of *Pseudomonas aureofaciens* (Tx-1). *Plant Dis.* 84:19–24.

Raturi, S., K.R. Islam, M.J. Carroll, and R.L. Hill. 2004. Thatch and soil characteristics in cool-and warm-season turfgrasses. *Commun. Soil Sci. Plant Anal.* 35:2161–2176.

Reicher, Z.J., and C.S. Throssell. 1997. Effect of repeated fungicide applications on creeping bentgrass turf. *Crop Sci.* 37:910–915.

Sanders, P.L. 1984. Failure of metalaxyl to control Pythium blight on turfgrass in Pennsylvania. *Plant Dis.* 68:776–777.

Sanders, P.L., W.J. Houser, P.J. Parish, and H. Cole, Jr. 1985. Reduced-rate fungicide mixtures to delay fungicide resistance and to control selected turfgrass diseases. *Plant Dis.* 69:939–943.

Schwinn, F., and T. Staub. 1995. Phenylamides and other fungicides against *Oomycetes*. In *Modern Selective Fungicides: Properties, Applications, Mechanisms of Action*, H. Lyr, Ed., pp. 323–346. Gustav Fischer Verlag, New York.

Settle, D.M., J.D. Fry, and N.A. Tisserat. 2001. Development of brown patch and Pythium blight in tall fescue as affected by irrigation frequency, clipping removal and fungicide application. *Plant Dis.* 85:543–546.

Shane, W.W., and T.K. Danneberger. 1989. First report of field resistance of *Colletotrichum graminicola* on turf to benzimidazole fungicides in the United States. *Plant Dis.* 73:775.

Shepard, D., M. Agnew, M. Fidanza, J. Kaminski, and L. Dant. 2006. Selecting nozzles for fungicide spray applications. *Golf Course Management* 74(6):83–88.

Sigler, W.V., Z. Reicher. C. Throssell, M. Bischoff, and R.F. Turco. 2003. Sorption and degradation of selected fungicides in the turfgrass canopy. *Water, Air, Soil Pollution* 142:311–326.

Sigler, W.V., C.P. Taylor, C.S. Throssell, M. Bishop, and R.F. Turco. 2000. Environmental fate of fungicides in the turfgrass environment: A minireview. In: *Fate and Management of Turfgrass Chemicals*, J.M. Clark and M.P. Kenna, Eds., pp. 127–149. American Chemical Society, Washington, DC.

Smiley, R.W., M. Craven-Fowler, R.T. Kane, A.M. Petrovic, and R.A. White. 1985. Fungicide effects on thatch depth, thatch decomposition rate, and growth of Kentucky bluegrass. *Agron. J.* 77:597–602.

Smiley, R.W., P.H. Dernoeden, and B.B. Clarke. 2005. *Compendium of Turfgrass Diseases*, 3rd ed. APS, St. Paul, MN.

Smith, J.D., N. Jackson, and A.R. Woolhouse. 1989. *Fungal Diseases of Amenity Turf Grasses*. E. & F.N. Spon., London and New York.

Tredway, L.P., and J.P. Kerns. 2005. Defining the nature of creeping bentgrass root diseases. *U.S. Golf Association Green Section Record* 43(4):7–9.

U.S. Environmental Protection Agency (EPA). 2008. Pesticides: Topical and Chemical Fact Sheets. Available at http://epa.gov/pesticides/factsheets/chemical_fs.htm. United States Environmental Protection Agency, Washington, DC.

U.S. Environmental Protection Agency (EPA). 2010. Fact sheets on new active ingredients. Available at http://www.epa.gov/opprd001/factsheets. U.S. EPA, Washington, DC.

Vargas, J.M., Jr. 2005. *Management of Turfgrass Diseases*. Wiley, Hoboken, NJ.

Vincelli, P. 2004. Simulation of fungicide runoff following applications for turfgrass disease control. *Plant Dis.* 88:391–396.

Vincelli, P., and E. Dixon. 2002. Resistance to QoI (strobilurin-like) fungicides in isolates of *Pyricularia grisea* from perennial ryegrass. *Plant Dis.* 86:235–240.

Vincelli, P., and E. Dixon. 2007. Does spray coverage influence fungicide efficacy against dollar spot? *Applied Turfgrass Science* Online. doi:10.1094/ATS-2007-1218-01-RS.

Wong, F.P., and S.L. Midland. 2007. Sensitivity distributions of California populations of *Colletotrichum cereale* to the DMI fungicides propiconazole, myclobutanil, tebuconazole, and triadimefon. *Plant Dis.* 91:1547–1555.

Wong, F.P., S.L. Midland, and K.A. de la Cerda. 2007. Occurrence and distribution of QoI-resistant isolates of *Colletotrichum cereale* from annual bluegrass in California. *Plant Dis.* 91:1536–1546.

Wong, F.P., and S.L. Midland. 2007. Sensitivity distributions of California populations of Colletotrichum cereale to the DMI fungicides propiconazole, myclobutanil, tebuconazole, and triadimefon. Plant Dis. 91:1547–1555.

Wong, F.P., S.L. Midland, and K.A. de la Cerda. 2007. Occurrence and distribution of Qol-resistant isolates of Colletotrichum cereale from annual bluegrass in California. Plant Dis. 91:1536–1546.

7

Herbicides for Creeping Bentgrass

There is no faster way to kill creeping bentgrass than with the misapplication of an herbicide (Figures 7.1 and 7.2). Most commercially available herbicides are known to discolor or severely damage creeping bentgrass and annual bluegrass, particularly on golf greens. Herbicides that normally are tolerated by creeping bentgrass and annual bluegrass may lose their margin of safety when applied during periods of heat or drought stress or when applied to turf grown in shaded or poorly drained sites where roots are restricted to the upper few inches (5 to 6 cm) of soil. Excessive soil wetness can intensify the toxicity of some herbicides. Creeping bentgrass maintained under low nitrogen fertility, excessive thatch-mat, lower than normal mowing height, or treated with plant growth regulators also is more susceptible to herbicide injury. There are few herbicides registered for use on creeping bentgrass, and even fewer for golf greens. It is important to note that the relative safety of labeled herbicides applied to vegetatively propagated, and many of the new seeded creeping bentgrass cultivars, is unknown. Even clones within a green may react differently to an herbicide or plant growth regulator treatment. Furthermore, there is little information on the relative tolerance of colonial (*Agrostis capillaris*) and velvet bentgrass (*Agrostis canina*) cultivars to herbicides. Field observations by golf course superintendents, and extension and U.S. Golf Association (USGA) agronomists often are our best sources of information on the tolerance of newer cultivars to chemicals, especially herbicides.

All herbicides applied to golf greens should be field tested under local conditions and bentgrass cultivars or annual bluegrass biotypes before they are used on a wide scale. Some of the more commonly used herbicides (and original U.S. trade names) labeled for creeping bentgrass and annual bluegrass turfs are listed in Table 7.1. Of those listed, only bensulide, oxadiazon + bensulide, mecoprop (MCPP), siduron, carfentrazone, and specially formulated "bentgrass" mixtures of 2,4-D + MCPP + dicamba are labeled for use on golf greens. Herbicides such as bispyribac-sodium, fenoxaprop-ethyl, quinclorac, and ethofumesate are labeled for use on bentgrass tees and fairways. Use of herbicides on immature creeping bentgrass presents special problems and generally is discouraged. It is likely that any herbicide applied to low-cut turf causes some injury, although there may be no visual symptoms of phytotoxicity. Hence, it is important to apply herbicides when the turf is vigorous and actively growing. Herbicides generally are applied to creeping bentgrass in ≥44 gallons of water per acre (≥411 L ha).

FIGURE 7.1
Green destroyed by a misapplication of a broadleaf weed herbicide mixture.

FIGURE 7.2
For ill-advised reasons this creeping bentgrass practice area was treated in summer with a mix of a pre- and postemergence herbicide meant to target crabgrass and goosegrass.

Except when otherwise noted, do not apply herbicides to creeping bent-grass seedlings or even in the spring following an autumn seeding of bent-grass. There are no good definitions of what constitutes a seedling becoming a mature plant. In general, after a turf has been mowed three or more times, usually a period of 4 to 6 weeks after seedling emergence, it is considered a turf if density is good. Use of herbicides on seedlings can be justified if weeds are dominating and the bentgrass is below the weed canopy. In this situation

TABLE 7.1

Some Herbicides Labeled for Use on Creeping Bentgrass in the United States

Trade Name(s)	Chemical Name	Creeping Bentgrass Safety Comments/Uses[a,b,c]
Broadleaf Herbicides		
Bentgrass formulations of Three-Way®, Trimec®, others	2,4-D + MCPP + Dicamba	May be used on greens, tees, fairways Likely to cause yellowing These are effective prepackaged blends that have been used for decades
Banvel®	Dicamba	May be used only on tees and fairways May stunt or discolor bentgrass
Confront®	Triclopyr + Clopyralid	Triclopyr + clopyralid is labeled for bentgrass fairways (not greens or tees) but causes objectionable yellowing High rates of triclopyr can kill bentgrass
Lontrel®	Clopyralid	Clopyralid is labeled for use on bentgrass fairways for white clover control and discolors bentgrass less than triclopyr + clopyralid
Mecomec®, others	MCPP	Good safety, generally weak on broadleaf weeds when applied alone and can cause turf yellowing
Quicksilver®	Carfentrazone	Used to control pearlwort and moss in greens and tees Do not apply in hot or cold weather May injure annual bluegrass
Postemergence Grass Herbicides		
Acclaim Extra®	Fenoxaprop	Multiple low rates used for crabgrass or goosegrass control on bentgrass mowed above 0.25 inches (6.2 mm) Low rates of fenoxaprop can discolor bentgrass for 2 weeks or longer Can be used at higher rates to suppress or control bermudagrass in roughs and green surrounds
Drive®	Quinclorac	Labeled for postemergence control of crabgrass, white clover, and other weeds in turf May be applied to creeping bentgrass fairways but not greens or collars Quinclorac applied to bentgrass can cause objectionable mottling and yellowing for several weeks
Progress®	Ethofumesate	Used for annual bluegrass control in bentgrass fairways Likely to stunt and discolor bentgrass, especially when used in multiple applications Poorly rooted turf or bentgrass grown in shaded or wet sites may be severely injured by ethofumesate Level of control is unpredictable

(continued)

TABLE 7.1 (continued)

Some Herbicides Labeled for Use on Creeping Bentgrass in the United States

Trade Name(s)	Chemical Name	Creeping Bentgrass Safety Comments/Uses[a,b,c]
Velocity®	Bispyribac-sodium	For postemergence annual bluegrass control and suppression of roughstalk bluegrass in mature creeping bentgrass or perennial ryegrass tees and fairways Can elicit a chlorosis in bentgrass for 7 to 21 days Suppresses dollar spot and annual bluegrass seedheads At least two applications needed and expect surviving target weeds.
Corsair® Manor®	Chlorsulfuron Metsulfuron	Used for perennial ryegrass or tall fescue control in mature bentgrass tees or fairways May require two or more applications and is best used in autumn Controls some broadleaf weeds Has a long soil residual that can impact overseeding Yellowing of bentgrass is likely, especially at higher rates
Preemergence Herbicides[b]		
Betasan®, Bensumec®, others	Bensulide	For preemergence crabgrass and annual bluegrass control on greens, tees, or fairways Provides poor control of goosegrass
Dimension®	Dithiopyr	For preemergence crabgrass and goosegrass control on tees and fairways Do not apply to bentgrass turfs in poorly drained sites or where turf has short and chronically weak root systems
Gallery®	Isoxaben	Do not use on golf greens Apply late summer or early autumn to fairways for preemergence broadleaf weed control, especially winter annuals Can be applied in winter to target early spring germinating weeds like knotweed
Barricade®	Prodiamine	For preemergence crabgrass, goosegrass, and annual bluegrass control Do not apply to creeping bentgrass maintained below 0.5 inches (1.25 cm), or where annual bluegrass is considered desirable Bentgrass rate should not exceed 0.5 lb ai/A (0.56 kg ai/ha)
Pre-M®, Pendulum®, others	Pendimethalin	For preemergence control of crabgrass, goosegrass, and annual bluegrass May only be applied to bentgrass maintained above 0.5 inches (12.5 mm) Bentgrass rate should not exceed 2 lb ai/A (2.2 kg ai/ha) Can inhibit rooting from creeping bentgrass stolons

(continued)

TABLE 7.1 (continued)

Some Herbicides Labeled for Use on Creeping Bentgrass in the United States

Trade Name(s)	Chemical Name	Creeping Bentgrass Safety Comments/Uses[a,b,c]
Ronstar®	Oxadiazon	For preemergence control of crabgrass, goosegrass, and annual bluegrass Labeled for use on creeping bentgrass mowed above 0.375 inches (9.4 mm)
Goosegrass/ Crabgrass Preventer®	Bensulide + Oxadiazon	For preemergence goosegrass and crabgrass control on greens, tees, or fairways Especially effective in controlling goosegrass on bentgrass greens and collars May cause some yellowing on greens
Tupersan®	Siduron	May be applied to the seedbed or overtop creeping bentgrass seedlings grown on greens, tees, and fairways for preemergence and early postemergence control of crabgrass and yellow foxtail
Others		
Basagran®	Bentazon	For yellow nutsedge control in bentgrass tees and fairways During heat stress periods in summer, this herbicide can severely injure perennial ryegrass
Manage®, Sedgehammer®	Halosulfuron	For yellow and purple nutsedge control in bentgrass tees and fairways
Dismiss®	Sulfentrazone	For postemergence control of yellow nutsedge and purple nutsedge, and green and false green Kyllinga control in creeping bentgrass fairways and tees Has preemergence activity for partially controlling yellow nutsedge

Note: Except for siduron, it is safest to avoid herbicide use on creeping bentgrass that is less than 1 year old.

[a] The relative safety of these herbicides on vegetatively propagated and many of the newer creeping bentgrass cultivars is unknown. Superintendents should field evaluate herbicides on small areas prior to using them on a wide scale.

[b] All preemergence herbicides can potentially injure shallow-rooted bentgrass. Do not apply a preemergence herbicide in the spring prior to an autumn conversion from an existing turfgrass species to creeping bentgrass.

[c] Any mention of field observations or nonlabeled usage of these products is not endorsed nor does it constitute legal authority, guarantee, or warranty. Reference to trade names does not constitute endorsement, guarantee, or warranty. No discrimination is intended against products not mentioned. Always consult the most current manufacturer's label instructions as rates, timings, labeled species, and so forth are subject to change.

the weed canopy would be expected to intercept most of the herbicide. Even if the herbicide does not provide complete control, it may injure weeds enough to give the underlying creeping bentgrass a competitive advantage. It also is important to note that a preemergence herbicide should not be applied in the spring prior to converting greens, tees, or fairways to creeping bentgrass.

This is because creeping bentgrass seedlings can be intolerant of even small soil residues of most preemergence herbicides. These residuals can cause significant reseeding failures, and when in doubt if a residue exists, the overseeding program should be delayed while a simple assay is performed (see section entitled "Dealing with Soil Residues of Preemergence and Other Herbicides"). Except as otherwise noted, the discussion that follows applies only to mature stands of creeping bentgrass and annual bluegrass.

Any mention of field observations dealing with a nonlabeled use of an herbicide does not constitute a legal endorsement, guarantee, or warranty for any product. All herbicides must be used according to the most recent manufacturer's label instructions.

Crabgrass and Goosegrass Control

Preemergence Herbicides

Premergence herbicides are applied in early spring prior to the time that annual grass weed seed begins to germinate. These herbicides do not kill seed in soil. Most (except oxadiazon) are mitotic inhibitors that must be taken up by the immature roots system of susceptible weeds. Once in roots, these herbicides prevent root cell division and elongation and seedlings succumb. Oxadiazon is shoot absorbed and interferes with enzymes involved in photosynthesis. Crabgrass (*Digitaria* spp.) and goosegrass (*Eleusine indica*) are the most common spring germinating annual grass weeds in turf in the United States. Yellow foxtail (*Setaria glauca*) is a common annual grass weed in higher-cut, low-maintenance turf sites, but it does not persist in close-cut creeping bentgrass or annual bluegrass turfs. Some superintendents use preemergence herbicides in late summer to target annual bluegrass as discussed below in the "Annual Bluegrass (*Poa annua*) Control" section. Crabgrass seedlings begin to emerge when soil temperatures average 60°F (15.6°C) for a 7-day period in spring (Fidanza et al., 1996). Goosegrass seedlings emerge shortly thereafter when soil temperatures rise to about 68°F (20°C) for a 7-day period. Both species can emerge throughout the summer, especially when there are frequent rain events.

When targeting crabgrass or goosegrass, labeled preemergence herbicides should be applied to golf greens in early spring about 1 to 2 weeks before these herbicides are applied to the rest of the golf course. The earlier timing does not affect the level of weed control, while minimizing potential turf discoloration or soil residuals that might impact overseeding. Preemergence herbicides should be applied in spring following coring and watered-in as soon as possible thereafter. If it is unavoidable to core prior to applying a preemergence herbicide some effectiveness can be lost, especially if cores are not removed. Avoid application of these herbicides if daytime temperatures

FIGURE 7.3
Granules of an herbicide collected in low areas following a heavy rain before they could be properly watered-in. (Photo courtesy of D. Settle.)

are expected to exceed 85°F (29°C), because the additional heat can cause damage to both creeping bentgrass and annual bluegrass. Granular forms of preemergence herbicides are subject to movement in heavy rains. In the event of a deluge, granules may collect in drainage patterns and low areas and cause extensive injury (Figure 7.3).

Bensulide, bensulide + oxadiazon, and siduron are the only preemergence herbicides labeled for use on golf greens. Bensulide has been used for decades for the purpose of preemergence crabgrass and annual bluegrass control on greens. Bensulide provides good levels of crabgrass control in some regions but fails to provide commercially acceptable goosegrass control. Research indicates that perennial types of *P. annua* are not controlled by bensulide. Mature stands of annual bluegrass may turn yellow for a few weeks following treatment with bensulide. Bensulide plus oxadiazon is a prepackaged mixture, which was first used on golf greens in the early 1980s. This product provides effective goosegrass control and good crabgrass control. Some bentgrass yellowing may occur within 30 days of application. Turf regains color shortly, but this is an occurrence worth noting so that it will not be confused with other factors that may cause turf to lose color. Siduron is generally used at the time of seeding creeping bentgrass and is discussed below. Siduron is ineffective when applied to mature turf, unless multiple applications are made targeting crabgrass only.

Dithiopyr was labeled for use on creeping bentgrass golf greens, and most other turfgrasses in 1991. In 2004, use of dithiopyr on greens was deleted from the manufacturers' label but was retained on some distributor labels. Dithiopyr effectively controls goosegrass preemergence and crabgrass both pre- and early postemergence. Dithiopyr does not effectively control goose-grass when applied postemergence, and as noted below, it should not be used in a postemergence timing for crabgrass control in creeping bentgrass. Because of its residual effectiveness, dithiopyr can be applied in winter or very early spring in northern regions of North America without loss of activity. Dithiopyr injury to bentgrass in poorly rooted, low, and wet areas appears as a purpling and thinning of the stand, which may not become apparent for up to 30 days or longer after treatment. Dithiopyr also can yellow bentgrass if applied during warm periods.

Prodiamine and pendimethalin are labeled for use on creeping bentgrass maintained above a height of 0.5 inches (1.25 cm). Oxadiazon may be used on bentgrass maintained above 0.375 inches (9.4 mm). Only a granular form of oxadiazon is labeled for cool-season grasses, and it should be applied to a dry canopy and watered-in as soon as possible. There are concerns that higher label rates and overlaps with the aforementioned products could injure bentgrass. For example, pendimethalin applied to a bentgrass fairway can inhibit rooting from stolons and slow or prevent filling-in of open areas before summer. Isoxaben is a preemergence herbicide labeled for use on fairways. Isoxaben's greatest strength is its ability to control broadleaf winter annuals such as common chickweed (*Stellaria media*) when applied preemergence in the autumn. Isoxaben, however, is seldom used to control these weeds on golf courses.

On newly established creeping bentgrass or when spring seeding greens, tees, or fairways, siduron can be applied in the seedbed or overtop of young creeping bentgrass seedlings. Sidurons' primary use is for preemergence crabgrass and yellow foxtail control in seedbeds. It has little or no effect on broadleaf weeds, sedges, or goosegrass. Vegetative (i.e., C-series) creeping bentgrass cultivars may be injured by siduron. Siduron injury appears as a yellowing of sensitive cultivars, which may not appear for 30 or more days after application. Even though there is not a large database, it is likely that most of the new creeping bentgrass cultivars tolerate siduron. Regardless, superintendents should assess the safety of siduron on bentgrass seedlings in a small area before treating larger areas with the herbicide. Apply 3 to 6 lb ai/A (3.4 to 6.7 kg ai/ha) of siduron to the seedbed or seedlings. Use a higher than normal seeding rate (i.e., ≥ 2 lb bentgrass seed/1000 ft^2; ≥ 98 kg/ha) when using siduron because some seedlings may be injured. A higher seeding rate will speed establishment in the spring. Siduron should be applied twice on a 21- to 28-day interval for best results. Siduron has pre- and early postemergence (one- to two-leaf seedling stage) activity on crabgrass but does not effectively control goosegrass. Crabgrass plants with three or more leaves are not controlled by siduron at rates recommended for use overtop of bentgrass seedlings. Except for siduron, other preemergence herbicides should not be used on creeping bentgrass sites that are less than 1 year old.

Postemergence Herbicides

Postemergence annual grass herbicides include disodium methane arsonate (DSMA), monosodium methane arsonate (MSMA), fenoxaprop-ethyl, and quinclorac. In summer, hand weeding is the safest and most effective means of weed control on golf greens. Historically, the organic arsenicals MSMA and DSMA were used for early postemergence crabgrass control. These herbicides are being phased out by the U.S. Environmental Protection Agency (EPA) and by 2012 no more will be produced. The organic arsenicals were ineffective in controlling goosegrass at rates safe for use on cool-season grasses. MSMA is currently labeled for use on bentgrass fairways and tees. MSMA requires multiple applications on a 7- to 10-day interval and should be applied in high volumes of water (>100 gallons/A; >936 L/ha) to reduce injury as it can be phytotoxic, especially when applied in hot weather. Fenoxaprop-ethyl and quinclorac are safer to use on bentgrass tees and fairways than organic arsenicals.

There currently are no postemergence herbicides labeled for crabgrass and goosegrass control on bentgrass golf greens. Postemergence crabgrass and goosegrass control can be effectively achieved in creeping bentgrass fairways and tees by applying multiple low-rate applications of fenoxaprop-ethyl. Quinclorac is labeled for use on bentgrass fairways and effectively controls crabgrass and white clover but not goosegrass. Both fenoxaprop and quinclorac can discolor or damage bentgrass fairways.

Only fenoxaprop-ethyl controls crabgrass, goosegrass, and yellow foxtail. For fenoxaprop, the best combination of creeping bentgrass safety and weed control is achieved using 0.016 lb ai/A (0.018 kg ai/ha). Fenoxaprop-ethyl can severely injure bentgrass if improperly applied. Ironically, fenoxaprop-ethyl is more likely to discolor creeping bentgrass during cool (70 to 80°F; 21 to 27°C) and moist periods than during relatively warm (85 to 92°F; 29 to 33°C) periods. Annual bluegrass may be more tolerant of fenoxaprop-ethyl than bentgrass. Applications should be initiated when weeds are in the one- to three-leaf stage. In northern regions of the United States, crabgrass seed begin germinating between mid-April and mid-May, whereas in the south germination begins much earlier depending on the state and sometimes regions within a state. Goosegrass seed begin to germinate shortly after the first crabgrass seedlings appear or about the time air temperatures consistently exceed 80°F (27°C). Large populations of goosegrass can emerge throughout summer (Figure 7.4). Should both crabgrass and goosegrass be a problem, fenoxaprop-ethyl sprays are initiated when crabgrass begins germinating, and they should continue throughout the summer to control the later germinating plants. Fenoxaprop-ethyl should be applied on a 10- to 14-day interval. Missing an application or failing to reapply the product within a maximum of 18 days of the last spray, will likely result in poor control and objectionable levels of escaping crabgrass and goosegrass plants. Furthermore, applications of most broadleaf weed-targeted herbicides (especially 2,4-D + MCPP + dicamba) within 14 days *prior* to applying fenoxaprop

FIGURE 7.4
Goosegrass in a creeping bentgrass tee.

will result in poor crabgrass and goosegrass control. Many broadleaf and potentially other herbicides block uptake sites for fenoxaprop-ethyl in weeds, rendering fenoxaprop-ethyl less effective. This unique form of antagonism does not occur when the broadleaf weed herbicide is applied *after* fenoxa-prop. Because of the greater likelihood for phytotoxcity in warmer weather, broadleaf herbicides generally are not used on creeping bentgrass or annual bluegrass in the summertime.

At low rates, fenoxaprop-ethyl injury to creeping bentgrass is normally superficial, but at times the herbicide can cause a reddening and thinning of the stand. Discoloration can be partially masked by tank-mixing fenoxaprop-ethyl with either a chelated iron product or slow-release liquid nitrogen (≤0.25 lb N/1000 ft^2; ≤12.5 kg N/ha) product. Flowable formulations of fungicides tank-mixed with fenoxaprop-ethyl, particularly chlorothalonil, also reduce discoloration. Conversely, fosetyl-aluminum and probably other *Pythium*-targeted fungicide tank-mixes with fenoxaprop elicit an objectionable level of yellowing that can persist for more than 2 weeks. Most importantly, never mix fenoxaprop-ethyl with another herbicide and apply the tank-mix combination to creeping bentgrass or annual bluegrass. Whenever in doubt, do not tank-mix any other products (except those noted on the label) with fenoxaprop-ethyl for use on bentgrass or annual bluegrass. Fenoxaprop-ethyl effectiveness is reduced greatly when applied to weeds growing in dry soil, despite an outward green and vigorous appearance of weeds. Inadequate soil moisture at the time fenoxaprop-ethyl is applied is the primary reason for erratic levels of weed control with this herbicide. This is particularly true on slopes, mounds, and other areas prone to wilt. Therefore, in the absence of rain, sites may have to be irrigated prior to applying this herbicide. Finally, fenoxaprop-ethyl needs to be applied through nozzles that atomize droplets, such as flat-fan nozzles. Avoid using floodjet or other nozzles that produce a large droplet when applying this herbicide.

In 1999, quinclorac was marketed for use as a postemergence herbicide on most turfgrasses, with the exception of bentgrass golf greens and collars and fine leaf fescues. Quinclorac controls crabgrass, white clover (*Trifolium repens*), and a few other broadleaf weeds but has no activity on goosegrass and is weak against yellow foxtail. Quinclorac provides best crabgrass control when applied mid-season and prior to the time that crabgrass plants have produced numerous tillers. Apply quinclorac when crabgrass can be seen in the canopy and is in the four-leaf to two-tiller stage. Once there are numerous tillers, more frequent applications may be needed. Quinclorac also controls or suppresses kikuyugrass (*Pennisetum clandestinum*) and torpedagrass (*Panium repens*), which are important weeds in California and Florida.

The application rate for quinclorac is 0.75 lb ai/A (0.84 kg ai/ha), and a single application is often effective when applied in the aforementioned timing and when crabgrass pressure is low to moderately severe. Where crabgrass is multitillered and pressure is severe, it is best to apply 0.5 lb ai/A (0.56 kg ai/ha) twice on a 14-day interval. In very heavily populated and dense crabgrass sites, 0.33 lb ai/A (0.37 kg ai/ha) applied three times on a 14-day interval may be required. Quinclorac should be tank-mixed (1% v/v) with methylated seed oil or other approved adjuvant noted on the label to obtain maximum weed control. Quinclorac should be used with caution on creeping bentgrass because it can cause several weeks of objectionable mottling or yellowing, even when applied once (Figure 7.5). Test quinclorac locally by applying it to a small area to ensure no unacceptable injury will occur. Multiple quinclorac applications can discolor creeping bentgrass for

FIGURE 7.5
Yellowing elicited by quinclorac in fairway height creeping bentgrass.

several weeks. Quinclorac-injured bentgrass appears as a general chlorosis, mottling, or yellow or tan blotches. Density may decline, but no exposure of bare ground should be expected when used at a label rate. Like fenoxaprop, quinclorac should be tank-mixed with a chelated iron product to help mask the yellowing. For best results, chelated iron should be reapplied on 2-week intervals after the final quinclorac application until the bentgrass has fully recovered.

Broadleaf Weed Control

White clover (*T. repens*), common (*S. media*) and mouseear chickweed (*Cerastium vulgatum*), and birdseye pearlwort (*Sagina decumbens*) are among the most common broadleaf weeds found in creeping bentgrass in North America (Figure 7.6). Pearlwort is a stolonifeous perennial that forms dense turf-like mats, which often are mistaken for annual bluegrass (Figure 7.7). Nearly all broadleaf weed herbicides will yellow or discolor creeping bentgrass. Carfentrazone has good safety and is used to control pearlwort and moss in creeping bentgrass golf greens. MCPP is generally safe for use on bentgrass, but it also will cause yellowing. MCPP has activity on the aforementioned weeds, but alone it would not be expected to provide effective control of these or most other weeds. When several different broadleaf weed species colonize creeping bentgrass turf, specially formulated mixes known as "bentgrass formulas" of 2,4-D + MCPP + dicamba are recommended.

FIGURE 7.6
Mouseear chickweed and white clover are among the most common broadleaf weeds found in creeping bentgrass.

FIGURE 7.7
Birdseye pearlwort is a weed with stolons and fleshy grass-like leaves, and it can be mistaken for annual bluegrass in golf greens.

These special mixes contain about 80% less 2,4-D than "regular" prepackaged mixtures. This is because 2,4-D can be extremely phytotoxic to creeping bentgrass (Figure 7.1). Use of these "bentgrass formulations," however, is associated with some bentgrass yellowing and possibly severe damage if not properly applied.

Triclopyr + clopyralid is a prepackaged mixture labeled for use on creeping bentgrass fairways (not greens or tees), but it usually causes objectionable yellowing and should be site tested before being used on large areas. High rates of triclopyr ester (1 lb ai/A; 1.1 kg ai/ha) alone are used to kill bentgrass selectively in some other cool-season grasses. Hence, it is unwise to use products containing triclopyr on creeping bentgrass. Clopyralid alone is extremely effective in controlling white clover, while causing only minor yellowing in creeping bentgrass fairways. Quinclorac is extremely effective in controlling white clover and some other broadleaf weeds. As previously noted, however, quinclorac can discolor bentgrass. Spurges (*Euphorbia* spp.), knotweed (*Polygonium aviculare*), and other broadleaf weeds can be very difficult to control in the summer. Broadleaf herbicides labeled for use on creeping bentgrass are likely to provide poor or erratic levels of control of mature spurge and knotweed plants in the summer and often cause severe yellowing or thinning in bentgrass when applied in hot and humid weather. Sites where spurge, knotweed, and other broadleaf weeds are troublesome should be mapped and treated between late autumn and late winter with isoxaben. As noted previously, isoxaben is a preemergence herbicide that targets broadleaf weeds. It also has fairly good activity on crabgrass. Do not overload a bentgrass site with more than one preemergence herbicide annually.

Summary of Key Points

- Misapplication of an herbicide is the fastest way to kill creeping bentgrass.

- Conduct small field tests on all new herbicides before blanket applying.

- Bentgrass formulas of 2,4-D, MCPP, and dicamba contain about 80% less 2,4-D than formulations used on roughs and lawns.

- Clopyralid is safe and effective for controlling white clover in creeping bentgrass tees and fairways.

- Carfentrazone can be used on creeping bentgrass golf greens to control pearlwort and moss.

- Avoid applying herbicides to creeping bentgrass in summer when daytime temperatures are expected to exceed 80°F (27°C).

- Bensulide, bensulide + oxadiazon, and siduron are the only preemergence herbicides labeled for use on golf greens. Some distributors may label dithiopyr. Apply preemergence herbicides 2 weeks earlier to greens than the rest of the golf course.

- Water-in preemergence herbicides as soon as possible. Granulars can move in heavy rain and collect in low areas and discolor or severely injure creeping bentgrass.

- Avoid using a preemergence herbicide (except siduron) on creeping bentgrass less than 1 year old.

- There currently are no postemergence annual grass weed control herbicides labeled for golf greens, but fenoxaprop-ethyl and quinclorac may be used on bentgrass tees and fairways.

- Only fenoxaprop-ethyl controls crabgrass, goosegrass, and yellow foxtail.

- Fenoxaprop-ethyl and quinclorac should be tank-mixed with a chelated iron product to mask potential discoloration in creeping bentgrass.

- Do not mix more than one herbicide at a time. Carefully read the label to know what products can and cannot be mixed with the herbicide you select.

Annual Bluegrass (*Poa annua*) Control

Annual bluegrass (known as annual meadowgrass in the United Kingdom, *Poa annua* or simply as Poa) is a highly diverse species with annual (*P. annua* ssp. *annua*) and perennial (*P. annua* ssp. *reptans*) biotypes. Its genetic diversity is its strength and is why annual bluegrass is such a difficult weed to control.

Cultural management practices can be effective in reducing the competitiveness of annual bluegrass. These practices, however, are less likely to be successful in poor growing environments. For example, creeping bentgrass does not compete with annual bluegrass in shade or in compacted, poorly drained soils. Hence, it is imperative that the growing environment be improved in these situations. This would involve aggressive tree removal programs and improvement of surface water drainage throughout the golf course. Unfortunately, in some regions where long periods of overcast weather are common, annual bluegrass likely will be able to maintain a consistent competitive advantage.

Cultural Management

Golf course managers have attempted to introduce creeping bentgrass into existing annual bluegrass dominated golf greens by intraseeding or plugging creeping bentgrass. Neither approach is effective because creeping bentgrass will not be able to compete in dense stands of annual bluegrass (Kendrick and Danneberger, 2005). Creeping bentgrass maintained at fairway height of about 0.5 inches (1.25 cm) competes better with annual bluegrass than does bentgrass grown on golf greens. Regardless, collecting clippings, alleviating soil compaction, improving drainage, reducing shade, core aerating at times when annual bluegrass seed is not germinating heavily, using lightweight mowers, and applying selected plant growth regulators reduce the competitiveness of annual bluegrass on greens, tees, and fairways (Figure 7.8). Late-season and winter applications of nitrogen promote the competitiveness of annual bluegrass. Hence, late autumn or

FIGURE 7.8
This green was cored in autumn when annual bluegrass seed was germinating in large numbers and thus dominates in coring holes.

dormant winter applications of nitrogen should be avoided where creeping bentgrass is being promoted in mixed creeping bentgrass and annual bluegrass stands. Conversely, applications of low rates of nitrogen (i.e., spoonfeeding) during summer improve the ability of creeping bentgrass to compete with annual bluegrass. Creeping bentgrass achieves its greatest competitive advantage versus annual bluegrass during summer when environmental conditions are warm and dry. This is because high-temperature stress limits annual bluegrass growth more than it does creeping bentgrass. During summer, creeping bentgrass should be irrigated deeply and infrequently and generally maintained as dry as possible. The combination of spoonfeeding and maintaining relatively dry soils enables creeping bentgrass populations to gradually increase.

Postemergence Herbicides for Use on Mature Creeping Bentgrass

Once established in creeping bentgrass fairways, annual types of *Poa annua* can be partially controlled with bispyribac-sodium or ethofumesate. The aforementioned herbicides can reduce the population of annual bluegrass, but the end result is seldom complete control. A challenge with these herbicides is that if they work well the result will be large dead areas that will require overseeding or sodding. Seeding into large dead areas must be accomplished and a bentgrass cover achieved prior to the time annual bluegrass germinates in late summer. Once annual bluegrass has emerged, and bentgrass seedlings are small and cover is poor, annual bluegrass is very likely to outcompete the creeping bentgrass. It is much more manageable to begin when annual bluegrass populations are low (e.g., <10% annual bluegrass). Eliminating annual bluegrass and spoonfeeding to encourage bentgrass recovery in early summer is the best approach when weed populations are small and there are no large dead areas to address. For reasons that will become apparent below, bispyribac-sodium is a better herbicide choice than ethofumesate for annual bluegrass control in creeping bentgrass.

Partial control or suppression can only be achieved with ethofumesate in creeping bentgrass because lower rates, when compared to the higher rates used on perennial ryegrass fairways, must be applied. Ethofumesate should not be applied to colonial or velvet bentgrass or to vegetatively propagated (i.e., C-series) creeping bentgrass cultivars. 'Pennlinks' and 'Penneagle' may be less tolerant of ethofumesate than most other seeded bentgrass cultivars. The label rate for ethofumesate use on creeping bentgrass fairways is 0.75 lb ai/A (0.84 kg ai/ha). Late autumn applications are most effective, whereas spring applications are generally ineffective. Frosts and low temperatures of winter appear to assist ethofumesate effectiveness in controlling annual bluegrass. For best results, ethofumesate should be applied twice (i.e., sequentially) in mid-to-late autumn when frosts begin to occur routinely. The second application should be made in about 21 days to a maximum of 30 days after the first spray. Two applications provide variable levels of control that often range from 30% to 60%, but sometimes 0 to >80% control may occur due to weather conditions and other unknown factors. Spring applications of ethofumesate are generally ineffective in controlling annual bluegrass, but they may suppress seedhead production.

Creeping bentgrass discoloration can be reduced by tank-mixing ethofumesate with a chelated iron product or a slow-release liquid nitrogen product. Poorly rooted creeping bentgrass is most severely damaged, especially bentgrass growing in shaded or wet sites, and even in areas where take-all patch (a root disease) is a problem. Extended areas on collars and greens where rooting is poor and in walk-on and walk-off areas where soil is compacted or where bentgrass is mechanically injured (e.g., where mowers turn on fairways) are also frequently damaged. Dry autumn and early winter weather conditions can be associated with increased ethofumesate injury in creeping bentgrass. This may be the result of poor creeping bentgrass root development associated with dry conditions in the autumn of some years. Overseeding fairways or tees with creeping bentgrass should be delayed for 6 weeks following use of ethofumesate. The erratic and unpredictable levels of annual bluegrass control provided, combined with the sometimes phytotoxic effects of ethofumesate, limit the utility of this herbicide in creeping bentgrass.

Bispyribac-sodium was labeled for the purpose of controlling annual bluegrass and suppression of roughstalk bluegrass (*Poa trivialis*) in established creeping bentgrass and perennial ryegrass fairways and tees in 2004. Bispyribac-sodium seldom kills roughstalk bluegrass but instead burns-back foliage. Bispyribac-sodium also controls some broadleaf weeds, suppresses annual bluegrass seedheads, and is labeled for suppression of dollar spot. Bispyribac-sodium inhibits the enzyme acetolactate synthase that plants require to produce three essential amino acids. The herbicide is both foliar and root absorbed, but most appears to enter through leaves and sheaths. Bispyribac-sodium can move off foliage in heavy rain and run-off to adjacent sites, where it can injure sensitive species. Kentucky bluegrass is severely stunted and discolored for long periods. Tall fescue also is sensitive, but mature stands recover when contacted by this herbicide. Perennial ryegrass is discolored and seedlings of both perennial ryegrass and especially tall fescue will be damaged by bispyribac-sodium.

The major uses of bispyribac-sodium will be to control or suppress annual and roughstalk bluegrass in creeping bentgrass fairways and tees. Bispyribac-sodium often elicits a yellowing or chlorosis in creeping bentgrass, which is referred to as the *yellow flash*. The yellow flash appears about 3 days following treatment and can remain evident for 14 to 21 days. An intense yellow flash generally indicates that the herbicide has not properly translocated out of the foliage (Figure 7.9). It also may be an indicator that weed control will be less effective. No major tip burn, thinning, or long-term damage is likely, but bispyribac may render bentgrass more susceptible to wilt, brown patch, and Pythium blight. Bispyribac-sodium seldom provides effective control of annual bluegrass or suppression of roughstalk

FIGURE 7.9
The *yellow flash* elicited by bispyribac-sodium in a creeping bentgrass fairway study.

bluegrass following a single application. Two or more applications are required for maximum control. Bispyribac-sodium should be applied during sunny weather when air temperatures range between 65 and 85°F (16 and 29°C) and when turf is actively growing. Applying bispyribac during cool to cold periods in spring or autumn can be phytotoxic to creeping bentgrass. A heavy, preapplication irrigation to moisten soil thoroughly should improve bispyribac performance. Mowing within 24 hours of application, and applications of bispyribac-sodium when rain is in the forecast should be avoided. Bispyribac-sodium has a short soil residual and treated sites can be overseeded 10 days following the final herbicide application.

Rate selection for bispyribac-sodium needs fine-turning on each individual golf course, but application frequency may be more important than rate. Superintendents should test bispyribac on a small scale to gain experience and local knowledge on the behavior of this herbicide. Two formulations of bispyribac-sodium were marketed, and given the low use rates of this herbicide there can be some confusion and users are cautioned to read the label carefully and to have their equipment properly calibrated. Rates in the range of 0.06 to 0.10 lb ai/A (0.067 to 0.112 kg ai/ha) of bispyribac-sodium are suggested. Bispyribac-sodium should be applied at least twice on a 14-day interval and the level of control assessed thereafter. In general, annual bluegrass and roughstalk bluegrass will show symptoms of severe herbicide injury within a week of the second application. A third or fourth application, depending on rate, may be required to achieve ≥90% control. When used repeatedly, bispyribac-sodium provides a significant reduction in bentgrass growth. Therefore, use rates of plant growth regulators may be reduced or suspended prior to and up to 14 days following the last bispyribac-sodium application. Where more than 15% to 20% of the stand is dominated by either annual or roughstalk bluegrass, you may want to take a more conservative approach. Lower rates applied weekly may be better suited to phytotoxically suppressing the targeted weeds without creating large areas of bare ground. Using bispyribac for annul bluegrass control should be completed before annual bluegrass seed begin to germinate in late summer (Figure 7.10). Creeping bentgrass does not effectively compete with annual bluegrass in the autumn. Thus, little may be achieved by using this herbicide if good creeping bentgrass cover is not achieved by early autumn as discussed below. As always, carefully read the label and familiarize yourself with the precautions stated when using this or any other herbicide.

Methiozolin is an herbicide under development for annual bluegrass control on golf greens. This herbicide is applied at rates ranging from 0.45 to 0.90 lb ai/A (0.5 to 1 kg ai/ha) shortly after spring green-up and again in about 21 days. Activity of this material is slow and takes about 60 days after the last spring application to have full effect. About 30 days after the first application, annual bluegrass turns an off-yellow color, but on greens with low populations, creeping bentgrass fills voids and dead spots are not usually noted. Creeping bentgrass is not discolored when methiozolin is applied in spring. Late summer preemergence applications of methiozolin, however, can cause objectional discoloration. Like bispyribac-sodium, there are annual bluegrass biotypes in all populations that are not sensitive to the herbicide.

In summary, bispyribac-sodium is a better choice than ethofumesate when targeting annual bluegrass in creeping bentgrass. Regardless of herbicide there are always surviving annual bluegrass plants. Bispyribac-sodium should not

FIGURE 7.10
Lighter-green annual bluegrass plants are reinfesting this overseeding following use of bispyribac-sodium. (Photo courtesy of S. Evans.)

FIGURE 7.11
This creeping bentgrass fairway was routinely treated with a plant growth regulator. Following an application of bispyribac–sodium in summer the fairway rapidly wilted and was injured by cart traffic. (Photo courtesy of S. Potter.)

be tank-mixed with surfactants or other adjuvants as severe bentgrass injury may occur. Unless otherwise stated on the label, do not tank-mix bispyribac with plant growth regulators, other herbicides, or products formulated as emulsifiable concentrates. Creeping bentgrass fairways sprayed with bispyribac-sodium and previously treated with plant growth regulators may be subjected to rapid wilt during periods of high-temperature stress (Figure 7.11). Wilting fairways should be syringed immediately and carts diverted to paths as long as wilt conditions persist. Also, there can be enhanced creeping bentgrass injury when bispyribac-sodium is applied in cold weather. Furthermore, you may not realize how much annual or roughstalk bluegrass you have until you use bispyribac-sodium. Superintendents should be prepared to overseed or sod sites when large losses of these weedy grasses occur. Finally, bispyribac-sodium should not be considered the "silver bullet" for annual bluegrass control. There always are survivors, and bispyribac-tolerant biotypes of annual bluegrass are known to exist in most populations. A substantial number of buds on roughstalk bluegrass stolons will survive treatment of bispyribac-sodium, and with the advent of cool and wet weather a resurgence of the weed will occur. Both annual bluegrass and roughstalk bluegrass can rapidly reinvade open areas from seed in soil. Bispyribac-sodium should be considered a valuable tool in the suppression of the weedy bluegrasses, but it has limitations that would preclude its use on an annual basis. Thus, golf course superintendents should continue to utilize other cultural, herbicide, and plant growth regulator management approaches to promote creeping bentgrass competition with these grassy weeds.

Bispyribac-Sodium and Ethofumesate Use on Creeping Bentgrass Seedlings

It is important to seed creeping bentgrass in late summer prior to annual bluegrass emergence. Should annual bluegrass emerge in seedling bentgrass stands, it can be very competitive and rapidly dominate. Therefore, it may be necessary to attempt to control annual bluegrass should it become competitive early in the establishment of creeping bentgrass. Bispyribac-sodium (0.06 lb ai/A; 0.67 kg/ha) can be safely applied overtop of creeping bentgrass seedlings 3 to 4 weeks after seedling emergence (Dernoeden et al., 2008; Branham and Sharp, 2011). In an Illinois study, it was concluded that bispyribac-sodium (0.06 lb ai/A; 0.67 kg/ha) safely and effectively controlled annual bluegrass when applied twice at 3 and 5 weeks after seedling emergence in late summer (Branham and Sharp, 2011). Both aforementioned studies noted that bispyribac-sodium was most phytotoxic to bentgrass seedlings when rainfall was plentiful and soils were saturated for long periods. In colder regions, bispyribac-sodium applied in autumn during cool to cold and wet weather likely will injure or severely damage creeping bentgrass seedlings.

In summary, bispyribac-sodium will yellow and stunt creeping bentgrass seedlings, but this is an acceptable risk because reducing annual bluegrass competitiveness is extremely important. Bispyribac-sodium is likely to severely injure creeping bentgrass seedlings when applied within 3 weeks following emergence. Saturated soil conditions following use of this herbicide can contribute to increased phytotoxicity. To be effective, bispyribac needs to be applied twice on a 2-week interval (about 3 and 5 weeks after seedling emergence) beginning in late summer, when temperatures are relatively warm. Ethofumesate may be safer to apply to creeping bentgrass seedlings (0.75 lb ai/A; 0.84 kg ai/ha) in colder regions, but applications should not be initiated until 4 weeks after bentgrass seedling emergence.

Preemergence Herbicides

Most preemergence herbicides reduce annual bluegrass populations when applied in late summer; however, they may fail to provide sufficiently high levels of control once annual bluegrass has become well established. Furthermore, there is a much greater risk of phytotoxicity when applying preemergence herbicides in late summer during hot weather. The use of pre-emergence herbicides in late summer also conflicts with overseeding practices. Hence, preemergence herbicides should be used with caution for the purpose of annual bluegrass control on creeping bentgrass greens, tees, or fairways. For example, consider the potential problems of having to over-seed herbicide-treated sites should large areas of bentgrass or preexisting annual bluegrass die during winter. Soil residues of preemergence herbicides may not dissipate completely during winter. These soil residues could result in either a failure of creeping bentgrass seed to emerge or an inability of seedlings to develop properly. They also could have an adverse effect on rooting from sod. Preemergence herbicides have a long residual and could delay or retard autumn root development. This type of injury would be most problematic during the following spring should unusually cool, wet, and overcast conditions prevail for extended periods. A general lack of sunlight, oxygen deprivation in excessively wet soils, and the presence of a preemergence herbicide could all contribute to poor spring creeping bent-grass root development. Blanket sprays of herbicides on golf greens less than 1 year old should be avoided. The bottom line is that superintendents should carefully consider the risks and benefits of using these types of herbicides in late summer on creeping bentgrass.

Successful preemergence control of annual bluegrass is contingent on proper application timing, which is just prior to the major emergence period in late summer or early autumn. Annual bluegrass seedlings emerge at constant temperatures ranging from 39 to 70°F (4 to 21°C) in an incubator (Vargas and Turgeon, 2004). Working in the field, Kaminski and Dernoeden (2007) found that most annual bluegrass seed emerged at daily average air temperatures ranging from 61 to 68°F (16 to 20°C), and that little germina-tion occurred when daily average air temperatures fell below 46°F (7.7°C). In Maryland, which is in the mid-Atlantic region of the United States, annual bluegrass typically begins emerging in large numbers following a major rain event in early September. Studies conducted in Maryland and California (Shem-Tov and Fennimore, 2003) showed that annual bluegrass seed germi-nate primarily in the autumn. In Maryland, 63% to 78% of all annual blue-grass seedlings emerge between early September and mid-October, and 90% of seedlings emerge by early December (Kaminski and Dernoeden, 2007). Annual bluegrass will germinate from early winter to spring as long as soils are not frozen, but the major emergence period is in the autumn. Substantial populations of annual bluegrass can emerge in spring, but usually this occurs where winter-kill of annual bluegrass is common. Disturbing soil to reseed winter-killed areas in the spring brings annual bluegrass seed to the surface and stimulates germination. In the absence of competition from other grasses, the plants that emerge in spring can provide an annual blue-grass cover by summer.

The best preemergence application timing varies from region to region and can be influenced by temperature, rainfall, photoperiod, and ecotype of annual bluegrass (Cline et al., 1993; Sweeney and Danneberger, 1995; McElroy et al., 2004). Depending on temperature, rainfall, and other fac-tors, germination will occur later in southern and earlier in northern U.S. regions and Canada. If unknown, consult with university extension special-ists or an USGA agronomist in your region for the likely annual bluegrass emergence time in your area. Mid-to-late autumn preemergence herbicide applications are less effective because many annual bluegrass seedlings will have emerged.

FIGURE 7.12
Green destroyed by a misapplication of bensulide to creeping bentgrass sod.

Bensulide is the most commonly used preemergence herbicides for annual bluegrass control on creeping bentgrass golf greens. Bensulide should only be applied to creeping bentgrass that is more than 1 year old. Avoid applications of bensulide to sodded greens within 1 year following installation (Figure 7.12). Maximum control is achieved with a single application of a high label rate about 1 or 2 weeks prior to the time annual bluegrass emergence is expected. To lower potential risk, the high rate of bensulide can be split, with the half rate applied just prior to germination and the other half rate 6 to 8 weeks later.

Use of bensulide, prodiamine, dithiopyr, pendimethalin, or oxadiazon in roughs and green surrounds will help reduce the potential for annual bluegrass seed to move into other areas in play. Sprayable formulations of prodiamine are among the most effective preemergence materials for roughs. Coarse granular prodiamine formulations tend to be erratic, with finer grades being preferred. Pendimethalin is a very effective herbicide for the combined preemergence control of annual bluegrass and winter annual broadleaf weeds. Do not use these herbicides, however, if overseeding of treated sites is anticipated or expected in the autumn or the following spring. In addition to herbicides, the plant growth regulators flurprimidol and paclobutrazol have preemergence activity when applied regularly into late autumn (Gaussoin and Branham, 1987; McElroy et al., 2004; Dernoeden, 2011). Control may be as high as 50%. The mechanism by which the aforementioned growth regulators suppress annual bluegrass populations is unknown but may be related to their long half-life in most soils (i.e., >100 days). It may be a true preemergence effect or they in some way render annual bluegrass seedlings unable to compete or survive winter.

Other Chemicals for Annual Bluegrass Control

For new golf greens, the best approach to annual bluegrass management is a vigilant hand-picking, plugging-out, or spot-treatment program with a nonselective herbicide. Dead spots resulting from a nonselective herbicide reduce aesthetic quality but not playability of putting surfaces. Glufosinate and glyphosate can be used for spot-treatment of annual bluegrass plants. Glufosinate is preferred because it kills only contacted tissues, whereas glyphosate can wick outwards killing a larger spot. This wicking was attributed to glyphosates' highly systemic behavior. Some superintendents mix the two herbicides and apply them with "bingo blotter"–type devices. A few plants in larger patches of annual bluegrass may survive treatment because it can be difficult to get good herbicide coverage to all tillers in larger patches. For small individual spots of annual bluegrass, start at the center and rotate the blotter in a clockwise fashion until the outer tillers are treated with herbicide. Nonselective herbicides should be mixed with a tracker so that applicators and golfers can see where the herbicides were applied. It is best to irrigate greens prior to mowing because glufosinate and glyphosate can track for up to 24 hours after application during humid weather.

Nonselective herbicides and desiccants like endothall, paraquat, and diquat have been used on golf greens, mostly with little or no success. All of these materials phytotoxically suppress annual bluegrass. Most annual bluegrass plants invariably recover. Lethal levels of the aforementioned herbicides may kill both annual bluegrass and creeping bentgrass. Hence, annual bluegrass-free golf greens should be kept weed free one plant at a time and on a routine schedule. The most effective and safest, albeit costly approach is vigilant hand-picking or spot-treating individual annual bluegrass plants with a nonselective herbicide.

Plant growth regulators such as paclobutrazol and flurprimidol are effective in reducing existing annual bluegrass populations in creeping bentgrass. Use of these materials for this purpose is described in Chapter 8. There is no single annual bluegrass control program for creeping bentgrass. Integrating proper fertility practices, reasonable mowing heights, careful irrigation, spot-treatment or hand-picking, use of plant growth regulators, and controlling this weed with herbicides in green surrounds, roughs, and so forth, together will hopefully provide an effective level of annual bluegrass suppression. Avoiding core aeration during peak annual bluegrass seed germination periods is very important in keeping annual bluegrass from invading golf greens (Figure 7.8). This also is true for renovated greens that were fumigated. Fumigants do not reach deep into soil, and core aeration will bring viable annual bluegrass seed to the surface.

Summary of Key Points

- Creeping bentgrass does not compete with annual bluegrass in shade or in wet and compacted soils.

- Annual bluegrass competiveness is reduced by improving sunlight exposure, increasing mowing height, irrigating deeply and infrequently, and spoonfeeding in summer. Avoid coring in early autumn when most annual bluegrass seed is germinating.

- Bispyribac-sodium is the preferred herbicide for controlling annual bluegrass postemergence in creeping bentgrass tees and fairways.

- Bispyribac-sodium is the preferred herbicide and should be used in mid-summer so that creeping bentgrass overseeding can take place prior to the emergence of annual bluegrass seedlings in late summer.

- Bispyribac-sodium must be applied at least two times about 14 days apart. Turf yellowing should be expected.

- Bispyribac-sodium is not a "silver bullet," and there always will be surviving annual bluegrass plants.

- When renovating fairways with new creeping bentgrass cultivars in late summer, there will be intense competition with annual bluegrass from the seed bank in soil. Bispyribac-sodium can be used overtop creeping bentgrass seedlings to control annual bluegrass.

- Bispyribac-sodium is applied two times about 3 and 5 weeks after emergence of creeping bentgrass seedlings. Injuries to bentgrass seedlings occur, but most plants normally recover, and the benefit of minimizing competition from annual bluegrass outweighs the injury.

- Bispyribac-sodium may become phytotoxic to creeping bentgrass seedlings if there are long periods of soil being saturated by rain.

- Annual bluegrass can be controlled preemergence using a late summer application of bensulide to greens and bensulide or dithiopyr in creeping bentgrass tees and fairways.

- Oxadiazon, pendimethalin, and prodiamine are good choices for controlling annual bluegrass preemergence in green surrounds and roughs. Pendimethalin provides the best combination of preemergence annual bluegrass and broadleaf weed control.

- There are risks associated with a late summer application of a preemergence herbicide to creeping bentgrass that need to be carefully considered.

- Timing of a preemergence herbicide is key to success. If you do not know when annual bluegrass seed germination begins and peaks in your area, consult a university or USGA agronomist in your region.

- Annual bluegrass can be spot-treated with a nonselective herbicide, which reduces aesthetic quality but not playability of greens.

- Flurprimidol and paclobutrazol are growth regulators that reduce competitiveness and have preemergence activity on annual bluegrass.
- There is no single program for annual bluegrass control.

Bacterial Wilt and Other Diseases Specific to Annual Bluegrass

Bacterial wilt of creeping bentgrass and annual bluegrass was the first recorded bacterial disease of turf in the United States; initially the casual agent was reported to be *Xanthomonas campestris*. Bacterial wilt in the United States was first recorded in 1981 on 'Toronto' (aka C-15) creeping bentgrass. Cause of the disease initially was unknown, and the malady was referred to as *C-15 Decline*. Other vegetatively propagated cultivars of creeping bentgrass such as 'Cohansey' and 'Nimislia' also proved to be susceptible. Since the original report of C-15 Decline in the early 1980s in the midwestern United States, there have been no other authenticated cases of bacterial wilt in creeping bentgrass turf. A new bacterial pathogen (i.e., *Acidovorax avenae*) of creeping bentgrass was reported in 2010. Acidovorax bacterial disease is discussed in Chapter 3. The distribution and importance of this new bacterial pathogen is currently unknown.

Summer patch (*Magnaporthe poae*) specifically attacks annual bluegrass in most regions; however, it was reported attacking creeping bentgrass greens in very warm climates such as in North Carolina and Florida. Mysteriously, anthracnose (*Colletotrichum cereale*) is often specific to either annual bluegrass or creeping bentgrass growing on greens of a particular golf course. In most cases, anthracnose attacks the annual bluegrass and not the creeping bentgrass in mixed stands. Summer patch and anthracnose were described and discussed in detail in Chapter 4.

Despite the possibilities, the greatest probability is that bacterial wilt, summer patch, and anthracnose will specifically attack annual bluegrass in mixed stands with creeping bentgrass on golf greens. Where these diseases are chronic, and annual bluegrass populations are low (i.e., <10% of the putting green), the aforementioned diseases should be allowed to go unchecked by plant protection chemicals. By allowing diseases to selectively control annual bluegrass, a true form of IPM is achieved. Spoonfeeding creeping bentgrass with 0.1 to 0.2 lb N/1000 ft^2 (5 to 10 kg N/ha) with water-soluble nitrogen on a 7- to 10-day interval during the time the aforementioned pathogens are actively infecting annual bluegrass will speed the spread of creeping bentgrass into voids created by the disease. These light applications of soluble nitrogen are an important approach to managing creeping bentgrass in summer. This is because creeping bentgrass is more competitive with annual bluegrass during warm and dry periods versus cool to cold and wet periods.

Bacterial Wilt of Annual Bluegrass

Plant pathogenic bacteria are single celled, usually rod shaped, and reproduce at an extraordinarily high rate by binary fission. Most have rigid cell walls and some are motile by means of flagella. Bacteria have no means of penetrating cells so they must enter plants through natural openings such as stomates and hydathodes, or through wounds. Bacteria are most likely to enter plants through cut leaf tips during mowing or through wounds created by sand topdressing. The organism that causes bacterial wilt in annual bluegrass can colonize the vascular system of plants but may fail to produce symptoms when conditions are not optimal for the disease (Mitkowski et al., 2005). Once inside plants, pathogenic bacteria cause damage by enzyme activity, toxin production, and vascular plugging. By occluding xylem vessels they interrupt water transport causing plants to wilt and eventually die.

Until 1995, many *Xanthomonas* pathogens were grouped into the species *Xanthomonas campestris*. Further differentiation among isolates was made by assigning them to pathovars, based on host specificity. The pathovar system allows for easy classification, but it is not a valid method of taxonomic separation. DNA and other analyses eventually were used to divide *X. campestris* into multiple species (Vauterin et al., 1995). According to a new taxonomic scheme, all *X. campestris* species that attack grasses are now considered to be *X. translucens*. Researchers confirmed that the organism responsible for bacterial wilt of annual bluegrass in Japan is *X. translucens* pv. *poae*. DNA analysis and host range assays from bacterial wilt isolates in the United States also have shown that *X. translucens* pv. *poae* is responsible for the disease in the northeastern United States (Mitkowski et al., 2005).

Symptoms in Annual Bluegrass

Bacterial wilt occurs on annual bluegrass golf greens and may also attack annual bluegrass in collars and approaches. Observations indicate that annual bluegrass turf is predisposed to bacterial wilt by various stresses such as intensive mowing and grooming, poor growing conditions (i.e., shade, poor drainage, soil compaction, etc.), and winter injury. The incidence of the disease appears to be related in large part to the trend for very low mowing heights, more frequent mowing, lower nitrogen fertility, and aggressive grooming practices to improve green speed.

Bacterial wilt in annual bluegrass was observed primarily in the Mid-Atlantic and Northern regions of the United States. In annual bluegrass grown on golf greens in the aforementioned regions of the United States, the disease generally first appears in late May or June but may remain active into September. Heavy rains and thunderstorm activity intensify the disease. Infected annual bluegrass plants may initially turn yellow, lime-green, or blue-gray. Soon, individual infected plants turn reddish-brown and die in white or tan-colored spots about the size of a dime (1.8 cm). From a standing position, badly infected greens will have a speckled or pitted appearance (Figure 7.13). There appear to be no consistent leaf or sheath symptoms that can be used reliably to diagnose bacterial wilt. When infected plants are incubated in a laboratory for about 2 days, the youngest leaf of some infected plants may become yellow and elongated (i.e., etiolated) (Figure 7.14).

FIGURE 7.13
Bacterial wilt (*Xanthomonas translucens*) in annual bluegrass. Note that the blue-green bentgrass is unaffected.

FIGURE 7.14
Etiolated growth of annual bluegrass infected with the bacterial wilt pathogen (*Xanthomonas translucens*) in a fairway.

Infected, etiolated leaves sometimes are bent in the shape of a shepherd's crook. Leaf etiolation, however, is not uncommon in apparently healthy turf and is referred to as etiolated tiller syndrome (see Chapter 3). Older leaves of infected plants can appear darker green, mottled, and stunted, but infected leaves usually are yellowed at the tip or the base of the lamina. Stem bases of infected plants may be variously discolored and water soaked. The variable leaf symptoms of bacterial wilt are not stable characteristics and are of little value in field diagnosis.

Bacterial wilt tends to develop first and more severely on pocketed or shaded greens. Damage appears in high and low areas but tends to be most severe on the outer periphery or clean-up areas, walk-on and walk-off areas, and in impeded surface water drainage patterns subject to ice formation during winter months. Large areas of the putting surface may become speckled with numerous spots or pits, creating uneven putting conditions. In severe cases, whitish-tan dead spots ranging from 0.5 to 1 inch (1.2 to 2.5 cm) in diameter coalesce and the pattern of injury appears as a general blighting, rather than a speckled appearance. Bacterial wilt symptoms can mimic active or residual anthracnose basal rot. To add to the confusion both diseases may occur simultaneously. It is unknown whether infection by one of these pathogens predisposes plants to infection by the other. Bacterial wilt may abate during dry periods and flair up following rainfall.

Bacterial wilt is often very difficult to identify. In annual bluegrass, the bacteria generally are found in leaves and crowns and less commonly in roots. In the laboratory, a diagnostician will place yellow or etiolated leaves or crowns of plants suspected of being infected into a drop of water on a microscope slide, sever them with a razor blade, and look for streaming of bacterial cells from vascular bundles. Slow oozes from senescent tissues are common, but rapid streaming of cells from vascular bundles of infected leaves, crowns, or roots is the best indicator of bacterial wilt. Unfortunately, it is difficult and time consuming to isolate and positively identify *Xanthomonas* spp.

Management

Bacteria are spread by moving water and mowers. Obviously, mowing creates wounds and contaminated equipment will distribute bacteria onto wounded tissue. Mowing turf when leaves are dry later in the morning may slow progression of the disease. Increasing mowing height reduces bacterial wilt severity, but this practice also slows the speed of golf greens. It is very important to avoid mowing when greens are excessively wet and soft and to mow every other day, particularly the perimeter clean-up cut. When the disease is restricted to one or a few greens, a mower should be dedicated to those greens. It is best to use a walk-behind greens mower, and dedicated mowers should be disinfected with a 10% sodium hypochlorite bleach (Clorox®) solution or other disinfectant after use. Grooved rollers should be replaced with solid rollers and grooming, aeration, and sand topdressing should be avoided when the disease is active. This is because grooming, aeration, and topdressing practices abrade and wound tissue and create openings for easy entry of the bacteria.

Antibiotics such as oxytetracycline and streptomycin suppress bacteria wilt. However, antibiotics are very expensive, difficult to handle, may sometimes be phytotoxic, and must be applied in very high water volumes at dusk every few weeks. Furthermore, there currently are no antibiotics labeled for use on turf. Products containing copper hydroxide and copper oxychloride plus copper sulfate are contact protectants. These products will not control bacteria in infected plants but may help to slow spread of the pathogen to healthy plants. Superintendents observed poor to good results with these copper-based products. There are, however, no research-based recommendations for using copper products for bacterial wilt suppression in turf. Copper-based products should be applied in at least 5 gallons of water/1000 ft^2 (19 liters/93 m^2) on 5- to 7-day intervals or following heavy rain as long as the disease is active. Using lower water volumes or applying copper products on warm or hot (>85°F; >29°C) days could intensify chlorosis or injury, especially in annual bluegrass. It therefore may be necessary during summer stress periods to apply copper products in the evening when temperatures are falling. Mature creeping bentgrass appears to be more tolerant of copper than annual bluegrass; however, bentgrass seedlings are likely to be injured or killed. Copper may accumulate in soils eliciting chlorosis and eventually may cause toxicity to the turf. Currently, it is unknown how much copper can be applied to golf greens before turf decline occurs. Observations by some superintendents suggest that hydrogen dioxide–based disinfectants may slow spread of the disease and perk up affected greens. The residual effectiveness of hydrogen dioxide is measured in seconds or minutes and could only be expected to impact bacteria on leaf surfaces at the moment it was applied. Potable and not pond water should be used with hydrogen dioxide. Unless otherwise stated on labels, do not tank-mix hydrogen dioxide or copper-based products with other chemicals or fertilizers. Acibenzolar is a chemical used on some crops to suppress bacterial diseases by stimulating natural plant defense mechanisms (aka "systemic acquired resistance"). Research with acibenzolar shows promise in reducing the severity of several turfgrass diseases and is available in a pre-packaged mixture with chlorothalonil (i.e., Daconil Action®) to suppress bacterial wilt in annual bluegrass.

The most important cultural practices for managing bacterial wilt include increase mowing height; reduce mowing frequency; replace grooved rollers with solid rollers; avoid mowing when greens are wet and soft; avoid using plant growth regulators; and cease all grooming, aeration, and sand topdressing practices when the disease is active. On greens where bentgrass is the dominant species, the disease is best managed by employing cultural practices that promote bentgrass growth. Golf greens with a good creeping bentgrass base will respond positively to weekly spray applications of about

0.125 lb N/1000 ft² (6.2 kg N/ha) from urea. Water management is important. Soils should not be allowed to become excessively wet. This necessitates frequent syringing and hand-watering of areas prone to rapid drying. Longer-range planning should include modifications to the growing environment. This may involve tree and brush removal, use of fans to promote air circulation, and improved water drainage. In situations where the disease is a chronic problem, greens composed primarily of annual bluegrass may have to be regrassed with creeping bentgrass.

Pathovars of *X. translucens* (formerly *X. campestris* pathovar *poannua*) were developed as biological agents for annual bluegrass control on golf courses but were not marketed because of erratic results. These bacteria are sensitive to ultraviolet (UV) light and thus need to be applied in the evening. They also require six to ten weekly applications. Mowing was needed after each application to create wounds. The reported levels of annual bluegrass control with some *X. translucens* biotypes (reported as *X. campestris* pathovar *poannua*) varied from 0 to 82% (Johnson, 1994; Zhou and Neal, 1995). It is likely that different *Poa annua* biotypes adapted to different regions will vary in their susceptibility to *X. translucens*. Biological agents also require specific environmental conditions in order to survive, reproduce, and incite disease, which are difficult to achieve.

Summary of Key Points

- Anthracnose, summer patch, and bacterial wilt are diseases of annual bluegrass; all three are seldom found attacking creeping bentgrass.

- If greens have less than 10% annual bluegrass, a true form of IPM would be achieved by allowing diseases specifically attacking annual bluegrass to go unchecked.

- *Xanthomonas translucens* causes bacterial wilt in annual bluegrass by plugging the vascular system of infected plants.

- Symptoms of bacterial wilt in annual bluegrass include small white to tan-colored spots about the size of a dime (1.8 mm). Leaves may appear mottled, yellow, or etiolated before dying.

- Bacterial wilt is diagnosed in a laboratory by cutting leaves and looking for rapid streaming of bacterial cells from vascular bundles.

- Acibenzolar is a chemical that triggers natural plant defense mechanisms and shows promise for the suppression of bacterial wilt.

- Bacteria wilt is best managed by increasing mowing height, reducing mowing frequency, using solid rollers, mow when the canopy is dry, and avoid mechanical injury from sand topdressing, coring, and so on.

Yellow and Purple Nutsedge and Kyllinga Control in Creeping Bentgrass

Yellow nutsedge (*Cyperus esculentus*), purple nutsedge (*C. rotundus*), green Kyllinga, and false green Kyllinga (*Kyllinga* spp.) are perennials and members of the sedge family. Sedges have grass-like leaves and three-angle stems. Leaves die in the winter and yellow and purple nutsedge survive as underground tubes (i.e., nutlets), stolons, and seed (Figure 7.15). Kyllinga species do not produce tubers, but both have stolons and rhizomes. Kyllinga often is found in clusters or turf-like patches. False green Kyllinga produces seed in the fall, but green Kyllinga produces seed all summer (Figure 7.16).

Yellow nutsedge is effectively controlled postemergence in creeping bentgrass tees and fairways with bentazon (Basagran®), halosulfuron (Manage®, Prosedge®, Sedgehammer®), and sulfentrazone (Dismiss®) (Table 7.1). Halosulfuron controls purple nutsedge and sulfentrazone controls green and false green Kyllinga and has activity on purple nutsedge. Sulfentrazone also has preemergence activity on sedges and Kyllinga. The aforementioned herbicides should be applied to sedges in the three- to five-leaf stage. Once yellow nutsedge plants develop seven or more leaves, plants begin to produce underground tubers. Older nutsedge plants require higher herbicide rates and sometimes more frequent treatment to be effectively controlled. Late summer applications of herbicides to nutsedge in low-cut fairways may not have enough leaf surface area for the herbicide to be taken-up in lethal concentrations. As a result, contacted leaves may die, but plants regenerate new leaves from stems or tubers. Because seed and underground tubers can

FIGURE 7.15
Yellow nutsedge is a perennial that turns brown after frost or herbicide treatment. (Photo courtesy of S. McDonald.)

FIGURE 7.16
Green Kyllinga grows in turf-like patches and flowers all summer.

germinate throughout the summer, two or more applications of an herbicide are usually needed.

Bentazon is a contact herbicide that has no preemergence activity on yellow nutsedge and generally requires three applications a season to provide complete control. Two applications of halosulfuron or sulfentrazone, however, often provide season-long nutsedge control because of their systemic activity. Sulfentrazone may be more effective in spring and autumn. Avoid using these herbicides when turf or sedges are under heat or drought stress. Herbicides may discolor creeping bentgrass but do not cause significant damage to turf if properly applied. Manufacturers recommend that their product be tank-mixed with a nonionic surfactant.

Perennial Ryegrass Control in Creeping Bentgrass

Perennial ryegrass (*Lolium perenne*) fairways, tees, collars, and approaches are sometimes converted to creeping bentgrass without first completely killing the perennial ryegrass. Perennial ryegrass seeded into roughs or green surrounds invariably finds its way into fairways. Perennial ryegrass growing in bentgrass fairways can be safely controlled in mature (i.e., >1 year old) creeping bentgrass with chlorsulfuron or metsulfuron (Table 7.1). These herbicides will severely damage creeping bentgrass seedlings. They should be applied in autumn to creeping bentgrass. Two or more applications on a 21-day interval may be required to control some of the more tolerant cultivars of perennial ryegrass. Both herbicides cause some yellowing in creeping bentgrass, especially when applied during warm periods in early autumn (Figure 7.17). They work very slowly and injury to perennial ryegrass may not be evident for 60 or more days. Overseeding bentgrass should be delayed for at least

FIGURE 7.17
Yellowing from spot-treating chlorsulfuron. Note greener bentgrass in sprayer skip.

8 weeks after applying chlorsulfuron or metsulfuron. These herbicides can move in water and should not be applied if rain is forecast. Bentazon will phytotoxically suppress perennial ryegrass in bentgrass fairways when applied during hot and humid periods in summer.

Roughstalk Bluegrass Control in Creeping Bentgrass

Roughstalk bluegrass (*Poa trivialis*) is an invasive weed grass in northern regions of the United States, but it is useful for overseeding bermudagrass greens in the south. Roughstalk bluegrass is often introduced as a contaminant in seed and once established it rapidly spreads by seed and stolons. The invasiveness of roughstalk bluegrass may be suppressed by some plant growth regulators (i.e., flurprimidol and paclobutrazol) as discussed in Chapter 8. During hot and dry periods, roughstalk bluegrass typically goes dormant and turns brown but invariably recovers with the advent of rainy weather (Figure 7.18). In frequently irrigated and syringed fairways, tees, and collars, however, roughstalk bluegrass can remain green throughout summer. It is best to keep tees and fairways as dry as possible in summer to induce dormancy in roughstalk bluegrass. Spoonfeeding and possible use of flurprimidol or paclobutrazol in summer allow creeping bentgrass to gain a competitive edge and push roughstalk bluegrass partially out of the stand. This approach reduces aesthetic quality by allowing the weed to become brown and drought dormant, but it is a viable cultural approach to reducing its invasiveness.

There are no highly effective selective herbicides for controlling roughstalk bluegrass. Bispyribac-sodium phytotoxically suppresses roughstalk bluegrass

FIGURE 7.18
Brown and summer dormant roughstalk bluegrass in a creeping bentgrass fairway.

in creeping bentgrass and perennial ryegrass fairways. Sulfosulfuron also targets roughstalk bluegrass but is no longer labeled for use on any cool-season grasses due to phytotoxicity issues. Substantial populations of roughstalk bluegrass will survive or will recolonize treated areas from surviving buds on stolons or from seed. Bispyribac-sodium should be applied at least twice on a 14- to 21-day interval beginning in late spring. This herbicide may elicit a brilliant yellowing that can persist for several weeks. See the "Annual Bluegrass (*Poa annua*) Control" section for more information about bispyribac-sodium.

Unfortunately, the best way to control roughstalk bluegrass if it has dominated more than 50% of the stand is to renovate with the nonselective herbicide glyphosate. This should be done on a large scale just prior to overseeding in late summer, assuming there are no lingering soil residues from a preemergence herbicide. Roughstalk bluegrass must be actively growing because glyphosate will not kill plants or stolons that are dormant. Escaping roughstalk bluegrass is a likely problem, and sufficient time (20 to 30 days) usually is not available to determine where surviving plants exist. Furthermore, if creeping bentgrass seed lots used in renovation are contaminated with roughstalk bluegrass, little will ultimately be achieved. Creeping bentgrass seed should always be tested by an independent laboratory to ensure that no contamination exists. It should be noted that even "Blue Tag" seed may contain agronomically significant levels of roughstalk bluegrass. Golf course superintendents should buy certified seed and insist on at least a 50 gram seed sample to better detect the potential presence of roughstalk bluegrass.

Bermudagrass Control in Creeping Bentgrass

Bermudagrass (*Cynodon* spp.) is a common weed in creeping bentgrass in transition and southern regions of the United States (Figure 7.19). There currently are no safe and effective herbicides for removing bermudagrass from creeping bentgrass. Multiple applications of bispyribac-sodium, ethofumesate, and siduron at rates labeled for use on creeping bentgrass discolor and suppress but do not eliminate bermudagrass. The most effective approach is to control bermudagrass in roughs and green surrounds, preventing it from encroaching into creeping bentgrass fairways, tees, collars, and greens. Fenoxaprop-ethyl and triclopyr are effective in phytotoxically suppressing bermudagrass, but not in creeping bentgrass. Bermudagrass, however, is seldom completely eliminated. Fenoxaprop-ethyl and triclopyr are safe to apply to most of the common cool-season grasses grown on golf course roughs and green surrounds in North America and elsewhere. Kentucky bluegrass, perennial ryegrass, tall fescue, and fine leaf fescues are very tolerant of both herbicides, but there can be some yellowing or growth suppression. Some superintendents tank-mix fenoxaprop-ethyl and triclopyr for more rapid results. Programs of suppression should begin at spring green-up of the bermudagrass. Three or four applications of fenoxaprop and triclopyr on a 30-day interval will provide more than 90% bermudagrass control. Note, however, that multiple applications of triclopyr would kill creeping bentgrass. Fenoxaprop-ethyl rates required for bermudagrass suppression are much higher than label use rates for crabgrass and goosegrass control in bentgrass fairways (Dernoeden, 1989). Hence, fenoxaprop-ethyl at rates used to control bermudagrass would severely injure creeping bentgrass.

FIGURE 7.19
Bermudagrass entering winter dormancy in creeping bentgrass green surround.

Where bermudagrass has become established in small pockets in creeping bentgrass, the only realistic approach is to spot-treat the bermudagrass with glyphosate, remove the dead turf, and re-sod. Removal of the dead turf with a sod cutter reduces the potential for bermudagrass survival from stolons and shallow rhizomes, which sometimes escape from glyphosate (Figure 7.20). For large areas the recommendation would be to use glyphosate or a fumigant and overseed with creeping bentgrass. See Chapter 9 for more information on renovation.

Creeping Bentgrass Control in Green Surrounds and Roughs

Creeping bentgrass becomes a serious weed problem when it escapes collars and fairways and colonizes green surrounds and roughs (Figure 7.21). One of the most dreaded lies (i.e., the bird's nest lie) in a green surround is when the ball finds itself immersed in creeping bentgrass (Figure 7.22). Herbicides that are available to control or suppress creeping bentgrass are listed in Table 7.2. Mesotrione and triclopyr can be used to control or significantly suppress creeping bentgrass in Kentucky bluegrass, perennial ryegrass, tall fescue, and fine leaf fescue roughs and green surrounds (Figure 7.23). Generally, three to four applications on a 30-day interval are required (Dernoeden et al., 2008). The most striking symptom of mesotrione activity is that it turns susceptible plants white about 1 week after application. Mesotrione can elicit a similar effect (usually whitening of leaf tips) in most other cool-season grasses. Perennial ryegrass and tall fescue are somewhat sensitive species

FIGURE 7.20
Bermudagrass cannot be selectively controlled in creeping bentgrass and is treated with a nonselective herbicide (usually glyphosate) and the dead sod removed.

FIGURE 7.21
Creeping bentgrass becomes a weed when it escapes collars and grows into green surrounds.

FIGURE 7.22
A golf ball nestled in a creeping bentgrass green surround is difficult to advance accurately.

to mesotrione, but they normally recover fairly quickly. In some unusual cases with mesotrione, there is significant yellowing and possibly death to perennial ryegrass and tall fescue plants when soils remain saturated with water for long periods. Should overseeding be necessary, there is no waiting period for mesotrione and about a 14-day period for triclopyr. Green surrounds totally dominated by creeping bentgrass should be renovated in late summer using glyphosate as described in Chapter 9.

TABLE 7.2

Nonselective and Other Herbicides Used to Control or Phytotoxically Suppress Creeping Bentgrass in the United States

Original Trade Name[a]	Chemical Name	Uses and Comments[b]
Finale®	Glufosinate	Used to spot treat annul bluegrass plants on greens
		It is nonselective and will kill all green tissues contacted
		Can be used for renovation but may not be as effective as glyphosate
Round-Up®	Glyphosate	Preferred nonselective (i.e., kills all green tissue contacted) herbicide for renovation because of its highly systemic nature
		May be used to spot treat annual bluegrass on greens
Tenacity®	Mesotrione	Selectively controls creeping bentgrass in fescue species, Kentucky bluegrass, and perennial ryegrass
		May discolor or injure desired grasses, but turf normally recovers
Turflon Ester®	Triclopyr	Selectively controls creeping bentgrass and bermudagrass safely in fescue species, Kentucky bluegrass, and perennial ryegrass
		May discolor desired grasses temporarily

[a] No discrimination is intended against other trade names or generic products not mentioned.
[b] Carefully read the most current manufacturer's label for information on rates, timing, and target plants because changes can and do occur.

FIGURE 7.23
Creeping bentgrass was selectively removed from tall fescue (*Festuca arundinacea*) with mesotrione and triclopyr ester in this study site.

Summary of Key Points

- Halosulfuron and sulfentrazone are safe and effective in controlling yellow and purple nutsedge in creeping bentgrass. Sulfosulfuron controls sedges, but at high rates it can discolor or injure creeping bentgrass.

- Sulfentrazone also effectively controls Kyllinga postemergence and has preemergence activity on yellow nutsedge sedge and Kyllinga.

- Yellow and purple nutsedge and Kyllinga should be controlled before they produce more than five leaves.

- Perennial ryegrass can be selectively controlled in creeping bentgrass fairways and tees with chlorsulfuron and metsulfuron.

- Chlorsulfuron and metsulfuron are best applied in the autumn, have long soil residuals, and can move in water drainage patterns.

- Roughstalk bluegrass can be phytotoxically suppressed by bispyribac-sodium. Leaves are killed, but plants invariably recover from buds on stolons or seed in soil.

- Use certified creeping bentgrass seed and insist on at least a 50 gram sample to increase the probability of detecting roughstalk bluegrass in seed lots.

- Bermudagrass cannot be selectively controlled in creeping bentgrass.

- Bermudagrass can be excluded by phytotoxically suppressing it with fenoxaprop-ethyl or triclopyr in Kentucky bluegrass, perennial ryegrass, tall fescue, and fine leaf fescue green surrounds and roughs.

- Creeping bentgrass becomes a weed when it escapes from collars into green surrounds and from fairways into roughs.

- Creeping bentgrass can be selectively controlled with mesotrione or triclopyr in Kentucky bluegrass, perennial ryegrass, tall fescue, and fine leaf fescue green surrounds and roughs.

Dealing with Soil Residues of Preemergence and Other Herbicides

Soil residues from some herbicides, particularly preemergence herbicides, can be among the most difficult problems for superintendents to manage. The lingering soil residues of preemergence herbicides can be affected by several factors as follows: excessive thatch-mat layers, cool and dry weather during the summer, soil type and organic matter content, and the herbicide and rate applied. Most injury is due to overlapping or poor calibration or an error in communication between the superintendent and applicator. Preemergence herbicides are relatively insoluble, and they bind very tightly to organic matter. Most of these herbicides persist at levels that may adversely affect overseeding for 6 to 16 weeks after application. Some herbicides can build-up in some soils after 3 or more years of use.

Sterilants are nonselective, have a very long soil residual, and have no place in turfgrass management. Should a sterilant be involved, the soil may have to be excavated and removed from the site. Among the most persistent herbicides used on cool-season grasses include bensulide, dithiopyr, prodiamine, pendimethalin, and oxadiazon. Chlorsulfuron and metsulfuron, which selectively remove perennial ryegrass from bentgrass, also can persist for very long periods in soil. Broadleaf weed and postemergence annual grass weed herbicides (including bispyribac-sodium, ethofumesate, and fenoxaprop-ethyl) persist for about 1 to 3 weeks in soil. Where there is excessive thatch-mat (i.e., >0.5 inch; >1.25 cm), stems and roots may become largely confined to the organic layer and therefore would be in constant exposure to persistent herbicides. In this situation, root tips may swell and appear "clubbed" in response to residues because the meristem located behind the root cap can pick up the herbicide, thus causing interference with normal mitosis (i.e., cell division) (Figure 7.24). Shortened roots with numerous branches are another herbicide-induced root deformity. Herbicides are degraded mostly by microorganisms and some are lost via volatilization (especially pendimethalin and dithiopyr EC) and by ultraviolet light degradation. The strong sorption of most preemergence herbicides by thatch results in very slow biodegradation. Biodegradation is enhanced in warm and moist soil; however, the process would be slowed by unusually cool or dry summer weather. Hence, normal biodegradation can be retarded by both

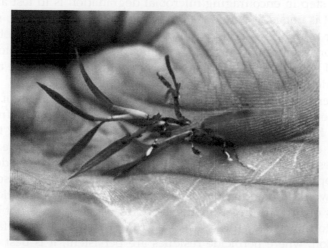

FIGURE 7.24
Preemergence herbicides interrupt cell division and can cause clubbing of roots on creeping bentgrass stolons.

high organic matter content in soil, mat or thatch, as well as by cool or dry soils that slow the activity of degrading microorganisms. There is little degradation of preemergence herbicides over winter months in northern and transition zone regions of North America, and little can be done to impact soil residues before spring.

If you suspect that herbicide residues may be present or that one could potentially adversely affect seed germination, there is a simple bioassay that you can perform. Collect several cup-cutter-sized plugs about 2 to 3 inches (5 to 7.5 cm) deep from affected areas. With a sharp knife, slice off the top layer of foliage and stems, leaving the underlying thatch-mat plus soil. Then slice off the top layer of thatch-mat and any soil lying just below the stems (about 0.5 in. soil; 1.25 cm). Scarify the top surface and seed the species of grass you plan to use into the upper 0.25 inch (6.2 mm) of thatch-mat plus soil. Also, plant seed into the top 1 inch (2.5 cm) layer of soil below the thatch-mat layer. As a check or standard for comparison, plant seed into soil collected from a nearby area that received no herbicide. Keep soil moist and in a sunny and warm location. Monitor seed germination from both layers and the non-herbicide-treated soil standard. If there is an herbicide residue it will appear in the form of poor germination or poor seedling vigor of plants from the top layer. Seedlings in herbicide-affected soils usually are stunted, yellow, or whitish in color, leaves tend to curl, and roots are short and stubby. The second layer usually will exhibit normal seedling development because the more persistent preemergence herbicides are insoluble and are not likely to leach at herbicidal levels below the top inch (2.5 cm) of thatch-mat plus soil. The ultimate standard, however, will be to compare the health of seedlings growing in herbicide-treated versus non-herbicide-treated soil.

The first step in encouraging microbial degradation is to test and adjust soil pH if it is outside the 6.5 ± 0.5 pH range. For a pH soil test, randomly select 10 or more 0.75 inch (1.9 cm) diameter cores from the root zones of three or more representative sites. Combine cores and send them to a laboratory for pH determination. If pH is low, apply ground limestone at the rate recommended based on the soil test.

Thatch removal is an important step in negating the effects of preemergence and other herbicides. The best way to deal with a residue from a preemergence herbicide is to strip to a depth of 0.5 to 1 inches (1.2 to 2.5 cm) of soil and discard the old sod. This should remove most of the herbicide because these chemicals are relatively insoluble and do not move deeply into soil. An application of activated charcoal to the soil prior to resodding or reseeding would add additional insurance to a successful renovation. Where large acres of turf are involved the stripping of sod would not be practical and rototilling may ruin a level and uniform green, tee, or fairway. In this situation, where herbicide residues are persistent, it is best to scarify thatch-mat and soil by a combination of core aeration and vertical cutting (Figure 7.25). Vertical cutting should be performed as deeply and in as many directions as possible. This will tear up a lot of turf and organic matter. Soil

FIGURE 7.25
Seeding failed and site was cored to minimize a herbicide residue. Inset shows bentgrass seedlings dying from an herbicide residue.

should be disturbed to the point where the site has a seed-bed-like quality. Organic matter should be raked or blown-off and the site then treated with an agricultural grade of activated charcoal at a rate of 10 lb/1000 ft² (488 kg/ha). Most preemergence herbicides are effectively tied up by activated charcoal. Organic herbicides bind to the carbon lattice of activated charcoal. Although binding renders the herbicide unavailable to microbial degradation, the herbicide will be bound so tightly that it will not desorb (i.e., move off the lattice) and therefore will not be available for root exposure or uptake.

Once seedlings have emerged in charcoal-treated areas in the autumn, apply a complete N + P + K fertilizer using a slow-release N-source before winter; if turf is mature and dense, core aerate the following spring if a significant thatch-mat layer remains; spot apply activated charcoal to areas where seedling development is weak, and overseed these thin areas within 48 hours of applying charcoal; do not apply any herbicides in the spring; protect young stands with preventive fungicides applications; spot-treat annual grass weeds in summer with postemergence herbicides and avoid blanket-spraying of herbicides; apply micronutrients and biostimulants; mow as high as possible in the spring and collect clippings; and apply 0.125 to 0.2 lb N/1000 ft² (6.2 to 10 kg N/ha) every 7 to 14 days to weakened areas that do not appear to experience a normal spring flush of growth.

Summary of Key Points

- Residues of preemergence herbicides can interfere with successful overseeding or renovation projects.

- Preemergence herbicides are relatively insoluble and can persist for a period of 6 to 16 weeks depending on product and rate. Clubbing of root tips or failure of bentgrass stolons to root in soil is a good indicator of a preemergence herbicide residue.

- Sterilants, chlorsulfuron, and metsulfuron have extremely long persistence, whereas most broadleaf weed herbicides persist for only about 2 weeks.

- Most herbicides are degraded by microbes, but some are lost to volatilization or photodecomposition.

- For a soil chemical analysis you need to tell the chemist what herbicide to search for in a sample, and getting test results may take several weeks.

- A simple bioassay may be the quickest way to detect a soil residue.

- Collect several cup-cutter plugs and slice off the top layer of green tissue leaving the underlying thatch and soil. Scarify the thatch plus soil sample and seed the desired turf species. As a standard for comparison, collect soil from an area where no herbicides were used.

- Place samples in a warm and sunny location and keep moist until seedlings appear.

- Inspect seedlings from the non-herbicide-treated soil and compare to those in the thatch layer and second soil layer.

- If there is a surface residue issue, strip off existing sod and either re-seed or sod.

- Deeper soil residues may require thorough mixing of soil and use of activated charcoal.

Controlling Primitive Plants: Blue-Green Algae and Moss

Blue-green algae (aka cyanobacteria) and moss contain chlorophyll and are very simple plants. These plants do not parasitize grasses, but they are highly invasive and often outcompete grasses for space, especially in wet or shaded environments. Hence, they are similar to weeds in terms of their effects on turf.

Blue-Green Algae or Cyanobacteria

Algae are described as primitive plants because they lack roots, stems, and leaves. The types of algae that grow in wet turf soils are primarily blue-green, filamentous, and have a mucus-like coating. Blue-green algae were reclassified as cyanobacteria (*cyano* is Greek for the color blue). The terms *algae, blue-green algae,* and *cyanobacteria* are used interchangeably by golf course superintendents and herein. Like bacteria, the organelles of blue-green algae are not bound by a membrane (i.e., prokaryotic), and they reproduce by fission. These algae contain chlorophyll and are capable of photosynthesis, and some continue to view these organisms as being more like primitive plants than bacteria. The true algae (i.e., eukaryotic) have membrane-bound nuclei and can be green or brown in color. Among true algae, only a few species of green algae are found in turf. The green algae that inhabit the turf microenvironment are unicellular and do not gather into colonies or form filaments. Their possible role in turf maladies is unknown (Hodges, 1993). Blue-green algae consist of filaments termed *trichomes* and reproduce by fission (i.e., splitting in two). Blue-green algae are common inhabitants of turf and their presence in high populations can result in the formation of scum, which can seal the surface. Their slimy filaments also can bind sand particles and impede water infiltration and percolation. Algae may be associated with the formation of some black-layers (Hodges, 1992).

Blue-green algae are capable of fixing atmospheric nitrogen, which may enable them to contribute some nitrogen to turf. This may seem beneficial, but blue-green algae can seal off soils or plug pore spaces, creating an anaerobic condition that is very harmful to the grass. Common genera of blue-green algae found in turf include *Nostoc* spp., *Phormidium* spp., and *Oscillatoria* spp. The filaments of these algae are long and are composed of a single cell. Individual cells may develop thick walls and behave like spores, enabling them to survive unfavorable environmental conditions. Blue-green algae secrete a gelatinous substance (i.e., mucilage) and eventually can form a black scum in thatch-mat or on the surface of wet soils. The mucilage protects cells by helping to maintain a moist microenvironment around individual filaments.

Blue-green algae can be a chronic problem on golf greens, especially those located in shade, other poor-growing environments, or where turf lacks density. Black algal blooms develop on greens during extended periods or warm, rainy, and overcast weather from spring to autumn in the United States (Figure 7.26). Black algal scums are particularly common on greens that are compacted, low lying, adjacent to ponds, and shrouded in trees. Algae are seldom a problem on higher mowed collars, tees, or fairways. Their growth is encouraged by extended periods of rainy, overcast, and warm weather in summer. Use of fertilizers containing phosphorus and some fungicides (primarily DMI/SI) can promote blue-green algae in greens in the summer. They are particularly a problem on greens because the very low mowing heights expose a much greater thatch-mat surface. Algal scums slow water infiltration into soil and keep thatch-mat wet for extended periods, further encouraging the proliferation of algae (Figure 7.27). Filaments can bind sand particles and plug pore spaces thereby impeding water percolation. Algal scums also impede oxygen and other gas diffusion into and out of soils. Algal scums can accumulate on older, lower leaves, thus further reducing turf density. In the United Kingdom,

FIGURE 7.26
Long periods of rainy, overcast weather promote blue-green algae growth in golf greens.
(Photo courtesy of J. Kaminski.)

FIGURE 7.27
Blue-green algal scums and crusts can seal the surface of golf greens and cause death of turf.

copious amounts of slimy algal scums built up on the surface of turfs during cold wet winter months. The blackish scums, known as *squidge*, are so massive that they can cause footwear to stick in the mucilaginous masses.

Management

Blue-green algae cannot be effectively controlled over the long run unless the conditions that predispose turf to their growth are corrected. Control is therefore aimed at alleviating wet soil conditions by improving surface and internal water and air drainage, sunlight exposure, and by employing proper irrigation practices. This ultimately may require removal of trees and shrubs; installation of fans; repairing, cleaning, or installing new drains; and possibly rebuilding of chronically wet greens. Increasing mowing height is recommended; persistent low mowing will continue to aggravate the condition.

Applications of ground agricultural limestone or iron sulfate may provide some short-term alleviation by desiccating algae. Hydrated lime (1 to 3 lb/1000 ft^2; 50 to 150 kg ha^{-1}) is more effective in algae suppression than ground limestone or iron sulfate, but it can scorch turf. Prior to application of any desiccant, scums should be manually broken up by handheld garden tools (e.g., "Weed Weasel") or by vertical cutting, spiking, or quadratine aeration and the debris removed. Simultaneously, mowing height should be increased. The higher turf canopy will shade algae, reducing their photosynthetic capacity and therefore their ability to grow and compete with turf. A light application of water-soluble nitrogen such as urea (i.e., 0.1 to 0.2 lb/1000 ft^2; 5 to 10 kg N/ha) in combination with iron sulfate (1 to 2 oz/1000 ft^2; 3 to 6 kg/ha) could be beneficial to golf greens that exhibit slow growth or poor vigor. Avoid complete (N + P + K) fertilizers, such as water-soluble 20-20-20, when algae are present because the phosphorus can promote algal growth. During autumn, nitrogen fertility should be increased to stimulate tillering, which improves stand density. Ammonium sulfate has been shown to be the most effective nitrogen source for suppressing blue-green algae. Ammonium sulfate has a high burn potential and should always be watered-in immediately following application.

Fatty acid soaps also reduce algal growth; however, they should be applied only when temperatures are below 80°F (27°C). When applied on hot and sunny days, these soaps dissolve the cuticle on grass blades and can cause a severe burning of turf. Some fungicides, such as fosetyl-aluminum, mancozeb, and particularly chlorothalonil, arrest blue-green algae. To be effective these fungicides need to be applied on 2-week intervals, and applications should commence prior to the time blue-green algae blackening appears. Preventive applications of chlorothalonil are most effective in controlling blue-green algae on greens. Higher rates and more frequent applications of chlorothalonil may be required to arrest blue-green algae once there is a conspicuous blackish bloom. Copper hydroxide and copper hydroxide plus mancozeb also help to reduce algal competitiveness. Copper-based products should be used with caution as they can yellow turf. Conversely, some DMI/SI fungicides (e.g., propiconazole, triadimefon, and others) and azoxystrobin can encourage growth of blue-green algae on greens in the summer.

Summary of Key Points

- The most common algae found growing in golf greens are blue-green and have been reclassified as cyanobacteria.
- Blue-green algae are filamentous, contain chlorophyll, and can fix nitrogen.
- Blue-green algae produce black mats or scums on the thatch-mat surface of golf greens, especially during overcast and rainy weather in summer.
- The presence of blue-green algae on greens is linked to very low mowing, low nitrogen fertility, shade, and excessively wet soils.
- Filaments can plug pore spaces or seal-off soils thus limiting gas exchange, which can lead to an anaerobic soil condition.
- Heavy black algal blooms compete for space and can reduce density of turf grown on golf greens.
- Some approaches to managing blue-green algae in golf greens include the following:
  - Increase mowing height and increase nitrogen fertility.
  - Ammonium sulfate is the preferred N-source for managing blue-green algae, but it has a high burn potential.
  - Avoid applying fertilizers containing phosphorous in the summer.
  - Avoid daily irrigation.
  - Improve drainage and soil aeration.
  - Improve sunlight penetration and air circulation by selective removal of trees and brush.
  - Desiccants, disinfectants, lime, iron sulfate, and soaps can suppress but do not eliminate blue-green algae.
- Chlorothalonil is the safest and the most effective chemical for controlling blue-green algae on golf greens.

Moss

Mosses are more advanced plants than algae, but like algae they have no vascular system. Common species found in turf include *Amblystegium* spp., *Brachythecium* spp., *Byrum* spp., *Funaria* spp., *Mnium* spp., and *Polytrichum* spp. *Byrum argentum* (silvery thread moss) is among the most common species found on golf greens in the United States. Colonies of *B. argentum* have a shiny or silvery appearance. They are composed of sheets of cells with leafy stems and root-like rhizoids. Rhizoids form at the base of shoots, anchor moss in soil, and function in water and nutrient uptake. Mosses have a complicated life cycle and water is required for reproduction to occur. They produce spores in a sporangium, which is borne on the top of a stalk. Although

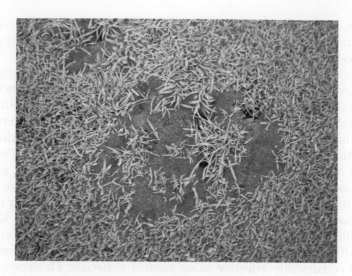

FIGURE 7.28
Moss colonies disrupt the uniformity of putting surfaces.

mosses usually grow in shady and moist places, they are capable of growing in full sun while surviving periods of drought. They also can grow in open areas during extended periods of overcast and wet weather. Moss often is found on slopes or knolls and other droughty areas of golf greens that are subjected to frequent scalping.

Moss colonies disrupt the appearance, surface uniformity, and ball roll on golf greens (Figure 7.28). The presence of moss is an indicator that greens are being cut too low, are underfertilized, or are over-watered. Mosses are seldom a problem on collars, tees, or fairways, demonstrating that low mowing is a major cause for moss invasion. Mosses are very aggressive and opportunistic, and they can eventually dominate large areas on putting greens. Mosses are especially invasive in the Pacific Northwest, Great Lakes, New England regions of the United States, and in the United Kingdom and northern Europe because weather conditions remain cool and overcast for long periods.

Management

Like blue-green algae, mosses gain a competitive advantage in poorly nourished turf or stands that are mowed very low. Wet conditions, aggravated by soil compaction, shade, and poor air circulation encourage the proliferation of moss. As is the case with blue-green algae, control is aimed at correcting those soil or cultural conditions that provide moss their competitive edge. Increasing mowing height while increasing nitrogen fertility are key cultural approaches to reducing the invasiveness of moss. Other cultural measures include improving water and air drainage, avoiding excessive irrigation, pruning trees and shrubs, maintaining proper soil fertility and soil pH levels, and alleviating soil compaction. Frequent sand topdressing and core aeration help to reduce moss competitiveness by disrupting its growth, injuring cells, and improving water and air movement into soil.

Iron sulfate is one of the oldest chemicals used for moss management. Even today in the United Kingdom, greenkeepers apply iron sulfate three or four times in late summer to burn-back or suppress moss. They follow with vertical cutting to physically remove moss and then apply some nitrogen to assist in turf recovery and to improve its competitiveness against reinvasion. Ammonium sulfate applications to supply 3 to 4 lb N/1000 ft^2 (150 to 200 kg N/ha) per year reduces moss. Higher rates of nitrogen, however, increase mowing frequency and reduce green speed. Furthermore, spoonfeeding at rates in excess of 0.3 lb N/1000 ft2 (15 kg N/ha) can increase moss (Thompson et al., 2001). Boesch and Mitkowski (2005) and Kennelly et al. (2010) provide good reviews of chemical moss control investigations. Carfentrazone, chlorothalonil, and sodium bicarbonate are probably the most consistently effective moss control materials. Older approaches to moss control such as the use of desiccants, soaps, iron, and copper are briefly reviewed here, but most have failed to provide any significant long-term control and are potentially phytotoxic.

Because moss have no vascular system, desiccants can be used to dry moss out, while giving turf some time to grow and compete better. Desiccants only temporarily arrest growth of mosses. Prior to applying a desiccant, existing moss should be physically injured or removed by raking or vertical cutting 2 to 3 days prior to applying the desiccant. When using desiccants, it is important that they be applied in enough water to thoroughly saturate the entire layer of moss. Desiccants should be applied in the evening, and tank-mixing with a wetting agent may improve their performance. Iron sulfate, fatty acid soaps, hydrogen dioxide, and copper compounds are common desiccants. Some golf course superintendents had success in safely reducing moss on greens in the summertime with Ultra Dawn® dishwashing detergent (Happ, 1998). Spot treatment is more effective than broadcast application. For spot drenching, mix only 1 to 3 fl. oz. Dawn per gallon of water (30 to 90 mL/3.8L) and spot-treat colonies of moss using a hand-pump or backpack sprayer. Dawn can be used as frequently as every 2 weeks. Soaps are erratic and do not work for everyone. Excessive applications of soaps are likely to run off and discolor turf. Blackened moss usually survives; however, if it develops a copper color it is more likely to die.

Copper hydroxide is effective but should be used with caution as it can burn or scorch bentgrass and annual bluegrass. Multiple applications on a 2-week interval during autumn and winter are most effective. Copper accumulation can occur over time and appears as a chlorosis as a result of inducing an iron deficiency. Chlorothalonil is perhaps the most effective material for suppressing moss in summer. Chlorothalonil does not kill moss but reduces the size of colonies. Chlorothalonil is most effective when applied at high label rates in the summer when air temperatures are high. Multiple applications made on 7- to 10-day intervals are required. Chlorothalonil generally is less effective in controlling moss in the autumn, spring, and winter. The fungicides maneb and zineb can be toxic to young moss; however, moss survives treatment with these materials.

Carfentrazone is safe and perhaps the most effective material available in the United States for controlling moss in golf greens. Carfentrazone is a herbicide that has other uses as described in Chapter 7. Carfentrazone is most effective when applied in the spring and autumn, but little or no control may be achieved with summer applications (Kennelly et al., 2010). Furthermore, summer applications of carfentrazone when air temperatures exceed 88°F (>31°C) may cause discoloration. The effective rate of carfentrazone is 0.10 lb ai/A (1.1 kg/ha). It should be tank-mixed with a surfactant and applied two or three times on a 14-day interval. It is helpful to spike or in some way injure the moss prior to applying carfentrazone. Spoonfeeding also improves the ability of creeping bentgrass to compete and fill-in areas where moss was injured or killed by carfentrazone.

Spring and autumn spot applications of sodium bicarbonate (i.e., baking soda) reduce moss populations (Kennelly et al., 2010). Sodium bicarbonate (6 oz/gallon water; 44 grams/L) should be spot applied twice on a 14-day interval to wet the moss colony. Some injury may be observed but is short lived. Spoonfeeding and increasing mowing height from 0.118 to 0.158 inches (3 to 4 mm) was shown to improve the level of moss control with sodium bicarbonate, chlorothalonil, and carfentrazone.

Whatever chemical approach you attempt, it is unlikely to provide long-term control unless the conditions that cause moss to develop are addressed. One product may work well in one year or at one location, but fail in other years or locations. Repeated treatments of the aforementioned materials often are needed, and most products are effective only when applied in the spring or autumn. As previously noted, the presence of moss is an indicator of poor growing conditions, and improving turf vigor is the best approach to minimizing a moss problem. Hence, a good nitrogen fertilizer and sand topdressing program, combined with proper water management, shade reduction, and core aeration, as well as increasing the mowing height are key to effective moss suppression.

Summary of Key Points

- Mosses are nonvascular plants.
- Moss disrupts the uniformity of putting surfaces and can dominate large areas of greens.
- Low mowing, scalping, poor drainage, and low nitrogen fertility promote moss invasion.
- Mosses proliferate in shady and moist places but also grow in open areas during extended periods of overcast and wet weather. Greens

in full sun with poor density, and kept excessively wet, also can develop moss problems.

- Moss can survive long periods of drought stress, even in sunny locations.
- Controlling moss involves similar approaches used to manage blue-green algae including:
  - Increase mowing height and nitrogen fertility.
  - Minimize moss invasiveness by applying ammonium sulfate at rates ranging from 3 to 4 lb N/1000 ft²/yr (150 to 200 kg N/ha/yr).
  - Avoid excessive or daily programmed irrigation.
  - Dry soils by improving water and air drainage, and prune trees to increase direct sunlight exposure.
  - Frequent topdressing and core aeration help to disrupt and reduce moss invasiveness.
  - Spring and autumn applications of carfentrazone and spot treatment with sodium bicarbonate are generally safe to apply to greens and can be effective.
  - Chlorothalonil can be used to safely manage moss in summer on golf greens but may only reduce the size of colonies.
  - It is helpful to spike or physically disrupt moss prior to applying a chemical treatment and to simultaneously spoonfeed with nitrogen.
  - Chemicals alone seldom provide for long-term moss control.

Bibliography

Beard, J.B. 2002. *Turf Management for Golf Courses*, 2nd ed. Ann Arbor Press, Chelsea, MI.

Boesch, B.P., and N.A. Mitkowski. 2005. Chemical methods of moss control on golf course putting greens. *Applied Turfgrass Science* Online. doi:10.1094/ATS-2005-1006-01-RV.

Branham, B.E., and W. Sharp. 2011. Annual bluegrass (*Poa annua*) control in seedling creeping bentgrass (*Agrostis stolonifera*) with bispyribac-sodium. *Applied Turfgrass Science* Online. doi: 10.1094/ATS-2011-1103-01.

Callahan, L.M., and E.R. McDonald. 1992. Effectiveness of bensulide in controlling two annual bluegrass (*Poa annua*) subspecies. *Weed Technol.* 6:97–103.

Cline, V.W., D.B. White, and H. Kaerwer. 1993. Observations of population dynamics on selected annual bluegrass-creeping bentgrass golf greens in Minnesota. *Intern. Turfgrass Res. Soc. J.* 7:839–844.

Dernoeden, P.H. 1989. Bermudagrass suppression and zoysiagrass tolerance to fenoxaprop. In *Proceedings of the Sixth International Turfgrass Research Conference*, H. Takatoh, Ed., pp. 285–290. Tokyo, Japan.

Dernoeden, P.H. 1998. Use of prodiamine as a preemergence herbicide to control annual bluegrass in Kentucky bluegrass. *HortScience* 33:845–846.

Dernoeden, P.H. 2011. Preemergence annual bluegrass control in fairway height zoysia. *Proc. Northeastern Weed Sci. Soc.* 65:75.

Dernoeden, P.H., C.A. Bigelow, J.E. Kaminski, and J.M. Krouse. 2003. Smooth crabgrass control in perennial ryegrass and creeping bentgrass tolerance to quinclorac. *HortScience* 38:609–612.

Dernoeden, P.H., J.E. Kaminski, and J. Fu. 2008. Selective creeping bentgrass control in Kentucky bluegrass and tall fescue with mesotrione and triclopyr. *HortScience* 43:509–513.

Dernoeden, P.H., S.J. McDonald, and J.E. Kaminski 2008. Creeping bentgrass and perennial ryegrass seedling tolerance to bispyribac-sodium. *HortScience* 43:2186–2190.

Dernoeden, P.H., and T.R. Turner. 1998. Annual bluegrass control tolerance of Kentucky bluegrass and perennial ryegrass to ethofumesate. *HortScience* 23:565–567.

Fidanza, M.A., P.H. Dernoeden, and M. Zhang. 1996. Degree days for predicting smooth crabgrass emergence in cool-season turfgrasses. *Crop Sci.* 36:990–996.

Gaussoin, R.E., and B.E. Branham. 1987. Annual bluegrass and creeping bentgrass germination response to flurprimidol. *HortScience* 22:441–442.

Happ, K.A. 1998. Moss eradication in putting green turf. *U.S. Golf Association Green Section Record* 36(5):1–5.

Hodges, C.F. 1992. Interaction of cyanobacteria and sulfate-reducing bacteria in subsurface black layer formation in high-sand content golf greens. *Soil Biol. Biochem.* 24:15–20.

Hodges, C.F. 1993. The biology of algae in turf. *Golf Course Management* 61(8):48,52,54,56.

Johnson, B.J. 1994. Biological control of annual bluegrass with *Xanthomonas campestris* pv. *poannua* in bermudagrass. *HortScience* 29:659–662.

Kaminski, J.E., and P.H. Dernoeden. 2007. Seasonal annual bluegrass seedling emergence patterns in Maryland. *Crop Sci.* 47:773–779.

Kaminski, J.E., and A.I. Putman. 2009. Safety of bispyribac-sodium on colonial bentgrass and influence on brown patch severity. *Int'l Turfgrass Soc. Res. J.* 11:219–226.

Kaminski, J.E., P.H. Dernoeden, and C.A. Bigelow. 2004. Creeping bentgrass seedling tolerance to herbicides and paclobutrazol. *HortScience* 39:1126–1129.

Kendrick, D., and K. Danneberger. 2005. Intraseeding to convert creeping bentgrass putting greens to new cultivars. *Golf Course Management* 73(1):143–146.

Kennelly, M.M., T.C. Todd, D.M. Settle, and J.D. Fry. 2010. Moss control on creeping bentgrass greens with standard and alternative approaches. *HortScience* 45:654–659.

McCarty, L.B. 2010. *Best Golf Course Management Practices*. Prentice-Hall, Upper Saddle River, NJ.

McCarty, L.B., J.W. Everest, J.W. Everest, D.W. Hall, T.R. Murphy, and F. Yelverton. 2001. *Color Atlas of Turfgrass Weeds*. Ann Arbor Press, Chelsea, MI.

McDonald, S.J., P.H. Dernoeden, and J.E. Kaminski. 2006. Creeping bentgrass tolerance and annual bluegrass control with bispyribac-sodium tank-mixed with iron and nitrogen. *Applied Turfgrass Science* Online. doi:10.1094/ATS-2006-0811-01-RS.

McElroy, J.S., R.H. Walker, and G.R. Wehtje. 2004. Annual bluegrass (*Poa annua*) populations exhibit variation in germination response to temperature, photoperiod, and fenarimol. *Weed Sci.* 52:47–52.

Mitkowski, N.A., M. Browning, C. Basu, K. Jordan, and N. Jackson. 2005 Pathogenicity of *Xanthomonas translucens* from annual bluegrass on golf course putting greens. *Plant Dis.* 89:469–473.

Shem-Tov, S., and S.A. Fennimore. 2003. Seasonal changes in annual bluegrass (*Poa annua*) germinability and emergence. *Weed Sci.* 51:690–695.

Sweeney, P.M., and T.K. Danneberger. 1995. RAPD characterization of *Poa annua* L. populations in golf course greens and fairways. *Crop Sci.* 35:1676–1680.

Thompson, C., M. Kennelly and J. Fry. 2011. Effect of nitrogen source on silvery-thread moss on a creeping bentgrass putting green. *Applied Turfgrass Science.* doi: 10.1094/ATS-2011-1018-02-RS.

Vargas, J.M., Jr., and A.J. Turgeon. 2004. Poa annua: *Physiology, Culture and Control of Annual Bluegrass.* Wiley, Hoboken, NJ.

Vauterin, L., B. Hoste, K. Kersters, and J. Swings. 1995. Reclassification of *Xanthomonas. Int. J. Syst. Bacteriol.* 45:472–489.

Zhou, T., and J.C. Neal. 1995. Annual bluegrass (*Poa annua*) control with *Xanthomonas campestris* pv. *poannua* in New York State. *Weed Technol.* 9:173–177.

8

Use of Plant Growth Regulators on Creeping Bentgrass

Two major factors influencing turfgrass maintenance budgets are the high costs of fuel and labor. Among the most time-consuming, labor-intensive cultural practice is mowing and clipping removal. Plant growth regulators (PGRs), which suppress the vertical shoot growth of turfgrasses, have the potential to significantly reduce expenditures allocated to fuel and labor used in mowing operations. Some products improve color, density, environmental stress tolerance, and playability of golf course turfs. Some PGRs are effective in slowing turf growth under fences, on steep banks, and other difficult to mow areas. Unfortunately, most PGRs will not uniformly suppress all grass or weed species in mixed stands.

Mefluidide (Embark T&O®), flurprimidol (Cutless®), paclobutrazol (Trimmit®), ethephon (Proxy®), and trinexapac-ethyl (Primo MAXX®) are the primary growth regulators used on fine, cool-season turf (Table 8.1). Older materials like maleic hydrazide and amidiochlor are too phytotoxic for use on fine golf course turfs and are seldom used. Most of these chemicals are absorbed by leaves and then are translocated to actively growing (i.e., meristematic) tissues. Flurprimidol and paclobutrazol are primarily root absorbed, although some of the chemical may be taken up by foliage. Once these chemicals reach growing points, they inhibit or slow down cell division or cell elongation by affecting hormone levels in plants. Mefluidide inhibits cell division, whereas flurprimidol, paclobutrazol, and trinexapac-ethyl inhibit cell elongation by interfering with gibberellic acid synthesis. Ethephon is a third type of growth regulator that inhibits cell elongation by interfering with the hormone ethylene.

Ethephon is marketed for use on cool-season grasses grown on greens, tees, fairways, and roughs. Multiple applications of ethephon to greens, tees, and fairways with excess thatch may render creeping bentgrass more prone to scalping due to the elongation of stems. Ethephon commonly is tank-mixed with trinexapac-ethyl and applied to creeping bentgrass greens and fairways to reduce clippings and for annual bluegrass seedhead suppression. Mefluidide primarily is used to suppress annual bluegrass seedheads on golf courses.

Flurprimidol, paclobutrazol, and trinexapac-ethyl were developed for the golf course market. Flurprimidol and paclobutrazol are primarily used on creeping bentgrass greens, tees, and fairways as a means of reducing annual

TABLE 8.1

Plant Growth Regulators Used on Creeping Bentgrass in the United States

Trade Name(s)	Chemical Name[a]	Formulation(s) and Use Rates[b]	Remarks
Proxy®	Ethephon	2S 3.4 lb ai/A (3.8 kg ai/ha) for greens, tees, and fairways	Suppresses seedheads and should be applied when annual bluegrass is in the boot stage. Tank-mix with trinexapac-ethyl to improve consistency and minimize discoloration and scalping. Bentgrass may develop a lime-green color and may show signs of scalping about 30 days after the second application of ethephon.
Cutless®	Flurprimidol[c]	50 W and 1.3 MEC Rates for bentgrass greens range from 0.06 to 0.25 lb ai/A (0.07 to 0.28 kg ai/ha) and tees and fairways 0.25 to 1 lb ai/A (0.28 to 1.1 kg/ha)	Used to suppress the competitiveness of annual bluegrass on golf greens, tees, and fairways as well as to reduce mowing frequency and clippings. May cause bentgrass foliage to develop a red or bronze color when applied during cold weather. Root absorbed and should be watered-in. Avoid use with high rates of dimethylene inhibitor/sterol inhibitor (DMI/SI) fungicides on greens.
Embark Turf & Ornamental®	Mefluidide	0.2S Rates for creeping bentgrass tees and fairways range 0.05 to 0.075 lb ai/A (0.056 to 0.084 kg ai/ha)	Suppresses foliar growth and annual bluegrass seedheads. Apply when annual bluegrass is in the boot stage. Can discolor and thin turf if applied when air temperatures are above 85°F (29°C). May only be applied to creeping bentgrass maintained above 0.375 in. (9.4 mm).
Trimmit®	Paclobutrazol[c]	2SC Rates for bentgrass greens range from 0.1 to 0.25 lb ai/A (0.11 to 0.28 kg ai/ha) and tees and fairways 0.25 to 0.5 lb ai/A (0.28 to 0.63 kg ai/ha)	Used to suppress the competitiveness of annual bluegrass on golf greens, tees, and fairways, as well as to reduce mowing frequency and clippings. May cause bentgrass foliage to develop a red or bronze color when applied during cold weather. Root absorbed and should be watered-in. Avoid use with high rates of DMI/SI fungicides on greens.

(continued)

TABLE 8.1 (continued)

Plant Growth Regulators Used on Creeping Bentgrass in the United States

Trade Name(s)	Chemical Name[a]	Formulation(s) and Use Rates[b]	Remarks
Primo MAXX®	Trinexapac-ethyl	ME Rates for bentgrass greens range from 0.047 to 0.055 lb ai/A (0.053 to 0.062 kg ai/ha) and tees and fairways 0.086 to 0.17 lb ai/A (0.096 to 0.19 kg ai/ha)	Primarily used on golf greens, tees, and fairways to reduce clippings and improve ball roll. Trinexapac elicits color enhancement, improves summer stress tolerance of bentgrass and annual bluegrass, and improves turf quality in shaded environments. Can injure annual bluegrass when applied in cold weather. Foliar absorbed, but is rain-safe once it dries on leaves.
Legacy®	Flurprimidol and Trinexapac-ethyl	1.5 MEC Rates for bentgrass greens range from 0.06 to 0.12 lb ai/A (0.067 to 0.13 kg ai/ha) and tees and fairways from 0.12 to 0.24 lb ai/A (0.13 to 0.27 kg ai/ha)	This prepackaged mix is used to suppress the competitiveness of annual bluegrass on golf greens, tees, and fairways, as well as to reduce mowing frequency and clippings. May produce less rebound and better color than flurprimidol alone. Allow to dry on leaves before watering-in.

Note: See the most current manufacturer's label for important information regarding proper use of growth regulators and disposal of containers.

[a] Reference to trade names does not constitute an endorsement, guarantee, or warranty. No discrimination is intended against products not mentioned.

[b] Consult label for the proper use rate as dosage is greatly dependent on turfgrass species to be treated as well as fertility level, time of year, environmental conditions, and other factors. Any mention of field observations or nonlabeled usage of these products is not endorsed nor does it constitute legal authority, guarantee, or warranty.

[c] Consult labels for specific information regarding the use of flurprimidol and paclobutrazol for annual bluegrass suppression in creeping bentgrass.

bluegrass competition and increasing ball roll speed of putting surfaces. All three PGRs are used to reduce clippings and can cause creeping bentgrass to develop a darker green color. Trinexapac-ethyl is used at lower rates and with greater frequency, and has a lower potential to injure turf than other PGRs. The length of time that PGRs suppress growth is dependent on product, rate, and weather conditions. In general, normal use rates of most PGRs would be expected to provide about 2 to 4 weeks of some level of growth suppression. Turf should be mowed prior to a PGR application. Periodical mowing will be necessary when using PRGs to maintain a uniform appearance.

Growth regulators can have harmful side effects. These include discoloration of or phytotoxicity to grass tissues; inhibition of cell division in all actively growing tissues (i.e., meristems), rather than selective inhibition of meristems affecting vertical shoot growth (e.g., mefluidide); postinhibition growth stimulation (i.e., a growth rate exceeding normal rates of growth after the regulatory effects have dissipated; aka rebound); inhibition of root development; reduction in recuperative capacity of turf; deterioration of stand density; and ethephon may predispose some grass species to scalping. Rebound or postinhibition growth stimulation is a problem with most PGRs other than ethephon. Basically, when PRG effects dissipate, turfgrasses will grow more rapidly (i.e., rebound) than if they had not been treated with a PGR. Problems with rebound include scalping, heavy clipping deposits, and slower green speed. Reduced recuperative potential occurs because there is less leaf, tiller, or stolon development when turf is being regulated. This factor would preclude the use of these compounds on summer-stressed and weakened golf turfs whose recuperative potential must be high. For obvious reasons, it is important to avoid scalping and other forms of mechanical injury to PGR-treated turfs. Wilt, heat stress, and pest activity should be monitored closely. Discoloration and turf injury associated with PGRs are described below. Plant growth regulators should be applied in ≥44 gallons of water per acre (411 L/ha) to creeping bentgrass.

Plant growth regulator usage for clipping management offers the greatest benefits when applied in early spring. Spring applications of a PGR can reduce the "spring flush" growth of turfgrasses. This is particularly important in spring because frequent rainstorms create difficult mowing conditions, especially on sloped areas. Furthermore, part-time laborers needed for mowing operations on golf courses generally are not available during early spring. Aside from suppressing foliar growth, PGR treatment may enable turfgrass plants to use water and nitrogen fertilizers more efficiently during summer months. Gibberellic acid inhibitors such as flurprimidol, paclobutrazol, and trinexapac-ethyl darken and enhance turf color and improve density. Gibberellic acid inhibitors may stunt annual bluegrass seedheads, but they do not dramatically reduce the numbers of seedheads produced.

Annual Bluegrass Suppression and Maintenance with Plant Growth Regulators

Paclobutrazol and flurprimidol are used for annual bluegrass suppression. The key to success is never to allow annual bluegrass to have a growth advantage over creeping bentgrass. This means beginning PGR treatments in early spring and continuing applications into late autumn. During mild winters, annual bluegrass should be closely monitored for growth. Once annual bluegrass leaves begin to grow above the bentgrass canopy, the PGR should be reapplied, even in late autumn or early spring. Frosts coinciding with applications of these PGRs, however, may cause objectionable levels of reddish-brown-purple-blue or bronze discoloration in creeping bentgrass (Figure 8.1). Both flurprimidol and paclobutrazol will elicit a yellow or brownish-green color in annual bluegrass, and creeping bentgrass may develop a blue-green color during warmer periods (Figure 8.2). Both suppress annual bluegrass growth for a longer period than for creeping bentgrass. As the PGR effect

FIGURE 8.1
Flurprimidol and paclobutrazol can discolor creeping bentgrass clones when applied in cold weather. (Photo courtesy of S. McDonald.)

FIGURE 8.2
Flurprimidol and paclobutrazol elicit a blue-green color in creeping bentgrass and a yellow color in annual bluegrass. (Photo courtesy of S. McDonald.)

FIGURE 8.3
Creeping bentgrass encroaching in a winter-purple annual bluegrass clone in response to a paclobutrazol program.

on creeping bentgrass dissipates, stolon growth is stimulated (i.e., rebound). Vigorously growing bentgrass stolons physically encroach into surrounding annual bluegrass turf, essentially pushing annual bluegrass gradually out of the stand (Figure 8.3). In addition, fewer annual bluegrass seedlings emerge and survive with root-absorbed flurprimidol and paclobutrazol (Gaussion and Branham, 1987; McElroy et al., 2004; Dernoeden, 2011). To achieve long-term progress in annual bluegrass reduction, a PGR should be applied on a 2- to 4-week interval throughout the growing season. Some golf course managers apply low PGR rates weekly and believe that this approach provides them better control in regulating growth, while maintaining the beneficial effects of the PGR. During warm to hot and rainy periods, paclobutrazol and flurprimidol may dissipate rapidly. Thus, during warm to hot and rainy weather periods of summer higher rates of both paclobutrazol and flurprimidol may be needed to achieve the same PGR effect that was provided by lower rates used in spring and autumn. Flurprimidol and paclobutrazol applications are normally suspended during hot and dry and otherwise stressful summer periods when annual bluegrass growth is naturally suppressed by the environment. Superintendents who develop a disciplined program can achieve 20% or greater reduction in annual bluegrass per year on greens. An even greater reduction in annual bluegrass can be obtained annually in bentgrass fairways. For example, flurprimidol applied monthly throughout the golf season over a 2-year period reduced annual bluegrass populations in a creeping bentgrass fairway by 75% to 85% (Bigelow et al., 2007). Similar results were achieved using paclobutrazol (McCullough et al., 2005). These marked reductions in annual bluegrass, however, will be short lived if the PGR program is suspended or stopped for long periods. It is important to note that biostimulants containing gibberellic acid (GA) should not be used when applying PGRs to suppress annual bluegrass. These PGRs are GA inhibitors, and applications of even small amounts of GA can partially or totally counteract their effects.

Paclobutrazol and flurprimidol generally are applied to golf greens and fairways every 2 to 4 weeks beginning in early spring and ending in late autumn. Most superintendents develop their own comfort levels with flurprimidol and paclobutrazol by experimenting with different rates and timings. These products are root absorbed and should be watered-in within 24 to 72 hours of application. Areas treated with these materials should not be mowed prior to watering-in or clippings should be returned. Mid-to-late summer applications of flurprimidol or paclobutrazol may be deferred because high temperatures naturally suppress annual bluegrass. Because it is less potentially injurious, trinexapac-ethyl is used on creeping bentgrass and mixed bentgrass–annual bluegrass greens and fairways during summer stress periods instead of flurprimidol or paclobutrazol. If summer weather conditions are not very stressful, flurprimidol and paclobutrazol applications may be continued on an as-needed basis throughout summer. These PGRs should not be tank-mixed with or applied to mixed annual bluegrass-creeping bentgrass greens within 2 weeks of a DMI/SI fungicide (e.g., propiconazole, triadimefon, others). This is because DMI/SI fungicides have growth regulator additive effects that may cause a more rapid loss of annual bluegrass than may be desired, when used in conjunction with flurprimidol or paclobutrazol. The aforementioned tank-mix problem is most likely to occur when high rates of DMI/SI fungicides are used to target root pathogens. Frequent applications of flurprimidol and paclobutrazol darken turf color and can provide significant levels of dollar spot suppression. Both PGRs cause discoloration when applied in cold weather. In most cases turf cover usually is maintained, but there are exceptions. Turf loss can occur in weak or poorly rooted areas on golf greens (Figure 8.4). These PGRs can discolor or severely damage sod on greens and fairways where there is poor rooting between the sod layer and underlying soil (Figure 8.5). Due to potential soil residues, it is best to delay any overseeding for 6 to 8 weeks following an application of flurprimidol or paclobutrazol.

Trinexapac-ethyl is not effective in reducing the competitiveness of annual bluegrass. Conversely, frequent applications of trinexapac-ethyl improve the summer stress tolerance of annual bluegrass. Trinexapac-ethyl is used extensively on creeping bentgrass and annual bluegrass in summer because it not only is unlikely to discolor turf, but it darkens foliage, thus improving aesthetic quality. Early spring applications of this PGR in cold weather, however, can yellow and injure annual bluegrass. Trinexapac-ethyl helps to improve green speed, reduces clippings, may tighten or improve turf density, can improve the residual effectiveness of some fungicides, and may improve the environmental stress tolerance of turf. Trinexapac-ethyl also improves the quality of turfgrasses grown in shaded environments (Steinke

FIGURE 8.4
Weak, poorly rooted areas treated with flurprimidol or paclobutrazol can redden and thin creeping bentgrass in cold weather.

FIGURE 8.5
Paclobutrazol injury in 1-year-old creeping bentgrass sod on an approach. (Photo courtesy of S. Potter.)

and Stier, 2003). Evidently, inhibited GA synthesis helps to reduce excessive growth, and therefore slows carbohydrate usage by plants in shaded environments. Trinexapac-ethyl rates vary depending on formulation, site (i.e., greens vs. fairways), turf species, and desired effect (i.e., green speed improvement, clipping reduction, etc.). Label recommended rates and timings serve as a starting point for using this and other PGRs. Turf managers need to experiment with rates and application frequencies to determine the most efficacious way to use the product. Because trinexapac-ethyl is more rapidly degraded as air temperatures rise, managers may need to increase trinexapac-ethyl use rates or application frequency in the summer. High-density creeping bentgrass cultivars (e.g., 'Declaration', Penn A and G-series; Tyee, others) may require higher rates to be effectively suppressed.

Flurprimidol and trinexapac-ethyl are available in a prepackaged mixture (i.e., Legacy®). Flurprimidol blocks GA late in the biosynthetic pathway, and trinexapac-ethyl blocks GA synthesis early in the pathway. It was suggested that blocking GA biosynthesis both early and late creates more efficiency. Trinexapac-ethyl provides rapid growth suppression because it is foliar absorbed, whereas flurprimidol in soil provides longer-lasting regulation via root uptake. There are anecdotal reports of reduced rebound effect toward the end of application intervals, reduced bronzing effect and more consistent foliar growth with this combination. This combination, however, may not be as effective as flurprimidol alone in suppressing annual bluegrass on greens (Baldwin and Brede, 2011). Trinexapac-ethyl is leaf absorbed and should not be watered-in, whereas flurprimidol is root absorbed and should be watered-in. Thus, it is best to allow this product to dry on leaves and then water-in within 24 to 72 hours of application. Areas treated with these materials should not be mowed prior to watering-in or clippings should be returned.

Annual Bluegrass Seedhead Suppression

Annual bluegrass seedheads create bumpy putting surfaces in the spring and detract from the uniformity of turf. These seedheads often are recognized by golfers and can be a subject of criticism. Grooming removes many seedheads, and slightly lowering mowing height in spring may help mask their appearance. Many superintendents attempt to chemically control these seedheads, but success often is elusive. Mefluidide, ethephon, and tank-mixes of ethephon and trinexapac-ethyl are used to suppress annual bluegrass seedheads in creeping bentgrass and other cool-season grasses (Figure 8.6). Achieving effective seedhead suppression can be difficult, and timing is everything. The two methods used to time a PGR for annual bluegrass seedheads suppression are to scout for the "boot stage" of seedhead development or use a growing degree day model. Growing degree day models tend to be less effective for reasons explained below.

To be effective, PGRs need to be applied when annual bluegrass seedheads are in the "boot stage." The boot stage is scouted by examining stem bases for swelling (Figure 8.7). The swelling or bulging indicates stem elongation has begun as the apical meristem (i.e., main growing point of each tiller) is differentiating and transforming into an inflorescence (i.e., seedhead)

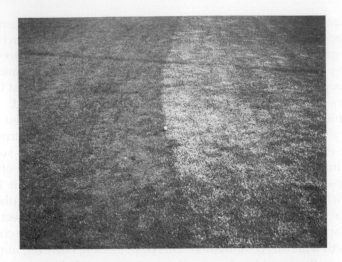

FIGURE 8.6
Annual bluegrass seedhead control with mefluidide. (Photo courtesy of S. Zontek.)

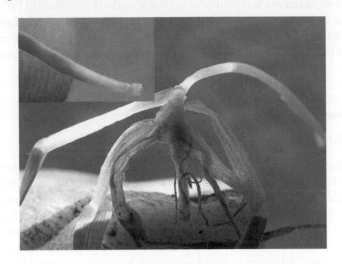

FIGURE 8.7
Annual bluegrass seedheads in the boot stage. (Photos courtesy of J. Borger.)

(Figure 8.7). As the inflorescence develops it elongates between leaf sheaths. Growth regulators are most effectively applied at the time swelling (i.e., boot stage of seedhead) is evident in stem bases. Once seedheads appear in large numbers effective control is unattainable.

Searching for the boot stage is tedious, and growing degree day (GDD) models tend to be unreliable in most regions. The problem with GDD models is that there are many microenvironmental niches on golf courses. Temperatures associated with open versus shaded and northern versus southern exposures greatly affect plant development. Hence, not all greens or fairways are ready for treatment on the same day or even the same week. These microenvironmental niches are what stifle efforts to develop reliable GDD models for predicting annual bluegrass seedhead development. Growing degree day models also can be unreliable because they are strictly temperature driven, and data are collected in one location. Large fluctuations in temperature and variation in microclimate (i.e., shade versus full sun, exposure, etc.) often render GDD models ineffective for this purpose. Hence, managers should use a GDD model while scouting for the "boot stage" and be prepared for variable results in different microclimates.

Mefluidide may be more consistently effective in northern and colder regions where temperatures remain fairly constant in spring. Ethephon is believed to be more consistent in southern California, where spring temperatures also are more consistent. In transition and southern regions there can be brief periods of warm and sunny weather in late winter. These early season fluctuations in temperature often result in early and more sporadic seedhead development versus what generally occurs in regions where temperatures remain seasonally consistent in winter and spring. Late winter and early spring heat-ups followed by cool-downs result in fickle annual bluegrass seedhead development, making effective seedhead suppression with PGRs difficult and erratic from year to year in many regions.

Mefluidide is labeled for use on creeping bentgrass mowed above 0.375 inches (9.4 mm). It is prudent to tank-mix mefluidide with a chelated iron product to mask most of the discoloration elicited by this PGR. Chelated iron products, however, reduce the level of seedhead suppression. Moderate rates of a wetting agent tank-mixed with mefluidide and chelated iron can slightly boost seedhead suppression (Watschke and Borger, 1998). Depending on rate, mefluidide is applied twice on a 2- to 4-week interval. Rain or irrigation within 8 hours of application can reduce efficacy. Levels of seedhead suppression vary, but 80% control would be considered excellent for mefluidide. Injury in the form of yellowing or browning of leaves may appear within 1 week of treatment and persist for 2 or more weeks (Figure 8.8). Severe leaf injury in bentgrass can occur with mefluidide if there are sprayer overlaps or if applied when air temperatures are above 85°F (29°C).

Ethephon alone and ethephon plus trinexapac-ethyl tank-mixes also yield inconsistent or variable levels of annual bluegrass seedhead suppression. The combination of ethephon and trinexapac-ethyl can improve bentgrass quality and the level of seedhead suppression compared to ethephon alone. As noted previously, it is believed that more consistent and moderate weather conditions contribute to greater consistency with ethephon. For maximum effectiveness, ethephon should be applied in the boot stage prior to the emergence of seedheads. A second application of ethephon alone or tank-mixed with trinexapac-ethyl may be made 2 to 3 weeks following the initial application. Ethephon is rain-safe within 2 hours of application. Growth suppression in creeping bentgrass will last a few weeks, but unlike other

FIGURE 8.8
Injury to a golf green caused by a misapplication of mefluidide.

PGRs, there is little or no postinhibition growth stimulation (i.e., rebound) with ethephon. Again, results can be highly variable, but excellent seedhead suppression with ethephon usually is in the range of 70% to 80%. Seedheads produced in ethephon-treated surfaces remain stunted and may not affect ball roll significantly.

Ethephon elicits a lime-green color in creeping bentgrass and can predispose creeping bentgrass to scalping. One study identified a potential problem with ethephon use on creeping bentgrass maintained as a golf green (Dernoeden and Pigati, 2009). In that study, ethephon and trinexapac-ethyl were applied alone or in tank-mix combination either once or twice beginning in spring (i.e., April) in two separate years. Scalping in the Providence creeping bentgrass green first became evident on a consistent basis in early June between 30 and 37 days following the second ethephon application (Figure 8.9). Two applications of ethephon alone or tank-mixed with trinexapac-ethyl resulted in significant scalping that persisted from 36 to 55 days. Plots treated twice with ethephon had unacceptable quality for 28 to 35 days in each year. Five applications of trinexapac-ethyl partially mitigated scalping in plots treated twice with ethephon. Close visual examination of plants revealed that ethephon caused a distortion in normal shoot development as a result of abnormal elongation of the apical meristem of individual axillary buds. That is, normal shoots were replaced with what appeared to be upright stolons with nodes (Figure 8.9). Two ethephon applications were required to elicit the scalping. Even though this phenomenon is not known to be widespread, golf course superintendents should be aware of the potential problem. Managers are cautioned to leave untreated areas for comparison to monitor for potential scalping beginning about 30 days following their second application of ethephon. Annual bluegrass is not known to be rendered susceptible to scalping by ethephon.

FIGURE 8.9
In some creeping bentgrass cultivars scalping and deformed plants may develop about 30 days following two ethephon applications. Inset shows distortion of bentgrass tillers caused by ethephon.

High rates of some wetting agents alone also reduce or delay annual bluegrass seedhead emergence. Wetting agents alone provide erratic levels of seedhead suppression and are generally ineffective. Late autumn applications of ethofumesate and bispyribac-sodium also may reduce annual bluegrass seedhead production the following spring. Some observed appreciable levels of annual bluegrass seedhead suppression on golf greens using the high label rate of the fungicide propiconazole. In this program, propiconazole is applied twice on a 21- to 28-day interval in early spring prior to the anticipated appearance of annual bluegrass seedheads. Flurprimidol, paclobutrazol, and trinexapac-ethyl do not significantly reduce the number of annual bluegrass seedheads produced, but they reduce their height.

Flurprimidol, paclobutrazol, and trinexapac-ethyl may provide some additional benefits as follows: improved green speed at mowing heights greater than 0.125 inch (3.1 mm), color enhancement, improved turf density, possibly increased rooting, and possibly reduced water and nitrogen requirements. These growth regulators often provide for smoother, less bumpy ball roll and generally provide more consistent green speed from morning to afternoon. Flurprimidol and paclobutrazol have fungicidal activity and can be very effective in reducing dollar spot severity (Putman and Kaminski, 2011). Trinexapac-ethyl is capable of extending the residual efficiency of some fungicides. This beneficial effect from trinexapac-ethyl is attributed to reduced clipping removal, thus keeping fungicides on or in plants for longer periods. Flurprimidol and paclobutrazol do not appear to enhance disease severity, but there is some concern that their use over time may lead to reduced sensitivity and even resistance of DMI/SI fungicides targeting dollar spot. Trinexapac-ethyl and probably other gibberellic acid synthesis PGRs (e.g., flurprimidol and paclobutrazol) improve turf quality in shaded environments.

Summary

- To suppress annual bluegrass seedheads, apply mefluidide, ethephon, or ethephon + trinexapac-ethyl in early spring when seedheads are in the boot stage. Once numerous seedheads are evident these PGRs will be ineffective.

- Mefluidide is labeled for use on creeping bentgrass maintained above 0.375 in. (9.4 mm), whereas ethephon may be used on greens, tees, and fairways.

- Chelated iron products can be mixed with mefluidide to mask or reduce discoloration.

- Flurprimidol, paclobutrazol, and trinexapac-ethyl do not reduce seedhead production.

- Label rate and application intervals provide important guidelines for using flurprimidol and paclobutrazol to suppress annual bluegrass competitiveness. Many turf managers adjust rates and application intervals to determine the most efficacious approach to using a PGR.

- Trinexapac-ethyl improves the summer stress tolerance of annual bluegrass and improves the ability of turfgrasses to tolerate shade.

- Use the correct concentration in 40 to 60 gallons of water per acre (374 to 561 L/ha), apply uniformly, and avoid overlaps.

- Rainfall or irrigation within 8 hours of applying mefluidide may reduce effectiveness. Trinexapac-ethyl and ethephon are rain-safe within 2 hours or when chemical has dried on foliage. Water-in flurprimidol and paclobutrazol following application.

- Delay spraying if clippings or other debris are present as they will prevent uniform PGR distribution.

- Avoid mowing within 12 hours of a PGR application.

- Mow PGR-treated turf as needed to improve uniformity and appearance.

- Do not apply a PGR to scalped turf.

- Do not scalp PGR-treated turf, unless attempting to convert to a different species.

- Do not apply a PGR to weak or stressed turf.

- Avoid use of flurprimidol, paclobutrazol, or trinexapac-ethyl with biostimulants containing gibberellic acid as they may counteract each other.

- Apply gibberellic acid to counteract any perceived deleterious effects of flurprimidol, paclobutrazol, or trinexapac-ethyl.

- Avoid applying flurprimidol or paclobutrazol and high rates of DMI/SI fungicides to golf greens within 2 weeks of one another during periods of high temperature stress.

- Application of flurprimidol or paclobutrazol during cold periods may result in objectionable levels of reddish-brown, purple-blue, or bronze discoloration in creeping bentgrass.

- Application of trinexapac-ethyl during cold periods may yellow or injure annual bluegrass.

- Ethephon elicits a lime-green color in creeping bentgrass and annual bluegrass, and multiple applications may render creeping bentgrass prone to scalping.

- Avoid drought stress and excessive traffic on PGR-treated turf.

- Closely monitor weed, disease, and insect pest encroachment.

- Nitrogen fertilizer used in conjunction with PGRs will improve turf quality but may decrease the duration of growth inhibition.

Bibliography

Baldwin, C.M., and A.D. Brede. 2011. Plant growth regulator selection and application rate influence annual bluegrass control in creeping bentgrass greens. *Applied Turfgrass Science* Online. doi:10.1094/ATS-2011-0517-02-RS.

Bigelow, C.A., G.A. Hardebeck, and B.T. Bunnell. 2007. Monthly flurprimidol applications reduce annual bluegrass populations in a creeping bentgrass fairway. *Applied Turfgrass Science* Online. doi:10.1094/ATS-2007-0509-02-RS.

Dernoeden, P.H. 2011. Preemergence annual bluegrass control in fairway height zoysiagrass. *Proc. Northeastern Weed Sci. Soc.* 65:75, Abstract.

Dernoeden, P.H., and R.L. Pigati. 2009. Scalping and creeping bentgrass quality as influenced by ethephon and trinexapac-ethyl. *Applied Turfgrass Science* Online. doi:10.1094/ATS-2009-0601-01-RS.

Gaussion, R.E., and B.E. Branham. 1987. Annual bluegrass and creeping bentgrass germination response to flurprimidol. *HortScience* 22:441–442.

McCarty, L.B. 2010. *Best Golf Course Management Practices*. Prentice Hall, Upper Saddle River, NJ.

McCullough, P.E., S.E. Hart, and D.W. Lycan. 2005. Plant growth regulator regimens reduce *Poa annua* populations in creeping bentgrass. *J. Appl. Turfgrass Sci.* Online. (http://www.plantmanagementnetwork.org/pub/ats/research/2005/pgr/).

McElroy, J.S., R.H. Walker, and G.R. Wehtje. 2004. Annual bluegrass (*Poa annua*) popu-lations exhibit variation in germination response to temperature, photoperiod, and fenarimol. *Weed Sci.* 52:47–52.

Putman, A.I., and J.E. Kaminski. 2011. Mowing frequency and plant growth regulator effects on dollar spot severity and on duration of dollar spot control by fungi-cides. *Plant Dis.* 95:1433–1442.

Steinke, K., and J.C. Stier. 2003. Nitrogen selection and growth regulator application for improving shaded turf performance. *Crop Sci.*, 43:1399–1406.

Watschke, T.L., and J.A. Borger. 1998. Seedhead suppression of annual bluegrass on a putting green. *Proc. N.E. Weed Sci. Soc.* 52(44). Abstract.

9

Renovation

Greens Renovation with Fumigants

It is not possible to convert to creeping bentgrass greens or fairways by intraseeding into existing stands of annual bluegrass or other grasses (Kendrick and Danneberger, 2005). When converting greens to eliminate annual bluegrass or some other highly undesirable pest, it is best to use a fumigant such as dazomet (Basamid®) or methyl bromide. Methyl bromide is being phased out by the U.S. Environmental Protection Agency (EPA), but some turfgrass sites may be exempted. Fumigation with methyl bromide is the best approach to renovating greens because it is rapid, very effective, and there is no soil residual. Methyl bromide may only be handled by a licensed applicator because it is a colorless, odorless, and deadly gas. Greens to be renovated should first be treated with the nonselective herbicide glyphosate and cored as deeply and in as many directions as possible. The applicator will lay down tubes that will carry the methyl bromide, then cover greens with a plastic tarp, and anchor its ends with soil. Methyl bromide is packaged as a liquid under pressure, but when released under a tarp the liquid volatilizes and becomes a gas. Methyl bromide kills all plants and most seed (does not control seed of sedges or morning glory) and microorganisms it contacts. Tarps are removed in 24 to 48 hours and greens are allowed to air-out for a few days before being seeded.

Unlike methyl bromide, dazomet is applied in granular form. Dazomet is effective in reducing annual bluegrass seed survival, especially when treated areas are covered with plastic tarps (Landschoot and Park, 2004). Dazomet-treated sites do not require a plastic cover; however, annual bluegrass seed survival is enhanced if treated areas are not covered. The site to be treated should be cored as previously described. If large fairway areas are being fumigated it is best to not apply glyphosate because errors, skips, washes, or other surviving areas will not be evident after dazomet is watered-in. Dazomet requires watering-in for several days, and it is important to ensure that granules do not run-off to adjacent, desirable turf. It is especially important to take all measures necessary to avoid movement of granules into ponds or streams. In general, dazomet-treated sites can be seeded with creeping bentgrass between 3 and 6 days after application.

Using only glyphosate to renovate likely will result in reinfestation and probably severe competition with annual bluegrass. Late seeding or poor creeping bentgrass seed germination often results in more annual bluegrass encroachment. Hence, fumigation is the best approach to renovating golf greens. Methyl bromide is the preferred fumigant, but due to cost in handling and applying the material it usually is not economically feasible to use on large numbers of acres. Dazomet may be more erratic and potentially problematic should granules wash and kill adjacent healthy sites or worse find their way into ponds and streams where the chemical can kill fish. Dazomet also is expensive.

Fairway Renovation with Herbicides and Plant Growth Regulators

For large numbers of acres that compose fairways, glyphosate is the preferred nonselective herbicide for renovation. Depending on region, fairways can be made playable by late autumn if overseeding is successfully achieved in late summer. Glyphosate will kill all vegetation, and this may not be desired or practical for some golf courses. Hence, golf clubs may find it preferable to attempt to overseed creeping bentgrass into an existing cool-season turf and to keep fairways in play during the conversion. This approach involves phytotoxically suppressing the undesired cool-season grass with a high rate of mefluidide. Mefluidide is the best PGR for overseeding because high rates will suppress existing grass without leaving a soil residual that would adversely impact bentgrass germination. Phytotoxic suppression will not be successful if bermudagrass (*Cynodon* spp.) or other warm-season perennial grasses are dominant. This method is much less effective than using a fumigant or glyphosate, but at least people can play golf during this type of renovation.

Trinexapac-ethyl is preferred by some superintendents because there is less discoloration, but it is not nearly as effective in promoting bentgrass seedling competitiveness compared to mefluidide. Paclobutrazol and flurprimidol are not options due to their lengthy (6 to 8 weeks) soil residual (Kaminski et al., 2004). Creeping bentgrass seeded into areas treated with flurprimidol and paclobutrazol will exhibit delayed emergence or a stunted growth habit.

For overseeding creeping bentgrass into an existing cool-season grass fairway, apply mefluidide at a high label rate (about two to three times higher than that used for seedhead suppression). High mefluidide rates used in overseeding programs will cause an objectionable level of yellowing and browning, which will persist for several weeks. Existing turf is likely to be stunted and severely damaged or killed. This phytotoxic response is what will give creeping bentgrass seedlings a competitive chance to establish.

Scalping the PGR-treated turf prior to overseeding also helps provide bent-grass seedlings with a competitive advantage. An application of siduron at the time bentgrass is seeded should be considered if crabgrass (*Digitaria* spp.) or yellow foxtail (*Setaria glauca*) invasion is likely. The overseeding should be completed and the creeping bentgrass emerged several weeks prior to the time annual bluegrass germinates. As previously discussed, creeping bentgrass seedlings cannot compete with large numbers of annual bluegrass seedlings going into autumn. Postemergence annual bluegrass control with bispyribac-sodium may be warranted in late summer to early autumn (see Chapter 7). Obviously, proper timing for planting seed is the best option to giving creeping bentgrass a chance to become established.

As previously discussed, a preemergence herbicide (except siduron) should not be applied in the spring prior to or following an autumn seeding of creeping bentgrass. Except for siduron, preemergence herbicide residues can cause a failure in bentgrass seeding projects. Poor or sparse establishment of creeping bentgrass as a result of herbicide residues contributes to significant increases in annual bluegrass and provokes criticisms from golf course officials and golfers.

Summary

- It is not possible to convert to creeping bentgrass by interseeding into annual bluegrass stands.
- Methyl bromide is the preferred fumigant for renovating golf greens because it acts rapidly and has no soil residual.
- Methyl bromide is a colorless, odorless, and deadly gas that is injected under a tarp by a licensed applicator.
- Dazomet is a fumigant that is applied in granular form and works best when treated areas are covered with a plastic tarp.
- Dazomet granules must be watered-in for several days, and granules carried in rain or irrigation water to streams will likely kill or injure all vegetation they contact as well as fish.
- Given the large number of acres involved, fairways usually are renovated with the nonselective herbicide glyphosate.
- To avoid closing the golf course, some memberships prefer that fairways be phytotoxically suppressed with mefluidide before over-seeding with creeping bentgrass.
- Mefluidide is applied at a high rate that will stunt, discolor, and possibly kill turf. Fairways then are scalped and disk-seeded with creeping bentgrass.

- The phytotoxic suppression approach using mefluidide yields inferior results compared to using a fumigant or glyphosate.

- Timing is critical and creeping bentgrass cover needs to be achieved in late summer prior to the time annual bluegrass seedlings emerge.

- Have creeping bentgrass seed lots tested independently using 50 gram samples to increase the probability of detecting noxious weed seed.

Bibliography

Dernoeden, P.H. 1997. The transition from perennial ryegrass to creeping bentgrass fairways for the Mid-Atlantic Region. *U.S. Golf Association Green Section Record* 35(5):12–15.

Kaminski, J.E., P.H. Dernoeden, and C.A. Bigelow. 2004. Creeping bentgrass seedling tolerance to herbicides and paclobutrazol. *HortScience* 39:1126–1129.

Kendrick, D., and K. Danneberger. 2005. Intraseeding to convert creeping bentgrass putting greens to new cultivars. *Golf Course Management* 73(1):143–146.

Landschoot, P.J., and B.S. Park. 2004. Renovating putting greens without methyl bromide. *Golf Course Management* 72(2):127–131.

Park, B.S., and P.J. Landschoot. 2003. Effect of dazomet on annual bluegrass emergence and creeping bentgrass establishment in turf maintained as a golf course fairway. *Crop Sci.* 43:1387–1394.

10

Selected Invertebrate Pests

There are a few significant insect and nuisance pests of creeping bentgrass that warrant mention. As was the case with soil and soil fertility, a thorough review on invertebrate pest biology and research is not within the scope of this monograph, but a few are briefly considered here. In North America some of the most notable invertebrate pests of creeping bentgrass, which will be briefly discussed, include the following: annual bluegrass weevils (*Listronotus maculicolis*), turfgrass ants (*Lasius neoniger*), black cutworms (*Agrotis ipsilon*), cicada killer wasps (*Sphecius speciosus*), sod webworms (*Parapediasia teterrella*, *Pediasia trisecta*, others), cranberry girdler or subterranean webworm (*Chrysoteuchia topiaria*), and earthworms (*Lumbricus terrestris*, *Apporecodea* spp., others).

Other important insect pests of cool-season turfgrasses grown on golf courses, which will not be discussed, include the following: larval species of root-feeding beetles including black turfgrass ataenius (*Ataenius spretulus*), green June beetle (*Cotinis nitida*), Japanese beetle (*Popillia japonica*), northern and southern masked chafers (*Cyclocephala* spp.), and others. Larvae of European crane fly (*Tipula paludosa*) are a significant problem in the Pacific Northwest and New England regions of the United States and Canada, the United Kingdom, and Europe. These crane flies also feed on roots and crowns. Chinch bug (*Blissus leucopterus hirtus*) nymphs and adults are surface feeders and are becoming increasingly important pests in areas of native fescues and bunker surrounds. Excellent descriptions and photographs of these pests, their biology, and control strategies can be found in Potter (1998) and Vittum et al. (1999). Insecticides commonly used to target insect pests in the United States are listed in Table 10.1.

Annual Bluegrass Weevils

The annual bluegrass weevil (ABW) formerly was considered a species of *Hyperodes*. It is a beetle in the weevil family and is a pest of highly maintained, short-cut turfgrasses in northeastern, Mid-Atlantic, and eastern Canada regions and is moving southward and westward. It was believed for years that the destructive ability of ABW was restricted to annual bluegrass, including the perennial subspecies (i.e., *Poa annua* spp. *reptans*), but

TABLE 10.1

Common Annual Bluegrass and Creeping Bentgrass Insect Pests and Insecticides[a] Available in the United States

Insecticide Class	Common Name	Original Trade Name[b]	Common Targets in North America
Carbamate	Carbaryl	Sevin®	Surface feeding insects[c] and surface migrating green June beetle larvae
Organophosphate	Chlorpyrifos	Dursban™	ABW[d] adults, surface feeders
	Acephate	Orthene®	White grubs[e]
	Trichlorfon	Dylox®	White grubs[e]
Pyrethroid	Bifenthrin	Talstar®	Ants, ABW adults, surface feeders
	Cyfluthrin	Tempo®	
	Deltamethrin	DeltaGard®	
	Lambda-cyhalothrin	Scimitar®	
	Permethrin	Astro®	
Neonicotinoid	Dinotefuran	Zylam®	Generally used for white grub control, billbugs, some have limited activity on ABW
	Clothianidin	Arena™	
	Imidacloprid	Merit®	
	Thiamethoxam	Meridian™	
Spinosyn	Spinosad	Conserve™	Black cutworm, sod webworms, ABW larvae
Microbial	*Bacillus thuringiensis*	Dipel®	Young black cutworm and sod webworm larvae
Diacylhydrazine	Halofenozide	Mach 2™	Used mostly for white grub control
Hydramethylnon	Hydramethylnon	Maxforce®	Ant bait
Avermectin	Abamectin	Advance®	Ant bait
Oxadiazine	Indoxacarb	Provaunt™	Black cutworm, sod webworm, ABW larvae
Anthranilic diamide	Chlorantraniliprole	Acelepryn™	Black cutworm, sod webworm, white grubs, ABW larvae

[a] Adapted with permission from S. McDonald, President of Turfgrass Diseases Solutions, LLC.

[b] Original trade name is the original patented product name. Others may exist on the market with a different trade name but similar formulation and same active ingredient. These are referred to as postpatent products. See individual product labels for application instructions, rates, and timing.

[c] Surface feeders generally include black cutworm, sod webworm, cranberry girdler, and chinchbugs.

[d] Annual bluegrass weevil (ABW) larval stage inflicts damage to stems causing wilting; however, applications targeting adults may provide the best control option.

[e] Many species of root feeding white grubs (i.e., larvae) including: Japanese beetle, northern and southern masked chafers, black turfgrass Ataenius, European crane fly, and the green June beetle. Different species can dominate a specific geographical location.

ABW has become a significant pest in creeping bentgrass fairways and golf green collars in some regions (Figures 10.1 and 10.2). The biology and control of ABW were outlined by Cowles et al. (2008) and Vittum et al. (1999) and are reviewed below.

Annual bluegrass weevils have a complete life cycle with as many as two to four generations per year possible. The adults are small (0.125 inch; 3.1 mm), compact, and differ in color from black to gray. Adults have a characteristic

FIGURE 10.1
Annual bluegrass weevil damage to a mixed annual bluegrass and creeping bentgrass fairway. Inset shows larvae in thatch. Inset shows an older larva. (Photos courtesy of S. McDonald.)

FIGURE 10.2
Annual bluegrass weevil damage in mixed annual bluegrass and creeping bentgrass collars. (Photos courtesy of D. Bevard and S. McDonald.)

FIGURE 10.3
Larvae, callow adult (bottom right), and true annual (top left) bluegrass weevil adult. (Photo courtesy of S. McDonald.)

weevil snout. Their body is covered with fine hairs and scales that are easily observed under magnification. Newly emerged adults (aka callow adults) look similar to an older adult ABW but have a reddish-brown color that gradually darkens (Figure 10.3). The eggs of the ABW are very small and oblong but can be found in leaf sheaths of the grass plant using a dissecting microscope. Larvae are legless, with a creamy white body and dark-brown head (Figure 10.3). Some larvae have black inner markings on their back. The older larvae appear slightly curved, but not nearly as C-shaped as white grub beetle species (Figure 10.1). Pupae are cream-colored and have some of the features of the adult (e.g., snout, wings, legs). Weevils over-winter as adults in higher mown turf and litter in wood lines. They begin to migrate from the fine turf areas to these sites in early autumn. It is believed that they fly back to over-wintering sites, although a mixture of flight and walking has been observed.

Adults typically begin to migrate back into fine turf areas in early spring. Adult feeding usually causes little or no significant damage, especially in the spring. Adults often are visible in the spring on warm, sunny days walking across the turf. Females chew a small hole in the leaf sheath and insert two or three eggs inside the sheath. Each female can produce at least 40 eggs, and research shows that each larva can kill 10 to 12 plants. Larval formation takes about a month for all five instars, and development from egg to adult takes 2 months.

Annual bluegrass weevil damage often shows up in mid-spring and can occur throughout summer months. Damage is first noticed in the perimeter of greens and fairways that support a high population of annual bluegrass. Most of the damage is caused by the larval stage and can go unnoticed for some time. Damage to shortly mown annual bluegrass initially appears as yellowing, which mimics basal rot anthracnose symptoms. Damage to plants begins when adult females chew into the outer leaf sheaths and lay their eggs between the sheaths. This process may weaken the plant and cause yellowing, but rarely will it kill a plant. When eggs hatch, young larvae feed on the inner leaf sheaths during their first two instars. Sometime between the second and third instar, larvae move down sheaths and begin feeding on stem or crown tissue. As more eggs hatch, more severe damage can occur. Feeding damage associated with the fifth instar (in the crown) is the most destructive. At this point, turf will appear to be under severe drought stress due to damaged stems. Turf will appear purple or wilted before it turns brown and dies out. As the fifth instars molt to pupae and then to callow adults, damage may continue to worsen although most of the feeding has stopped for this generation. The problem is that there usually are multiple generations. During mid-summer, all developmental stages (i.e., adults, pupae, and early and late instar larvae) can be present in the same general area at the same time. This overlapping of life cycles complicates control measures.

Early in spring the most effective scouting method for finding adults is to use a soap flush. Mix 0.5 to 1 fluid ounce of a lemon-scented soap in one gallon of water (15 to 30 mL soap in 3.8 L water) in a watering can. This amount of soapy solution should be spread over an area of 4 to 6 ft^2 (4.8 to 10.8 m^2). Wait about 5 minutes, and if present, adults will become visible walking on leaves. To avoid discoloring turf, be sure to rinse the treated area later with additional water to dilute the soap. Use of linear pitfall traps may be useful in tracking migration patterns in a single location. The limitation to the pitfall trap method, however, is that it is a snapshot in a single location. First and second instar larvae are very small and difficult to detect; however, it can be accomplished by using a salt flush. To make a salt flush, take a cup cutter plug of turfgrass and remove all soil using a knife. Cut the plug into small pieces (about 1 inch [2.5 cm] on a side), and place into a solution of salt and water. This solution is made by placing 1 lb of salt into a gallon of water (454 grams salt in 3.8 L water) and mixing. Early instar larvae will float to the surface within 5 minutes and appear white with a small brown head capsule. Late instar larvae can be easily detected by cutting a wedge of turf with a knife or cup cutter and looking in the turf/thatch interface. The threshold for damage when turf is vigorously growing is 30 to 50 larvae per ft^2 (900 cm^2). This threshold significantly decreases as turf becomes stressed, especially in the summer. Plants damaged by ABW feeding can easily be pulled from the soil.

Cultural management options include proper nutrition and irrigation, which can mask symptoms. The most effective cultural control method is to reduce or eliminate annual bluegrass populations. Few natural enemies exist. Spinosad, a biological product derived from the bacterium *Saccharopolyspora spinosa*, is highly effective on larvae. Insect parasitic nematodes and other biological control agents were investigated; however, results are highly variable (McGraw et al., 2011). Where there is a chronic infestation of ABW, it would not be expected for cultural or biological options to provide acceptable levels of control by themselves.

Insecticides remain the most effective way to ensure that the turf stand is not severely damaged by the ABW. Some strategies suggest spot-treating areas that have high populations of adults or had past ABW infestations. Chemicals need to be applied before damage is seen and while adults are active in the early spring and before eggs are laid. Typically, chemical applications are made when large numbers of adult ABW are observed walking across golf course fairways, greens, and roughs, which may be associated with the blooming (i.e., not a full bloom, but later when plants are half green with leaves and half gold with flowers) of *Forsythia* spp. Sequential applications are needed in early spring if a long migration period is expected or the area has experienced damaging populations in previous seasons. In high pressure situations, up to six applications per season may be needed on a 30- to 45-day interval or when significant adult activity is observed. Some insecticides for control of ABW adults include pyrethroids like deltamethrin, bifenthrin, cyfluthrin, lambda-cyhalothrin, and the organophosphate chlorpyrifos. The aforementioned insecticides generally provide high levels of adult ABW control (>85% control) in situations where ABW has not developed resistance to pyrethroids.

The neonicotinoids are systemic, long-residual, preventively applied insecticides for white grub control, but they can be used in ABW management programs if timed properly. The neonicotinoids, which include imidacloprid, clothianidin, dinotefuran, and thiamethoxam, have been inconsistent in their ability to control ABW. Neonicotinoid insecticides should be applied just prior to egg laying, which targets early instar feeding, but proper timing is difficult to achieve. Neonicotinoids have been shown to provide between 20% and 100% control but probably should not be relied upon for highly effective control under significant pressure.

Chlorantraniliprole is a newer systemic insecticide that will generally provide more consistent ABW control when compared to the neonicotinoids. This insecticide is effective in controlling other white grub species (e.g., Japanese beetles and chafers) and is used to target ABW larvae in leaf sheaths. Chlorantraniliprole should be applied 2 to 3 weeks after the adult application (i.e., the time following use of pyrethroids or chlorpyrifos) but about 2 weeks before the larva application. When larvae are observed damaging turf there are three chemical options. These include trichlorfon, indoxacarb, and spinosad. When targeting larvae below the soil/thatch surface, obtaining greater than 80% control can be difficult due to materials being tied up in thatch. Rhododendron flowering (e.g., *Rhododendron catawbiensis*) in the northeastern United States can serve as a good phenological indicator for timing spring applications targeting larvae.

For preventative ABW control in chronically infested sites, multiple applications of various classes of insecticides may be needed to provide acceptable control. Resistance has been documented in the pyrethroids. Thus, proper rotation and even tank-mixing different classes of insecticides could provide a delay in the development of tolerant ABW populations as well as provide for more effective control. Control measures should focus on the management of adults early in spring or when they are observed walking on turfgrass surfaces. Subsequent applications should be based on scouting and determining which stage (e.g., medium or large larva, pupa, or adult) is dominant at a given time. Chemicals should be applied in the proper amount of water (typically ≥44 gallon/acre; ≥411 L/ha). It is important that insecticides targeting surface adults be allowed to dry or only receive less than 0.10 inch of irrigation (2.5 mm) following treatment. Applications using systemic insecticides or those targeting larvae in the thatch/soil should be watered-in lightly with one turn of the irrigation heads.

Turfgrass Ants

Biology and control strategies were outlined by Potter and Maier (2006), which are briefly reviewed here. Ants are among the most common insects found in turf. They are social insects that live in colonies that consist of many thousands of sterile female workers, but only one reproductive queen. The queen, eggs, and larvae remain below ground and are fed by workers who forage on the surface. The favored foods for ants include the eggs and small larvae of grubs, cutworms, and sod webworms as well as other insects. Their nests consist of interconnected chambers that are mostly found in the surface 10 to 15 inches (25 to 38 cm). Each passage to the surface is marked by a volcano-shaped mound (Figure 10.4). The number of mounds increases from spring to summer. The mounding activity of ants poses problems for turfgrass managers. Mounds left undisturbed or flattened by mowers kill creeping bentgrass in spots 1 to 2 inches (2.5 to 5 cm) in diameter. Soil mounds are destructive to mower blades and reels and can prevent good stand establishment in newly seeded areas.

Turfgrass ants are most troublesome on golf greens but can also cause problems on tees and fairways. Control is difficult because fast-acting insecticides only control a portion of the workers on the surface and often fail to eliminate queens in the colonies. Several insecticides suppress mound building for a few weeks, but colonies recover and new mounds often appear. Mound drenches are effective but labor intensive. Broadcast treatments are much less effective, and dusts do not work well. Whatever strategies are followed, success is dependent on getting started as soon as the first mounds appear in spring. For golf greens, insecticides should be directed onto the perimeter, about 6 feet (2 m) on either side to include the collar (Potter and Maier, 2006). This is because many main nests are located just outside collars in the native soil. As the number of colonies

FIGURE 10.4
Turfgrass ant mounds on a golf green.

grow over a season, workers expand foraging territories into golf greens where tunneling is made easier in sand-based rootzones. Some effective insecticidal materials include chlorpyrifos and pyrethroids. Multiple applications of chlorpyrifos and pyrethroids may be needed on a 30- to 45-day interval until about mid-summer. Baits such as abamectin and hydramethylon can be effective. Small amounts of the aforementioned baits are sprinkled around mounds. Granules are carried by workers to the nest, and the chemical eventually is carried to the brood chamber. Baits sometimes reduce mounding activity within a few days. Mounds, however, often redevelop later in the season. Avoid irrigation when using a conventional insecticide for at least 8 hours and withhold water for at least 24 hours if using a bait.

Black Cutworms

Black cutworms are larvae of night-flying moths. Moths lay individual eggs on the tips of leaf blades and after larvae emerge, they feed at night and hide in thatch or in burrows during the day. Egg-laying begins in early spring and continues throughout the summer. Eggs often are laid in green surrounds, and caterpillars then migrate (walk) to the greens. Cutworms are most problematic in golf greens, and they can often be found hiding in core aeration holes during the day. At night they chew grass blades at the base of plants, usually from around their aeration hole hiding places (Figure 10.5).

FIGURE 10.5
Black cutworms feed at night from their hiding places in coring holes.

This feeding activity appears initially as a bronzing and then tissues turn tan or brown in circles or streaks around their hiding place. Larvae are caterpillars, 1 to 1.5 inches (2.5 to 3.8 cm) long, gray to olive in color, and coil in a spiral when disturbed (Figure 10.6). Larvae have no distinctive markings other than prominent black spiracles and a pale stripe down the middle of the back. Frass (i.e., fecal matter), chewed basal sheaths, and bird probes are key signs of cutworm activity (Figure 10.6).

Black cutworm adults are monitored by black light and pheromone traps, and larvae are monitored by using a soap flushing solution. Because many eggs are removed during mowing, some reduction in populations can be

FIGURE 10.6
Black cutworm damage and bird foraging activity on a golf green.

achieved by disposing of clippings well away (at least 50 feet; 15 m) from golf greens. The biological agent *Bacillus thuringensis* is moderately effective, but only against very small larvae. Spinosad, a biological product derived from the bacterium *Saccharopolyspora spinosa*, is highly effective. Many insecticides are very effective including carbaryl, chlorantraniliprole, chlorpyrifos, indoxacarb, and pyrethroids.

Sod and Subterranean Webworms

As in the case of black cutworms, sod and subterranean (i.e., cranberry girdler) webworm adults are moths. The larvae (caterpillars) are the damaging stage. Adults are whitish-gray moths with snout-like projections. Unlike cutworms, adults can be seen at dusk flying in a zigzag manner over turf areas. Larvae range from 0.75 to 1 inch (1.9 to 2.5 cm) long and are green to grayish in color and have black spots. Females scatter eggs across the turf and emerging larvae burrow into thatch and line their tunnels with a silk-like material that they produce. Feeding occurs at night as larvae sever grass stems and drag leaves into their burrows for consumption. Damage appears in summer on creeping bentgrass golf greens as small whitish-tan or brown spots about 0.5 to 1 inch (1.2 to 2.5 cm) in diameter (Figure 10.7). Sod webworm adults are monitored with black light or pheromone traps and larvae by using a soapy water flush. Signs of sod webworm include bird activity, frass, and silk-lined tunnels. Control is the same as previously described for black cutworm.

FIGURE 10.7
White spots, not unlike dollar spot or hale damage, were caused by sod webworm larvae on this creeping bentgrass green.

Cicada Killer or Sand Wasps

These wasps resemble very large hornets or yellow jackets. They have rust-red-colored heads, yellow wings, and yellow-and-black-striped abdomens (Potter and Potter, 1999). They have a buzzing and menacing manner but rarely sting or otherwise injure humans. Only females sting, and it is considered mild, whereas the males will dive at golfers if they get too close to nests. They are most noticeable during the heat of summer and have aggressive mounding habits at this time. Cicada killer wasps prefer to burrow into sandy embankments of bunker faces and southwest sloping greens. They also will tunnel into native soils of fairways. Soil is pushed to the surface as females tunnel and create U-shaped mounds that are ≥3 to 6 inches (≥7.5 to 15 cm) in height and length (Figure 10.8). Mounds can smother grass in patches and damage mower blades and reels. Once burrows are prepared, females ambush adult cicadas and bring them to the brood chamber, where the wasp lays an egg. The female backs out and seals the cell. The emerging larva consumes the cicada. Females continue the process until late summer, when cicada populations naturally decline.

Cicada killer wasps are beneficial insects, and because they rarely sting, their control generally is not recommended. Some people are intimidated and may request that they be controlled; however, these wasps become most problematic when they burrow into greens. Superintendents seldom achieve effective control with insecticides and use insect nets or tennis rackets to capture or incapacitate flying adults.

FIGURE 10.8
Cicada killer wasp and mound. (Photos courtesy of D. Settle.)

Earthworms

Earthworms, or night crawlers, are found in abundance on golf course fairways. They are less common in sand-based rootzones, but their castings or their trails (probably related to mating activities at night) can be seen on the wet canopy of greens and tees on some summer mornings. Evidently, angular sand particles irritate earthworms and deter their tunneling activities in sand-based rootzones. It also is probable that the generally drier soil conditions in sand-based rootzones discourage their activities. Earthworms are beneficial because their tunneling and feeding activities improve soil structure; provide channels in soil to allow for easier infiltration of air and water; and increase the availability of nutrients by degrading organic matter. They construct vertical burrows that can extend 8 to 12 feet (2.4 to 3.6 m) deep. They migrate up and down their burrows in response to fluctuations of soil temperature and moisture. They are most active following periods of rain in spring and autumn. They tend to remain closer to the surface in fairways where there is regular irrigation and a good food supply in the form of clippings. Earthworms can produce large numbers of castings (i.e., excreted organic matter and soil) (Figure 10.9). These castings interfere with mowing, playability, and aesthetics. The main problem is that rollers on mowers and tires of maintenance equipment flatten, smear, and compact castings, causing muddy spots (Figure 10.10). Sometimes grass is smothered in spots where mud has compacted turf for long periods, which usually is associated with rainy weather that drives earthworms to the surface. The compacted

FIGURE 10.9
Earthworm castings on a fairway.

FIGURE 10.10
Earthworm casts were flattened and compacted by rollers and tires when the fairway was mowed.

casts also promote weed invasion because they contain weed seed brought to the surface by earthworms (Potter, 1998).

The activity of earthworms may be reduced by acidifying soil with ammonium sulfate. Clipping removal reduces their food supply, and this causes earthworms to move deeper in soil and thus become less of a problem. Once it was believed that topdressing with abrasive sand would discourage earthworm activity, but this has not been the case for most superintendents who topdress fairways. Following years of topdressing fairways, however, castings become more sandy, dry quickly, and are more easily dispersed. Cultural approaches, however, seldom eliminate the problem. Because earthworms are considered beneficial there are no pesticides registered for their control. Carbaryl, granular forms of pyrethroid insecticides as well as thiophanate-based fungicides can be toxic to earthworms. In Europe, irritants are used to force earthworms to the surface, where they are raked into piles, shoveled into wheelbarrows, and relocated. Research demonstrated that watering-in tea seed meal pellets (i.e., seeds of the Chinese tea oil plant, *Camellia oleifera*) is very effective in expelling earthworms from soil (Potter et al., 2011). These tea seeds contain soap-like substances (i.e., saponins) that irritate the membranes of earthworms causing them to surface. A drawback to using an expellant is that when large numbers of earthworms die on the surface, they temporarily cause an unpleasant odor and an unsightly appearance. There are plans to commercially develop this natural product.

Summary

- Annual bluegrass weevil (ABW), ants, black cutworm, sod web-worm, cicada killer wasps, and earthworms are among the most common invertebrate pests of creeping bentgrass in North America.

- The ABW selectively attacks annual bluegrass in mixed stands with creeping bentgrass, but they eventually will feed on bentgrass.

- All three ABW stages (larvae, pupae, and adults) can be simultaneously active in the same area, which makes their control with insecticides challenging.

- Control programs are complicated and involve different insecticides and timings. Targeting ABW adults in spring may be most effective, but different insecticides are used to target larvae feeding within leaf sheaths versus those that have migrated into thatch.

- Mounding activity of ants can kill bentgrass in golf greens and damages mower reels.

- Ants are targeted with insecticides in spring as soon as mounding appears. Mound drenches are temporarily effective. Insecticides should be applied to collars and green surrounds where many main nests are located.

- Adult black cutworm and sod webworms are moths, but the larvae (i.e., caterpillars) cause damage in summer by feeding at the base of leaves, creating bronze or white-colored spots of dead grass on golf greens. These insect pests are easy to control, but problems may recur throughout summer.

- Cicada killer wasps are a nuisance to golfers but become most troublesome when they burrow into golf greens. These wasps are not effectively controlled with insecticides.

- Earthworms are more of a problem in fairways than greens or tees. Their casting activity is most pronounced during rainy weather, especially in early autumn. Casts are flattened by rollers on mowers and kill grass in spots, which can be numerous and interfere with maintenance and reduce aesthetics and playability.

- Earthworm activity can be discouraged by acidifying soil with ammonium sulfate and collecting clippings.

- Because earthworms are beneficial, there are no chemicals labeled for their control, but tea seed pellets appear promising for use as an earthworm expellant.

Bibliography

Cowles, R.S., A. Koppenhofer, B. McGraw, S.R. Alm, D. Ramoutar, D.C. Peck, P. Vittum, P. Heller, and S. Swier. 2008. Insights into managing annual bluegrass weevils. *Golf Course Management* 76(8):86–92.

Fermanian, T.W., M.C. Shurtleff, R. Randell, and H.T. Wilkinson. 2003. *Controlling Turfgrass Pests*. Prentice Hall, Upper Saddle River, NJ.

McGraw, B.A., P.J. Vittum, R.S. Cowles, and A. M. Koppenhofer. 2011. Nematodes for control of annual bluegrass weevils. *Golf Course Management* 79(2):88–90,92,94.

Potter, D.A. 1998. *Destructive Turfgrass Pests: Biology, Diagnosis and Control*. Ann Arbor Press, Chelsea, MI.

Potter, D.A., M.C. Buxton, C.T. Redmond, C.G. Patterson, and A.J. Powell. 1990. Toxicity of pesticides to earthworms (*Oligochaeta:Lumbricidae*) and effects on thatch degradation in Kentucky bluegrass thatch. *J. Econ. Entomology* 83:2362–2369.

Potter, D.A., and R.M. Maier. 2006. Turfgrass ants: Biology dictates strategies for control. *Golf Course Management* 74(11):87–90.

Potter, D.A., and M.F. Potter. 1999. Removing stingers from golf courses. *Golf Course Management* 67(9):53–58.

Potter, D.A., C.T. Redmond, and D.W. Williams. 2011. The worm turns: Earthworm cast reduction on golf courses. *Golf Course Management* 79(9):86,87,88,90,92,94,96.

Vittum, P.J., M.G. Villani, and H. Tashiro. 1999. *Turfgrass Insects of the United States and Canada*, 2nd ed. Cornell University Press, Ithaca, NY.

Index

Printed and bound by CPI Group (UK) Ltd, Croydon, CR0 4YY

18/10/2024

01776208-0010